The ARRL
Antenna Compendium

Volume 2

Editor
   Gerald L. (Jerry) Hall, K1TD

Assistant Editors
   Rod Gray, WA1DVU
   Randy Henderson, WI5W
   Charles L. Hutchinson, K8CH
   Joel P. Kleinman, N1BKE

Production
   Leslie Bartoloth, KA1MJP
   Michelle Chrisjohn, WB1ENT
   Sandra Damato
   Jacqueline Hernandez
   Jodi Morin, KA1JPA
   Steffie Nelson, KA1IFB
   David Pingree
   Dianna Roy
   Hilary Vose
   Jean Wilson

Cover Design
   Sue Fagan

The Cover

   Theme: Radio signals propagate from the antenna, through the ionosphere, and to distant points on the globe. Sometimes the ionosphere displays visible phenomena, as discussed in the final paper of this volume.

   Feature photo: Aurora australis. Taken south of New Zealand looking toward Antarctica, by Col Robert Overmyer from Spacelab III (Challenger). The principal investigators on the Spacelab III auroral project were Dr. Thomas Hallinan and mission specialist Don Lind. The photo shows not only cloud layers, but also the airglow layer, stars, and the aurora. *(NASA photo)*

   Inset photo: Aurora borealis, showing the entire auroral oval. Taken September 1981 from a spacecraft altitude of approximately 12,962 miles. The sun illuminates the earth from the left-hand side of the picture. *(NASA photo)*

Foreword

In 1985 the ARRL added *The ARRL Antenna Compendium, Volume 1* to its library of publications. That book contains 176 pages of previously unpublished material, covering a wide range of antenna types and related topics. The foreword of *Volume 1* suggests that *Volume 2* of *The Compendium* might logically follow.

Like its predecessor, *Volume 2* contains all brand-new material. Because antennas are a topic of great interest among radio amateurs, ARRL Headquarters continues to receive many more papers on the subject than can possibly be published as articles in the League's journal, *QST*. So again, as with *Volume 1,* those papers have been collected here and combined with solicited material. None of this material has appeared in print before. Whether you have only a casual interest in antenna construction or a serious interest in understanding fundamental antenna theory, you'll most likely find something to stimulate your thinking.

Will a companion *Volume 3* appear in the future? Very likely. Amateurs' interest in antennas never ends.

David Sumner, K1ZZ
Executive Vice President
Newington, Connecticut
September 1989

Diskette Availability

Six papers in this book contain listings of BASIC programs suitable for use with an IBM PC or compatible computer. Some papers include more than one program listing. The programs aid in performing various design tasks associated with antennas, as described in those six papers.

The ARRL offers a 5¼-inch (360K) computer diskette for the IBM PC as an optional supplement to this volume. The diskette contains 11 BASIC programs in ASCII text format, and one compiled Pascal program allowing a more extensive series of calculations than its corresponding BASIC counterpart. The diskette is copyrighted but not copy protected.

The program files have been supplied by the authors of corresponding papers, and filenames have been assigned to resemble the authors' last names. The diskette is made available as a convenience to the purchaser. The ARRL has verified that the programs run properly, but does not warrant program operation or the results of any calculations.

Contents

Vertical and Inverted L Antenna Systems

2 Vertical Antennas: New Design and Construction Data
Archibald C. Doty, Jr, K8CFU, John A. Frey, W3ESU, and Harry J. Mills, K4HU
What length should a ¼-λ vertical radiator be for resonance? More than likely it is not 234/f(MHz). Capacitive bottom loading? Folded monopoles? Data from extensive tests are summarized in this comprehensive paper.

10 Steerable Arrays for the Low Bands
Bob Alexander, W5AH
You can aim your beam electronically with time-delay beam steering. The beauty of this system is the steering angle is independent of frequency. Use multiband radiators for beam steering on different bands.

16 A Steerable Array of Verticals
S. L. Seaton, K4OR
This array covers the 40, 20 and 15-meter bands. On 40 meters you can steer the beam in four directions. Band changing and beam steering are done without a complicated relay system, using a simple manual method to switch base loading coils.

19 The Robert Tail—A DX Antenna
Robert L. Brewster, W8HSK
Invert a bobtail curtain, feed it with coax, and you have an array that reaches out and grabs DX. Nest one inside another for 80- and 75-meter coverage.

25 The Simplest Phased Array Feed System . . . That Works
Roy Lewallen, W7EL
Feeding two array elements from a T connector with lines that differ electrically by ¼ λ will probably not give you 90° element phasing. Why not, and how you can get the proper phasing is explained here.

36 Unipole Antennas—Theory and Practical Applications
Ron Nott, K5YNR
Unipole antennas are sometimes likened to gamma-matched or shunt-fed antennas. Are they the same? A properly designed folded unipole is superior in some ways.

39 Magnetic Radiators—Low Profile Paired Verticals for HF
Russell E. Prack, K5RP
Would you believe you can have an efficient HF vertical system without radials? And it need be only a small fraction of a wavelength high!

42 A Multiband Loaded Counterpoise for Vertical Antennas
H. L. Ley, Jr, N3CDR
There's been a lot of talk about using lumped loading in a counterpoise system of radials. Here are the results of experiments doing just that, with information included for you to build your own system.

46 A Multiband Groundplane for 80-10 Meters
Richard C. Jaeger, K4IQJ
If lumped loading of counterpoise radials works, then why not use traps instead? And with a multiband radiator, you end up with a multiband groundplane antenna.

50 Tunable Vertical Antenna for Amateur Use
Kenneth L. Heitner, WB4AKK/AFA2PB
You can feed a vertical antenna with an omega match. And with this arrangement you can change bands with a relay. A simple remote tuner gives you the final touch—the ability to QSY up and down the band but yet adjust for a low SWR.

52 A 5/8-Wave VHF Antenna
Don Norman, AF8B
For either permanent or temporary mounting, this antenna accounts well for itself. The location of the ground-plane radials is optimized for minimum antenna current on the transmission line.

54 Some Experiments with HF and MF 5/8-Wave Antennas
Doug DeMaw, W1FB
A 5/8-wave vertically polarized antenna is a winner on 160 meters! Install it as an inverted L.

Yagi-Type Beam Antennas

58   New Techniques for Rotary Beam Construction

G. A. "Dick" Bird, G4ZU/F6IDC

Rotatable beam antennas don't always require aluminum tubing, nor do they need a large turning radius for reasonable gain and F/B ratio. Use these techniques to construct inexpensive beams.

61   The Attic Tri-Bander Antenna

Kirk Kleinschmidt, NT0Z

Efficient Yagi beams can be made of wire. This indoor Yagi antenna covers three HF bands!

64   Yagi Beam Pattern-Design Factors

Paul D. Frelich, W1ECO

Over the years, antenna theorists frequently calculated complex radiation patterns for multielement systems by combining array patterns. They considered the overall antenna as an array of arrays. Couple this concept with the ability of a computer, and you have a truly powerful analysis tool.

Quad and Loop Antennas

88   Half-Loop Antennas

Bob Alexander, W5AH

Half of a full-wave loop antenna can exist as wire in an inverted-U or inverted-V shape above ground. The other antenna half then exists as an image in the ground and radial system. Such an antenna offers quite respectable performance.

90   Coil Shortened Quads—A Half-Size Example on 40 Meters

Kris Merschrod, KA2OIG/TI2

Try a quad with electrically lengthened elements. This paper tells you how to design them, and includes information on using a computer in the design work.

Multiband and Broadband Antenna Systems

96   A 14-30 MHz LPDA for Limited Space

Fred Scholz, K6BXI

You can cover 10, 12, 15, 17 and 20 meters with this rotatable array. Being mostly wire, it weighs little and has low wind resistance. It's just the ticket if you live on a small city lot. It's great for portable use, too.

100   Antenna Trap Design Using a Home Computer

Larry V. East, W1HUE

You can make traps for multiband antennas from nothing but lengths of coaxial cable. But how much coax, and wound in what coil dimensions? Avoid the tedium of cut-and-try designs with this simple computer program.

103   The Suburban Multibander

Charles A. Lofgren, W6JJZ

A single antenna covers the 80- through 10-meter bands. Although the design resembles the G5RV antenna, its lineage and operation are different.

106   Fat Dipoles

Robert C. Wilson

Mention a "cage" dipole and you immediately have a hot conversational topic. This simplified technique offers some advantages of a cage dipole with a lot less wire.

108   Swallow Tail Antenna Tuner

Dave Guimont, WB6LLO

This unique adaptation of a fan dipole lets you remotely tune the antenna for 75 or 80 meters by mechanical means. It covers 40 meters, too.

110   The Coaxial Resonator Match

Frank Witt, AI1H

If you use a dipole on 80 meters, you'd really like to feed it with coax and have an SWR below 2:1 across the entire 80-meter band. Recent ideas have met with success. This one uses a unique combination of coax sections as resonators.

119   A Simple, Broadband 80-Meter Dipole Antenna

Reed E. Fisher, W2CQH

Several different schemes have been tried over the years for getting an SWR below 2:1 across the 80-meter band. This system uses multiple coaxial stubs at the feed point.

Portable, Mobile and Emergency Antennas

126 Emergency Antenna for ARES/RACES Operation

Ken Stuart, W3VVN

The familiar J-pole antenna has been constructed with many mounting arrangements. A camera tripod makes a freestanding support that can be used almost anywhere, even on a flight of stairs.

128 Portable 2-Meter Antenna

Michael C. Crowe, VE7MCC

This portable 5/8-λ groundplane offers significantly better results than the typical antenna you'll find on any hand-held transceiver. And yet it folds into a very small package, less than 2 inches diameter by 10 inches long.

130 The Half-Wave Handie Antenna

Ken L. Stuart, W3VVN

Build this ½-λ antenna for your hand-held transceiver. You'll need only a few parts, and the results will amaze you.

Controlled Current Distribution Antennas

132 The Controlled Current Distribution (CCD) Antenna

Stanley Kaplan, WB9RQR, and E. Joseph Bauer, W9WQ

What do you have when you break your antenna conductor up into many sections coupled with capacitors? If the section lengths and capacitor values are chosen properly, you have an efficient antenna with numerous advantages over a straight conductor. Here's how to design it.

137 The End-Coupled Resonator (ECR) Loop

Henry S. Keen, W5TRS

The controlled current distribution (CCD) technique can be applied to loop antennas, too. Here is basic information on why CCD works, as well.

Balloon and Kite Supported Antennas

142 Balloons as Antenna Supports

Stan Gibilisco, W1GV

Helium-filled balloons can serve well as supports for long end-fed wires at MF and HF. But there are pitfalls to avoid. Follow these tips for success.

145 Kite-Supported Long Wires

Stan Gibilisco, W1GV

If you've ever flown a kite, you've probably thought about using one to hold up an antenna. Give it a try. The results can be well worth the effort, aside from the fun you'll have.

Antenna Potpourri

152 Antenna Selection Guide

Eugene C. Sternke, K6AH

Your antenna type, the soil conditions in your area, and propagation radiation angles determine how much useful signal you have for working DX. Your antenna height plays a part, too. These results of a comprehensive computer analysis will help you choose the best antenna for your needs.

157 A Ham's Guide to Antenna Modeling

Steve Trapp, N4DG

Anyone with access to a computer can calculate numerous antenna and matching-network parameters. If you do your own programming, take advantage of the tips and math relationships in this paper.

162 A Window Slot Antenna for Apartment Dwellers

Ermi Roos, WA4EDV

Use the reinforcing steel and other metal in your building to enhance your VHF/UHF signal, rather than to shield it from radiating.

164 Polar Pattern Plotter for the C64

Steve Cerwin, WA5FRF

Let your Commodore 64™ plot antenna radiation patterns for you. You can view patterns on the screen or send them to the printer.

169 A VHF RF Sniffer

Don Norman, AF8B

An absorptive wavemeter is a very useful item of test equipment when you work with RF power circuits. This design is one of the simplest, and yet it is quite effective.

Baluns and Matching Networks

172   Some Additional Aspects of the Balun Problem

Albert A. Roehm, W2OBJ

You don't need a complicated or expensive balanced matching network to go from coax to a balanced line operating with high SWR. Just put a good balun ahead of your tuner.

175   A Servo-Controlled Antenna Tuner

John Svoboda, W6MIT

Go one step beyond a manually adjusted matching network. With this semiautomatic, remotely controlled tuner, you can change bands quickly and easily.

182   Remotely Controlled Antenna Coupler

Richard Z. Plasencia, WØRPV

This device lets you remotely switch bands and tune a vertical antenna for 1.8, 3.5 and 7 MHz. It uses a unique, low-cost homemade variable capacitor that works as well as a vacuum variable.

187   Phase-Shift Design of Pi, T and L Networks

Robert F. White, W6PY

Many amateurs use a matching network at one end of the transmission line. These networks, pi circuits in particular, are typically designed for a specific system Q. Designing for a specific phase shift offers a simpler design approach, as well as a better understanding of the relationships between various network types.

Solar Activity and Ionospheric Effects

198   Sunspots, Flares and HF Propagation

Richard W. Miller, VE3CIE

MINIMUF and similar computer programs are excellent for making predictions of HF propagation under normal solar conditions. But unusual solar activity can render those predictions useless for a time. Use these guidelines to get more reliable day-to-day predictions.

204   Visible Phenomena of the Ionosphere

Bradley Wells, KR7L

You probably know the ionosphere contains regions (layers) with different electrical characteristics. You may also know the ionosphere is a zone of dynamic activity. But do you know why it sometimes emits light?

Vertical and Inverted L Antenna Systems

Vertical Antennas: New Design and Construction Data

By Archibald C. Doty, Jr., K8CFU
347 Jackson Rd
Fletcher, NC 28732

John A. Frey, W3ESU
841 Greenwood Dr
Hendersonville, NC 28739

Harry J. Mills, K4HU
631 4th Ave W
Hendersonville, NC 28739

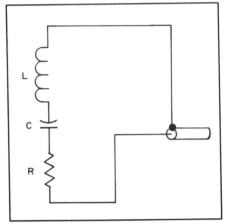

Fig 1—The equivalent circuit of a vertical antenna. The resistor, R, represents the several resistances inherent in an antenna, including the ohmic resistance value of the metallic antenna element, the radiation resistance of the antenna, etc. The inductor, L, is the self-inductance of the wire or tubing used as the radiating element, and the capacitor, C, is the capacitance between the radiating element and the ground or surrounding objects.

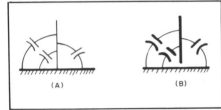

Fig 2—The capacitance of a vertical antenna to ground can be imagined as the result of many small capacitors, each having one end connected to the radiating element and the other end to ground. The drawing at A represents a thin radiator; at B, one of greater diameter, and one requiring a shorter element length in order to maintain resonance.

T he original objective of the test programs that were the basis of this paper was to determine the base-impedance transformation that can be accomplished by design variations of vertical folded-monopole antennas having an electrical height of ¼ λ. As folded monopoles consist of two ¼-λ vertical elements, the starting point for the program was to determine the correct heights for the various elements to be used. These elements had diameters ranging from 0.125 inch to 2.0 inches in diameter.

Vertical antennas are not too complicated. They can be drawn, looking from the feed point, as a series resistor, inductor and capacitor. See Fig 1. The resonant frequency of a vertical is "a frequency at which the input impedance of an antenna is nonreactive."[1] In other words, when the inductive reactance of the radiating element is exactly balanced by the capacitive reactance of the antenna system at a certain frequency, the antenna is resonant at that frequency.

Height-to-Diameter Ratio

To give an example of how this works, let's first look at a resonant vertical antenna. The vertical radiating wire has a certain natural value of inductive reactance. This value is balanced at resonance by the capacitive reactance resulting from the adjacency of the radiating element to ground or to nearby structures. For illustrative purposes this capacitance can be imagined as resulting from many small capacitors, each having one end attached to the antenna, and the other end attached to the ground. Fig 2A illustrates.

If the diameter of the radiator is increased, this increases the surface area of the element that is "looking" at the ground, as illustrated in Fig 2B. Thus the element will have greater capacitance to ground—just as making capacitor plates

larger increases their capacitance.

If the antenna system with the larger diameter element, Fig 2B, is to remain resonant at its original frequency, it must be shortened enough so that it once again provides the same capacitance to ground as did the original radiator, Fig 2A. This is referred to as the height- (or length-) to-diameter ratio, or the antenna-to-diameter effect on the resonant frequency of a vertical antenna. Simply stated, the "fatter" a radiating element is, the shorter it must be to maintain a fixed resonant frequency. This explains why the physical height of a ¼-λ vertical antenna can be considerably less than an electrical quarter wavelength.

End Effect

There is a second phenomenon, referred to as end effect, that also results in the physical length of a radiator being shorter than the electrical length. End effect results from the boundary condition at the end(s) of the radiator, ie, where the metallic element ends and a nonconducting medium begins (as air, or an insulator). In this area there is a concentration of electric lines of force, which implies a greater capacitance per unit length than is found along the rest of the radiator.

The result of this "extra capacitance" at

the end of an element is to load the element and thus reduce its resonant frequency. Thus it will be necessary to reduce the physical length of the element to maintain its original resonant frequency. In summary, anything that adds capacitance to a vertical antenna must be recognized when deriving a formula for the physical height of the radiating element.

As the ¼-λ vertical is one of the most commonly used antenna systems, we assumed that there would be no problem in finding a formula that would provide explicit dimensional data for the height of the test antennas that we wished to use. *Wrong!* Here is a small portion of what was found on this subject in an extended literature search. A number of references acknowledge that the height of a vertical antenna must be adjusted to compensate for the antenna height-to-diameter ratio, or for end effect. But most of them do not agree on how much compensation should be made for these important factors.

In previous publications ARRL provides a formula for various configurations of ¼-λ vertical antennas as[2]

$$\ell = \frac{234}{f_{MHz}}$$

This formula is apparently derived from

$$\ell = \frac{984}{4f} = \frac{246}{f}, \text{ less 5\% for end effect}$$
$$= \frac{234}{f}$$

This formula is incorrect, as it does not make provision for the important height-to-diameter factor discussed above. (The 1988

15th edition of the *ARRL Antenna Book* corrects this situation in part.) Moxon notes that, "Even when no insulators are used, the resonant length of a dipole is slightly less than the free-space half wavelength, depending on the ratio of length to diameter but typically 2.4% for wire elements rising to 4.7% for 1-inch diameter tubing at 14 MHz, or half this at 28 MHz."[3]

Another source of information disagrees with both the older ARRL and the Moxon data. In reporting on tests with Yagi-Uda dipole antennas, Green states, "During the experimental measurements, it was found that agreement between theory and practice was always obtained at a frequency one or two percent lower than the design value. This was not affected by the number of elements nor their spacings, but only by their diameter, suggesting a dipole 'end effect'."[4] Several other sources provide excellent or good data and mathematical analyses of end effect—but no specific formulas for designing a vertical antenna.[5-9]

Finally, however, two references provide explicit data describing the shortening effect on a 90° vertical antenna resulting from its height to diameter ratio. Brown and Woodward refer to the "shortening effect for zero reactance" (ie, resonance) in terms of percent of a quarter wavelength, and pro-

Fig 3—This photo shows the variation of heights found necessary to resonate elements of various diameters at 29.0 MHz. The length difference between the thinnest element (1/8 inch) and the thickest element (2 inches) is 5 inches.

Fig 5—Test equipment was located at the base of the antenna, beneath the counterpoise system. Equipment so located is virtually "invisible" to the antenna itself.

vide data on the basis of the "antenna/ diameter" ratio (height-to-diameter ratio expressed in degrees).[10] Laport takes another approach toward describing the height/diameter ratio effect—but comes out with almost identical dimensional conclusions.[11]

The importance of the height-to-diameter shortening factor is illustrated by Fig 3. A 5-inch variation of element heights was found necessary to resonate the thinnest and the thickest elements used in this test program at the same frequency—29.0 MHz.

In this test series, the objective was to determine correct antenna heights for the test frequency—and not to separately define each of the factors determining the correct height for the radiating element. Thus, the two element-shortening factors discussed above were combined into a single shortening factor, C, and the following formula derived.

$$h = \frac{246 \; C}{f}$$

where

h = height of a ¼-λ vertical monopole, feet

f = frequency, MHz

C = a factor to compensate for *all* aspects of the antenna system that cause the physical height of the radiating element to be less than its electrical length (ie, height-to-diameter ratio, end effect, etc)

Base Resistance and C Factor Tests

The antennas tested were located above the center of a 20- × 20-foot 64-radial counterpoise raised 5 feet above a brick terrace. The counterpoise is shown in Fig 4, A and B. Tests were conducted in the 27- to 30-MHz range. Extensive prior testing had proved that a counterpoise of this size would provide a fair approximation of the elusive "perfect ground," and should assure uniformity of testing without concern about anomalies in the ground system.[12] Test equipment was located at the base of the antenna, directly under the counterpoise, as shown in Fig 5. Previous experience had shown that equipment *under* a properly designed counterpoise is virtually "invisible" to an antenna located above the counterpoise radial wires.

When testing began, it was apparent that

(A)

(B)

Fig 4—Tests in the 27- to 30-MHz range were performed over this counterpoise system. It consists of 64 radial wires covering a 20- × 20-foot area.

the base impedances of the ¼-λ vertical monopoles tested were considerably lower than anticipated. Also, the height-to-diameter effects encountered were quite different from those which had been published for dipole antennas. Several weeks were spent rebuilding, revising and recalibrating the test setup, but the height anomaly persisted. As retest error with our venerable General Radio impedance bridge was consistently low, we finally concluded that our readings must be correct.

Fig 6 shows the shortening effect found for the physical length of an electrical ¼-λ vertical monopole antenna over a counterpoise. This graph takes into account *all* factors that cause a vertical antenna to have a physical height less than an electrical ¼ λ.

Further testing produced the base-resistance figures shown in Table 1. The lower than expected values for the "fatter" radiators were of concern until we noticed the recent work of Richmond, which indicates similar values for monopole antennas operating over a circular disk (ie, a solid counterpoise).[13]

Capacitive Bottom Loading of Vertical Antennas

The "final report" of the test program described so far in this paper was written, provided to participants, and filed away. We did not originally plan to publish this data. However, the nagging question remained: Had some factor been missed that would help to explain the unexpectedly short heights found for ¼-λ vertical monopoles?

The results of tests made by Brown and Woodward have been published in their highly respected paper.[14] They show that the formula for the height of a vertical antenna over normal ground (with all shortening factors included for an antenna having a height/diameter ratio of 100) was

$$h = \frac{230}{f}$$

where

h = height, feet
f = frequency, MHz

Our tests (using a counterpoise), however, showed figures in the range between 205 and 220. We also remembered an unexplained antenna height anomaly that occurred when we constructed the vertical antenna used in our extensive studies of counterpoises.[15] This antenna was carefully designed, using data from *The ARRL Antenna Book*, to be resonant at 1840 kHz. However, when tested, it was found to be resonant at approximately 1700 kHz (ie, more than 10 *feet* too high!).

Some months later the first clue to the unusually short antenna heights appeared when we built and tested a number of 440-MHz folded-monopole antennas over counterpoises. Fig 7 shows one of these antennas. Here is what we found from a series of tests on these antennas.

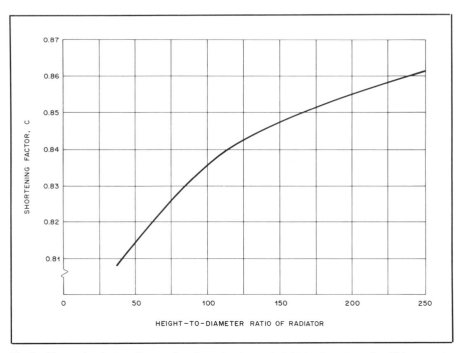

Fig 6—Shortening factor, C, as a function of antenna height-to-diameter ratio. This curve is based on measurements of ¼-λ vertical monopoles over a 64-radial counterpoise, approximately 0.6 λ square and approximately 0.15 λ above ground. A minor shift in the C-factor value will occur with different counterpoise sizes and height above ground.

Table 1

Base Resistance of ¼-Wavelength Vertical Antennas over a 64-Radial Counterpoise

Unipole Diameter	Actual Height	H/D Ratio	Base Resis.	Resonant Frequency	Theoretical Height	C Factor
0.125 inch	7.41 ft	712	33.5 ohms	29.1 MHz	8.45 ft	0.877
0.375	7.45	238	29.0	28.45	8.65	0.861
0.375	7.41	237	29.0	28.4	8.60	0.861
0.375	7.38	236	29.0	28.7	8.57	0.861
0.625	7.44	143	25.0	28.1	8.46	0.849
0.625	7.38	142	25.0	28.2	8.72	0.846
0.625	7.33	141	25.0	28.4	8.66	0.846
0.625	7.17	138	25.0	28.95	8.50	0.844
0.875	7.43	102	23.5	27.7	8.88	0.837
0.875	7.38	101	23.5	27.9	8.82	0.837
1.0625	7.26	82	22.5	28.0	8.79	0.826
1.3125	7.39	67	22.0	27.4	8.98	0.822
1.875	7.25	46	21.0	27.6	8.91	0.814
2.00	7.25	44	21.0	27.5	8.95	0.810

1) The resonant frequency of a ¼-λ vertical antenna used with a counterpoise varies as the size of the ground system (or plate, in the design tested) under the counterpoise. The larger the ground system, the lower the resonant frequency.

2) The resonant frequency of a vertical antenna used with a counterpoise varies with the distance between the counterpoise and the ground. The greater the distance, the higher the resonant frequency.

In other words, the capacitance between the counterpoise and ground is acting to *capacitively bottom load* the vertical radiator. The concept of loading vertical

antennas to reduce their physical height is certainly not new. Verticals are often inductively loaded, as with top, center, or base loading coils. Capacitive top loading (a top hat) is also common. As a matter of fact, the idea of capacitive bottom loading discussed here is mentioned in the literature, although it has apparently not been extensively tested or applied.[16]

Capacitive Bottom Loading Tests

A series of tests was conducted to better define and quantify the capacitive-bottom-loading effect. The emphasis in these tests was changed—instead of attempting to

Fig 7—One of the 440-MHz antennas used for testing. With these antennas it was learned that the "resonant" frequency of the vertical radiator depends on the size of the counterpoise and its distance from ground. This effect has been termed capacitive bottom loading.

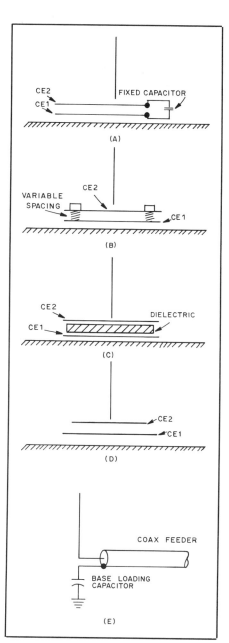

Fig 8—A through D, four test arrangements to determine the effects of capacitive bottom loading (see text). The feed method for all cases is shown at E. CE = capacitive element.

maintain a constant resonant frequency, we decided to determine what range of resonant frequencies could be attained with a fixed height of vertical radiator by varying the value of the capacitive bottom loading.

In deriving the test program, we considered that the capacitive-bottom-loading effect can be described as follows: As the frequency of operation is raised, an antenna of fixed length looks at its base feed point like an increasing resistance in series with a decreasing capacitance. The resulting inductive reactance at the feed point must be tuned out, which necessitates the use of capacitive reactance, which is provided by a capacitor.

In the antennas tested, capacitance was added between the feed point of the antenna and the ground system, in order to supply the required capacitive reactance. By varying the value of this capacitive reactance, the frequency at which the base impedance of the antenna is purely resistive, ie, its resonant frequency, was varied.

The following tests were conducted with two 11-inch-diameter copper-clad printed circuit board plates (capacitive elements 1 and 2 in Fig 8) adjacent to the base of the vertical radiator. A 24- × 36-inch metal plate was spaced under these capacitive elements to simulate a vehicle body or other ground system.

Test No. 1

The capacitance between capacitive elements 1 and 2 (CE1 and CE2 in Fig 8A) was varied by inserting fixed capacitors ranging in value from 0 to 1000 pF. The resonant frequency of the antenna system was found to vary from 27.8 to 30.7 MHz.

Test No. 2

See Fig 8B. The spacing between capacitive elements 1 and 2 was varied from 1/8 inch to 1-3/8 inch. The resonant frequency of the antenna system was found to vary from 28.7 to 32.1 MHz.

Test No. 3

Three different materials (sheet plastic, plywood and perfboard) were inserted between capacitive elements 1 and 2, Fig 8C. Each material had a different dielectric constant. The resonant frequency of the antenna system was found to vary from 28.7 to 29.2 MHz.

Test No. 4

As shown in Fig 8D, the relative diameters of capacitive elements 1 and 2 were varied from a ratio of 1:1 to approximately 1:5. The resonant frequency of the antenna system was found to vary from 25.2 to 36.4 MHz—more than 10 MHz!

Practical Antennas Utilizing Capacitive Bottom Loading

A number of antennas for use on 440, 145 and 29 MHz have been built to evaluate the practical aspects of capacitive bottom loading. The 29-MHz model was extensively tested, and found to be tunable from 25.9 to 31.5 MHz without changing the height of the vertical radiator. With 100 watts input, a small trimmer capacitor was found to be satisfactory. On-the-air performance on the amateur 10-meter band was excellent.

A and B of Fig 9 show 145-MHz mobile antennas constructed with the capacitive bottom-loading feature. In many months of on-the-air use this feature has allowed tuning of the antenna over a considerable frequency range while it was mounted at different locations on several different vehicles. Performance has been consistently excellent.

Impedance Transformation of Folded-Monopole Antennas

The folded-monopole antenna consists of two or more parallel, vertical elements approximately ¼-λ high, and all connected at the top. These elements are at right angles to, and fed against, a suitable counterpoise, ground plane or other artificial ground system. IEEE describes a folded monopole as being "a monopole antenna formed from half of a folded dipole with the unfed element(s) directly connected to the imaging plane."[17]

O. M. Woodward has provided the following description of the electrical characteristics of folded-monopole antennas.[18] He notes that this approach was first suggested by Walter van B. Roberts in 1947.[19] First, consider a folded monopole of height h, fed against an ideal ground of perfect conductivity and infinite extent, as shown in Fig 10A. The generator, e, has zero internal impedance. Next, the circuit can be modified as shown in Fig 10B, using three

(A)
(B)

Fig 9—The bases of mobile antennas constructed with the capacitive-bottom-loading feature. Performance is consistently excellent.

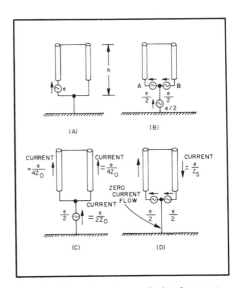

Fig 10—Illustrating the analysis of current flow within the boundary conditions of a folded monopole. See text.

generators having zero internal impedance. Instantaneous polarities are indicated by arrows.

At point A, the left-hand leg of the folded monopole is fed by two of the generators in series, ie, a total voltage of e. At point B (the right-hand leg) the two driving generators are bucking; hence point B is at zero potential with reference to ground. Thus it is seen that the boundary conditions of the two drawings are identical. We may now find the current that flows using superposition theory.

Consider first the case with only the generator that is joined to the ground turned on, Fig 10C. Since the other two generators have zero impedance, then the two lower ends of the monopole legs are tied in parallel and fed as a simple monopole by the

generator, e/2. The current from this generator is then

$$\text{Current} = \frac{e/2}{Z_0} = \frac{e}{2Z_0}$$

This current is the push-push mode, with equal currents flowing in the same directions on the parallel conductors. Z_0 is the radiation impedance of the "fat" monopole fed against ground.

Now, let us turn off this generator, and turn on the other two generators as shown in Fig 10D. We now have the push-pull mode, with the currents flowing in opposite directions in the two parallel legs. This circuit is now a nonradiating transmission-line section short circuited at the far end. The input reactance is then

$$Z_S = Z_C \tan (h/\lambda \times 368) \text{ ohms}$$

where
Z_S = input reactance at the base
Z_C = characteristic impedance of the transmission line
h = height, feet
λ = wavelength, feet
and the current flow is e/Z_S.

Finally, with all generators on, the current flow at point A of Fig 10B is the sum of the currents found in Figs 10C and 10D.

$$\text{Current} = \frac{e}{4Z_0} = \frac{e}{jZ_S}$$

The input impedance of the folded monopole is

$$Z_{in} = \cfrac{e}{\cfrac{e}{4Z_0} + \cfrac{e}{jZ_S}} = \cfrac{1}{\cfrac{1}{4Z_0} + \cfrac{1}{jZ_S}}$$

This equation, the reciprocal of the sum of reciprocals, is recognized as the impedance of two elements in parallel.

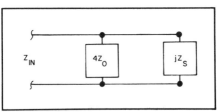

Fig 11—Equivalent circuit of the folded monopole.

Hence, the equivalent circuit of the folded monopole is that shown in Fig 11. Thus, the radiation impedance of the two legs fed in parallel (Fig 10C) is multiplied by four, and the resulting impedance is placed in parallel with the transmission line stub (Fig 10D).

If the stub is ¼-λ long, then the input impedance of the folded monopole is, theoretically, four times the radiation impedance of the push-push mode. Or, to put it another way, so long as the element diameters are equal, the input impedance of a folded monopole will be four times that of a unipole antenna. Our tests showed this to be true in practice, so long as the element diameters are relatively "fat" as compared to their height.

In practice it is of much more interest to determine what impedance transformations can be accomplished by changing the diameters of the folded-monopole elements and/or the spacing between the elements. Data of this type—which has not been previously available in the literature—will allow the design of folded monopoles having base impedances matching the feed line to the antenna. Thus separate impedance matching sections or devices are not required to feed the antenna. This reduces losses in system efficiency.

Common practice has been to feed most vertical antennas with coaxial cable that has a nominal impedance of 50 ohms. As the base impedance of the usual ¼-λ vertical antenna is only a fraction of 50 ohms, an impedance-matching device is required for efficient operation of the antenna system. Design data has not been previously available that will allow the construction of folded-monopole antennas with predetermined values of base impedance, as, for example, 50 ohms. Data from our tests, presented below, provides this design information.

Folded-Monopole Test Results

The center test frequency was 29.0 MHz. The artificial ground system was the same as described earlier, a 20- × 20-foot counterpoise elevated 5 feet above an extensive paved terrace (see Fig 4). Again, 64 radials were used, with a peripheral wire connecting their ends to eliminate the possibility of resonance in the array.

Equipment included a 5-watt, crystal-controlled RF source (with the output power monitored on a Bird model 43 RF wattmeter), a General Radio 916A RF impedance bridge, and a Drake model SSR-1 receiver. The test equipment was located immediately adjacent to the base of the test antennas so that it was not necessary to extend the lengths of the GR bridge leads, Fig 12. The operators also positioned themselves under the counterpoise, as shown in Fig 13. This arrangement was used because, as indicated earlier, extensive experience had shown that objects *under* a counterpoise are virtually "electrically invisible" to a vertical antenna *above* the counterpoise.

Table 2 shows the base impedances found for two-element folded-monopole antennas having element diameters from 1/8 inch to 2.0 inches, and with spacings between the elements of 2 to 10 inches. The "main element" is that which has its bottom end attached to the counterpoise. The "drop wire" is the fed element. In each case shown, the frequency was varied to achieve zero reactance. For reference, the base resistance for a ¼-λ unipole is included in the data for each basic main-element diameter.

From the theory of images, the input impedance of a folded monopole should be one half that of a folded dipole having the same element dimensions. It will be noted that this theoretical consideration does *not* apply to the values we found.

Since commonly used resistance values of 52 and 72 ohms are compatible with available coaxial feed lines, it is useful to note how these conditions can be obtained. As Table 2 shows, using the commonly available element lengths and diameters, there are many combinations resulting in base resistances at or near 72 ohms, and three at or near 52 ohms.

It is always interesting to check test results with similar work previously undertaken by others. This was done, and the data in Table 1 compared with the *only* other tests

Fig 12—To avoid long test leads, the General Radio impedance bridge is located right at the base of the antenna.

Fig 13—John Frey, W3ESU, at left, and Harry Mills, K4HU, discuss some of the test results.

of the same kind found in the literature—those conducted by Brown and Woodward more than 40 years ago.[20] After converting the Brown-Woodward data to 29.0 MHz, here is what was found. The Brown-Woodward data shows a resistance of approximately 115 ohms for a folded monopole having an A/D ratio (antenna height-to-diameter ratio) of 85. Our tests of an equivalent folded monopole (h/diam ratio of 82, with 1-1/16- × 7/8-inch elements spaced 4 inches) showed 120 ohms base resistance. Brown and Woodward's Graph 6 shows 70 ohms for another configuration of folded monopole. Our equivalent 1/8- × 5/8-inch folded monopole with 4-inch spacing showed 72 ohms.

Practical Application to Amateur Antennas

The impedance-transformation characteristics of folded-monopole vertical antennas offer several tempting possibilities to Amateur Radio operators. The folded-monopole vertical is inherently a broader antenna than a monopole, just as a folded dipole has broader bandwidth than a single-wire dipole. Thus the folded monopole offers significant operational advantage—particularly on the 80- and 160-meter bands.

As has been shown above, the folded monopole, if properly designed, can be constructed to offer base impedance values that will allow direct feed with coaxial line—without the need for an impedance matching device between the antenna and the transmitter. The removal of this matching device and its associated components, cables and fittings will improve overall system efficiency.

In order to tempt experimenters to try the folded-monopole type of antenna, provisional antenna dimensions for the 10- through 160-meter bands are presented in Table 3. It must be emphasized that these are *provisional* values for the reasons given in the note contained in the table. The authors would greatly appreciate comments and specific data from anyone who builds antennas to these dimensions.

Radiation Characteristics of Folded Monopoles

As a matter of curiosity, a number of tests were made to assess the radiation characteristics of folded-monopole antennas. Time was not available for an in-depth study. However, the results were both interesting and unexpected. Further study is certainly indicated.

We were aware, of course, that radiation from folded monopoles should be identical in strength and vertical angle with that from vertical unipoles of the same electrical height. Theory teaches that the radiation mode of a folded monopole is only from the push-push operation illustrated earlier; this is the same for the "fat" unipole and the folded monopole, ie, the relative current distributions are the same. Hence, they should have the same elevation-angle patterns and therefore the same directivity gains. This should hold for antennas above any kind of ground screen, counterpoise or other ground system that approaches the theoretical perfect ground.

In our tests, radiation-angle data was

Table 2
Base Impedance Characteristics of Two-Element Folded Monopoles

Notes on terminology: Drop wires are the fed elements. Main elements are connected to the counterpoise. Spacing was measured from outside of the element.

Main Element Diam	Drop Wire Diam	Diam Ratio	Spacing	Unipole Resistance	Folded Monopole Resistance	Resistance Ratio
1/8 inch	1/8 inch	1:1	4 inches	31.0 ohms	98 ohms	1:3.16
			7		102	1:3.29
			10		105	1:3.39
	3/8	1:3	4		—	—
			7		91	1:2.94
			10		97	1:3.13
	5/8	1:5	4		72	1:2.32
			7		75	1:2.42
			10		89	1:2.87
	7/8	1:7	4		67	1:2.16
			7		72	1:2.32
			10		84	1:2.71
	1-5/16	1:10.5	2		52	1:1.68
			4		60	1:1.94
			7		68	1:2.19
			10		73	1:2.35
	2	1:16	4		54	1:1.74
			7		59	1:1.90
			10		63	1:2.03
3/8	1/8	3:1	4	29.0	140	1:4.83
			7		142	1:4.90
			10		143	1:4.93
	3/8	1:1	4		104	1:3.59
			7		108	1:3.72
			10		111	1:3.82
	5/8	3:5	4		74	1:2.55
			7		94	1:3.24
			10		100	1:3.45
	7/8	3:7	4		73	1:2.52
			7		76	1:2.62
			10		92	1:3.17
	1-5/16	6:21	4		71	1:2.45
			7		73	1:2.52
			10		76	1:2.62
	2	3:16	4		54	1:1.86
			7		70	1:2.41
			10		72	1:2.48
5/8	1/8	5:1	4	28.0	164	1:5.86
			7		166	1:5.93
			10		178	1:6.36
	3/8	5:3	4		111	1:3.96
			7		123	1:4.39
			10		131	1:4.86
	5/8	1:1	4		98	1:3.50
			7		102	1:3.64
			10		105	1:3.75
	7/8	5:7	4		96	1:3.43
5/8	7/8	5:7	7 inches	28.0 ohms	98 ohms	1:3.50
			10		106	1:3.79
	1-15/16	10:21	4		82	1:2.93
			7		88	1:3.14
			10		93	1:3.32
	2	5:16	4		71	1:2.54
			7		73	1:2.61
			10		74	1:2.64
7/8	1/8	7:1	4	27.0	212	1:7.85
			7		218	1:8.07
			10		230	1:8.52
	3/8	7:3	4		137	1:5.07
			7		139	1:5.15
			10		140	1:5.19
	5/8	7:5	4		115	1:4.26
			7		117	1:4.33
			10		119	1:4.41
	7/8	1:1	4		106	1:3.93
			7		112	1:4.15
			10		114	1:4.22
	1-5/16	2:3	4		83	1:3.07
			7		90	1:3.33
			10		95	1:3.52
	2	7:16	4		72	1:2.60
			7		74	1:2.74
			10		75	1:2.89
1-5/16	3/8	7.2	4		150	1:6.00
			7		156	1:6.24
			10		185	1:7.42
	5/8	21:10	4		140	1:5.60
			7		141	1:5.64
			10		142	1:5.68
	7/8	3:2	4		120	1:4.80
			7		122	1:4.83
			10		124	1:4.96
			2		—	2:1.32
			4		89	1:3.56
			7		92	1:3.68
			10		96	1:3.84
2	7/8	16:7	4		—	—
			7		—	—
			10		202	1:8.42
	1-5/16	32:21	4		—	—
			7		—	—
			10		114	1:4.75
	1-7/8	16:15	4		—	—
			7		—	—
			10		92	1:3.83

Table 3

Provisional Folded-Monopole Dimensions

Operating Frequency	Desired Base Impedance	Height of Elements	Main Element Diameter	Fed Element Diameter	Spacing Between Elements
1.8 MHz	50 ohms	116.0 ft	2.0 in.	13.5 in.	23.5 in.
	75			9.75	109.75
3.8	50	56.5	1.0	6.75	11.5
	75			4.75	53.5
7.2	50	29.8	0.5	3.5	6.0
	75			2.5	28.25
14.2	50	15.1	0.25	1.75	3.0
	75			1.25	14.25
21.3	50	10.1	0.20	1.25	2.0
	75			0.75	9.5
29.0	50	7.4	0.125	0.875	1.5
	75			0.625	7.0

Note: These dimensions were derived from the results of tests on 29.0-MHz monopole antennas, using a 64-radial counterpoise as the ground system. Dimensions will vary if monopoles are used with other types of ground systems. Thus it may be necessary to vary the spacings shown to some degree to obtain the base impedances indicated if a monopole is used over some other type of ground.

measured by the use of ½-λ dipoles and associated detectors per Bry.[21] We constructed, tested and calibrated the detectors, and they proved to be well suited for use in our application. Our vertical dipoles, resonant at 29 MHz, were mounted 200 feet from the test antennas at vertical angles of 0 degrees and 17 degrees. Little, but noticeable difference in field strength was measured. The variation was in the range of 20% in favor of the 0° angle. In comparing the relative field strength of a unipole, as compared to several folded monopoles, we found gains as high as 30% in favor of the folded-monopole configuration.

Summary and Acknowledgments

The test programs described in this paper provide a considerable amount of new data on vertical antennas—with important practical implications. Through the use of Fig 6, proper dimensions can now be derived for ¼-λ vertical monopoles over a counterpoise or equivalent ground systems.

This data has not been readily available in the past, so far as we can determine.

The concept of capacitive bottom loading is explained. Illustrations are presented for the practical application of this convenient and efficient method of resonating vertical antennas. The final part of the paper provides—for the first time —data that allows the base impedance of folded monopoles to be selected by design. This data will permit the choice of element sizes and spacings to achieve a wide variety of impedances for matching, phasing and other purposes.

This paper resulted from a privately funded research effort to investigate several unresolved aspects of vertical monopoles and folded monopoles. The material was originally presented at the Technical Symposium of the Radio Club of America in New York City on November 20, 1987. Greatly appreciated data and encouragement for this program were extended by Dr. George Brown, Barry Boothe, W9UCW, Richard Frey, K4XU, John Furr and, particularly, by O. M. Woodward, Jr. All tests were performed by the authors.

Notes

[1] *IEEE Standard Dictionary of Electrical and Electronics Terms*, 3rd ed. (New York: IEEE, 1984).

[2] G. L. Hall, Ed., *The ARRL Antenna Book*, 14th ed. (Newington: American Radio Relay League, 1982), pp 2-2, 2-3, 8-10.

[3] L. A. Moxon, *HF Antennas for All Locations* (Potters Bar, Herts: Radio Society of Great Britain, 1982), pp 109-111.

[4] H. E. Green, "Design Data for Short and Medium Length Yagi-Uda Arrays," *Trans IE Australia*, Vol EE-2, No. 1, Mar 1966.

[5] S. A. Schelkunoff and H. T. Friis, *Antennas Theory and Practice* (New York: John Wiley & Sons, 1952), pp 244-246, 339-343, 386, 401.

[6] J. D. Kraus, *Antennas*, 1st ed. (New York: McGraw-Hill Book Co, 1950), p 314.

[7] H. Jasik, *Antenna Engineering Handbook*, 1st ed. (New York: McGraw-Hill Book Co, 1961), pp 3-21 to 3-24.

[8] G. B. Welch, *Wave Propagation and Antennas* (New York: D. Van Nostrand Co, 1958), pp 180-182.

[9] P. H. Lee, *The Amateur Radio Vertical Antenna Handbook*, 2nd ed. (Port Washington, NY: Cowen Publishing Co., 1984) pp 30, 31, 56, 57, 75-80, and 115-117.

[10] G. H. Brown and O. M. Woodward, Jr., "Experimentally Determined Impedance Characteristics of Cylindrical Antennas," *Proc IRE*, April 1945.

[11] E. A. Laport, *Radio Antenna Engineering* (New York: McGraw-Hill Book Co, 1952), pp 109-111.

[12] A. C. Doty, J. A. Frey and H. J. Mills, "Characteristics of the Counterpoise and Elevated Ground Screen," Professional Program, Session 9, Southcon '83 (IEEE), Atlanta, GA, Jan 1983.

[13] J. H. Richmond, "Monopole Antenna on Circular Disc," *IEEE Trans on Antennas and Propagation*, Vol AP-32, No. 12, Dec 1984.

[14] See note 10.

[15] A. C. Doty, J. A. Frey and H. J. Mills, "Efficient Ground Systems for Vertical Antennas," *QST*, Feb 1983, pp 20-25.

[16] K. Henney, *Principles of Radio* (New York: John Wiley and Sons, 1938), p 462.

[17] See note 1.

[18] O. M. Woodward, Jr., private communication, 1985.

[19] W. van B. Roberts, "Input Impedance of a Folded Dipole," *RCA Review*, Jun 1947.

[20] See note 10.

[21] A. Bry, "Beam Antenna Pattern Measurement," *QST*, Mar 1985.

Steerable Arrays for the Low Bands

By Bob Alexander, W5AH
2720 Posey Dr
Irving, TX 75062

The use of vertical arrays on the low bands is not uncommon. Generally two elements are used, and the phase shift between elements is adjusted by means of a phasing line to provide the desired radiation pattern. Electrically steerable phased arrays that permit placement of a beam peak or null where desired are less common, and generally consist of more than two elements. A steerable, multiband phased array is rare. However, if you think in terms of time delay rather than phase shift, a two-element, multiband, beam-steerable array becomes a relatively simple project.

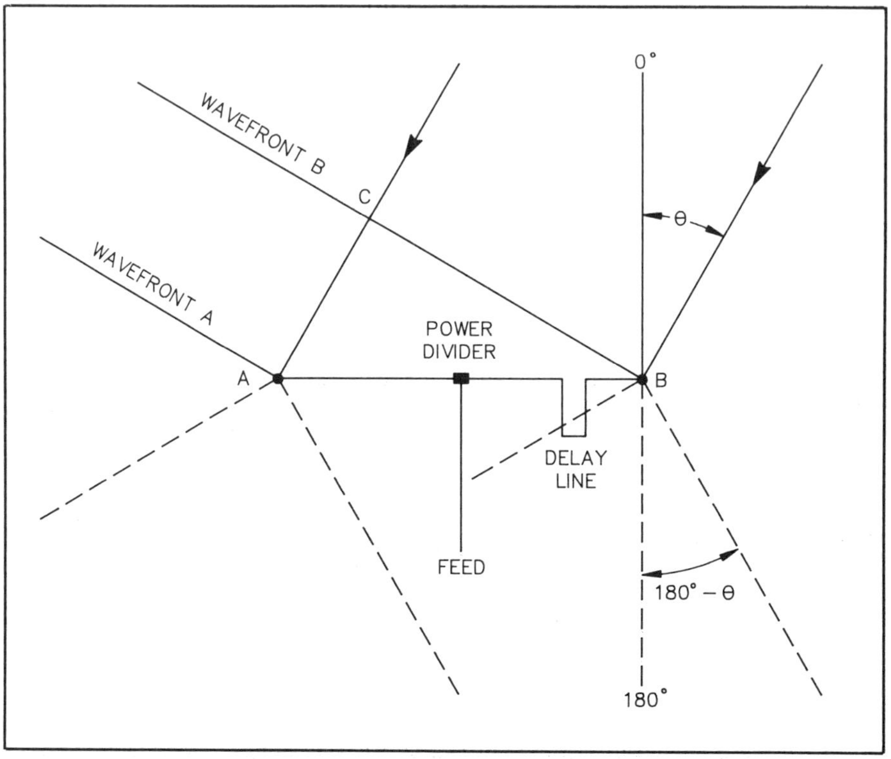

Time-Delay Beam Steering

Time-delay beam steering has been previously described by Fenwick and Schell,[1] and only a simplified explanation is given here. Fig 1 illustrates two verticals with spacing AB. When elements A and B are fed with equal power through equal length transmission lines, radiation is in the broadside direction. In the case of an incoming signal arriving at angle θ, the signal will arrive at element B prior to element A. CA is the extra distance the signal must travel before arriving at element A. When a delay line electrically equal to CA is added to the transmission line to element B, the signals will add in the directions of $0° + \theta$ and $180° - \theta$. (Zero and 180° are arbitrary designations for the broadside direction.) Typical patterns for element spacings between ¼ and 2 λ are shown in Fig 2.[2]

Delay-line length is a function of element spacing and angle θ, and is given by

$$\ell = VF\ S\ \sin\ \theta$$

where
   VF = velocity factor of the cable in use
   S = spacing between elements (same units as ℓ)
   θ = steered angle (less than or equal to 90°)

Since the delay line, and therefore the steered angle, is independent of frequency, operation on more than one band with the

[1]Notes appear on page 15.

Fig 1—Array using time-delay beam steering.

same steered angle can be achieved through the use of multiband verticals and a single delay line. If provision is made to select the amount of delay and the element to which it is applied, 360° time-delay beam steering is obtained.

Circuit Description

The beam-steering system described here is based on the Omega-T 2000C Beam Steering Combiner[3], which is no longer in production. The system is rated for 1500 watts over a frequency range of 1.8 to 7.3 MHz, provided certain antenna requirements are met. These requirements are discussed later.

Equal power to the antenna elements is provided by a broadband power divider, and the amount of delay is adjusted through the use of relay switching. See Fig 3. T1 is a 2:1 transformer used to match the 25-Ω input impedance of the hybrid (T2)

to 50 Ω. C1 serves to cancel the inductive reactance of the input wiring and transformer windings. C2 cancels the reactance in the relay wiring.

The hybrid difference port is terminated with a noninductive resistor, R1. When identical impedances are connected to the hybrid output ports, as in broadside operation or with ideally matched antennas, no current flows through the resistor. If a mismatch exists between the output ports, current will flow through the resistor.

Three delay lines with a combined electrical length equal to slightly less than the element spacing are switched in and out of the signal path by relays K1, K2, and K3. The element to which the delay lines are connected is selected by K4.

W4 on the schematic is 50-Ω coaxial cable. Its electrical length is equal to that of the signal path from the hybrid output through the delay-line switching relays

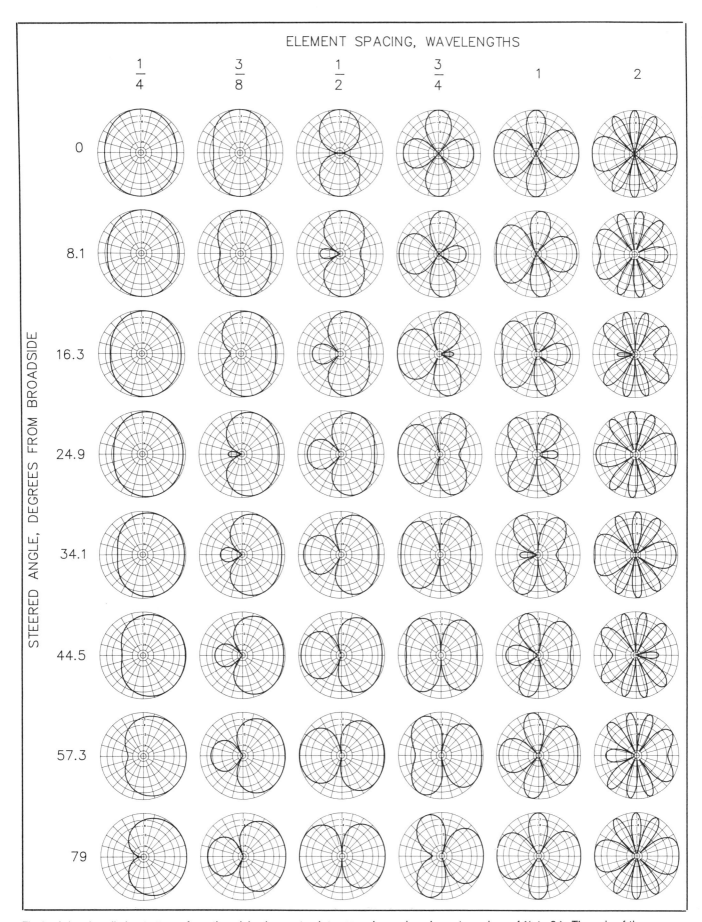

Fig 2—Azimuth radiation patterns for a time-delay beam-steering array when using element spacings of ¼ to 2 λ. The axis of the elements is along the 90-270°line. Responses are relative, shown in decibels. The array gain will vary with spacing and steering angle. Also see Table 1 and note 2.

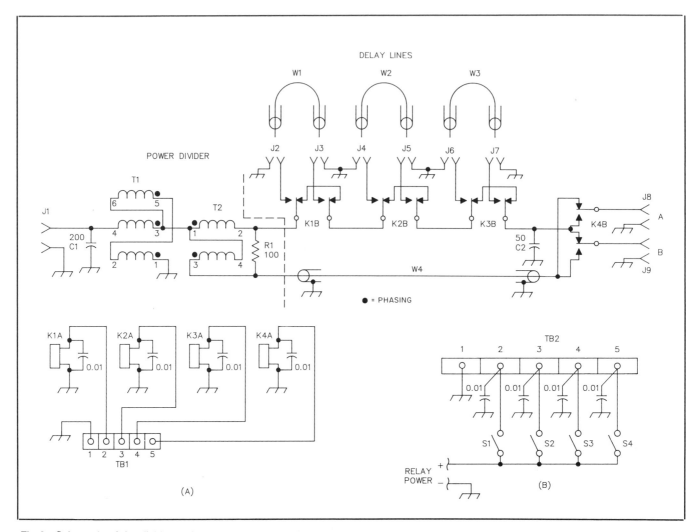

Fig 3—Schematic of the divider and switch box at A. The control unit is shown at B.

C1—200 pF transmitting-type doorknob
  capacitor.
C2—50 pF transmitting-type doorknob
  capacitor.
J1-J9—SO-239 connectors.
K1-K4—DPDT relay (contact rating and insulation must handle 750 watts).

R1—100 Ω, 100-watt noninductive resistor
  with mounting clips, Carborundum
  886-SP101K or equiv.
S1-S4—SPST switch.
T1, T2—F240-61 toroid cores, 4 req'd
  (Palomar Engineers). See text and Fig 6

for winding information.
TB1, TB2—5-position terminal block.
W1-W3—50-Ω coaxial cable (RG-213 or
  similar; see text).
W4—50-Ω coaxial cable (RG-316 or RG-58;
  see text).

(relays not energized) to the antenna relay. It serves to provide an equal time delay from the two hybrid outputs to the antenna jacks. If a direct connection is made, a built-in time delay sufficient to skew the broadside pattern and the steered angles by a few degrees exists.

The control unit has been made as simple as possible, consisting of the relay power supply and four switches, as shown in Fig 3B. S1, S2 and S3 are used in a binary manner to select the desired delay line(s), as shown in Table 1. S1 selects W1 for a steered angle of 8.1° off broadside. Switching in W2 while switching out W1 steers the beam to 16.3°, and so on until all lines are selected, providing 79° beam steering. A steered angle of 90° is not needed since end-fire radiation patterns are quite broad. S4 switches the delay lines

from one element to the other to reverse pattern directions.

The choice of 8.1° for the minimum steered angle is dictated by several factors. The delay-line segment lengths must be related to each other in a binary manner in order to produce smoothly increasing amounts of delay. Since the equation for delay-line length contains a sine function, then making W2 = 2 × W1 and W3 = 4 × W1 to obtain the required binary relationship results in angles greater than 2 or 4 times that chosen for determination of W1. At less than 8.1° end-fire radiation is not obtained. Therefore, the formula for determining line lengths becomes

$$W1 = VF \ S \sin 8.1$$
$$W2 = 2 \times W1$$
$$W3 = 4 \times W1$$

Table 1
Switch Position Versus Steered Angle

S4	S3	S2	S1	Steered Angle
0	0	0	0	Broadside
0	0	0	1	8.1 and 171.9
0	0	1	0	16.3 and 163.7
0	0	1	1	24.9 and 155.1
0	1	0	0	34.1 and 145.9
0	1	0	1	44.5 and 135.5
0	1	1	0	57.3 and 122.7
0	1	1	1	79 to 111 (end-fire)
1	0	0	0	Broadside
1	0	0	1	351.9 and 188.1
1	0	1	0	343.7 and 196.3
1	0	1	1	335.1 and 204.9
1	1	0	0	325.9 and 214.1
1	1	0	1	315.5 and 224.5
1	1	1	0	302.7 and 237.3
1	1	1	1	281 to 259 (end-fire)

where the W numbers = the cable length in the same units as S, VF = velocity factor and S = element spacing.

Construction

The divider and switching relays are housed in a 3 × 8 × 12-inch aluminum box. Component layout has a definite effect on the bandwidth of the power divider. The parts arrangement shown in Fig 4 should be adhered to as closely as possible.

The autotransformer and hybrid are each wound on stacked F240-61 cores prepared as shown in Fig 5. A coating of thermal grease is applied between the cores and aluminum bracket. The cores are centered on the 1.4-inch hole in the bracket and secured with tie wraps. One side of the toroids is then double wrapped with Scotch no. 61 TFE or similar tape. It should be noted that the cores will overlap the edge of the bracket by 0.2 inch on each side.

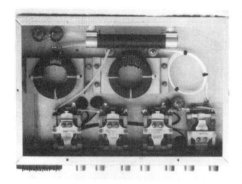

Fig 4—Interior view of the divider and switch box.

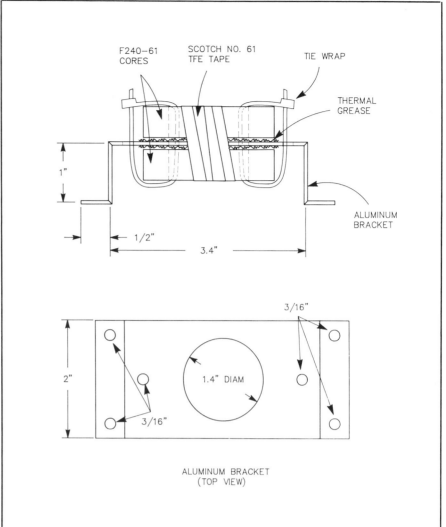

Fig 5—Transformer and hybrid assembly.

The transformer and hybrid are wound as shown in Fig 6. The transformer is trifilar wound with 5 turns in each of the outside windings and 4 turns in the center winding. The hybrid has 10 bifilar turns of twisted no. 18 enameled wires. The wires are twisted 3 times per inch prior to winding on the cores. In each case windings should be tight around the cores and bracket, and as close as possible to adjacent turns.

Transmitting-type doorknob capacitors are used at C1 and C2. Two such capacitors were connected in parallel at C1 to obtain the required capacitance. C1 should be mounted as close as possible to T1. Connections are made via solder lugs secured to the tops of the capacitors. Braid (from RG-58) with Teflon sleeving over it is used for the connection between J1 and T1-C1. Likewise, braid is used between T1 and T2.

Two ceramic standoffs, mounted adjacent to T2, are used for the hybrid output connections. The mounting clips for R1 are affixed to ceramic standoffs secured to the

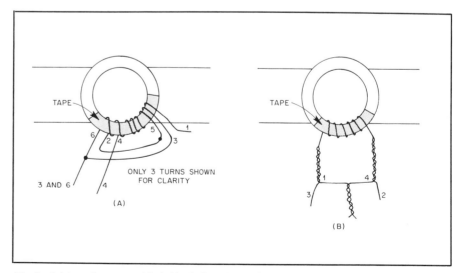

Fig 6—2:1 transformer and hybrid winding detail. Core and brackets are assembled as shown in Fig 5. The transformer (A) is 5 trifilar turns of no. 18 enameled wire, with 1 turn removed from lead no. 4. Lead 1 is ground, lead 4 is the 50-Ω input, and leads 3 and 6 are the 25-Ω output. The hybrid (B) consists of 10 bifilar turns (twisted 3 times per inch). See text and Fig 3.

Fig 7—Power divider and switch box (left) and control box (right).

Fig 9—The array installation for 80- and 40-meter operation using Butternut HF2V verticals.

Fig 8—Power divider and switch box with all cables connected.

side of the box opposite the relays. They should be positioned so the leads to T2 are short and of equal length.

The relays are mounted as close as possible to the coaxial connectors, as the connections between them must be short and direct. C2 is mounted between K3 and K4. The connection from T2 to K1 is made by using braid with Teflon sleeving over it. All connections between relays are made using braid.

W4 is 50-Ω coax cable. Its length is determined by measuring the length of the leads used from T2 to K1 to K2 to K3 to K4. The length of the signal path through K1, K2, and K3 is also added in (relays not energized). Cable length is the measured length of the signal path times the velocity factor of the cable. RG-316 was used here because of its small size and high power-handling ability. However, RG-58 can be used.

TB1 is mounted on the same side of the divider box as the coax connectors. All connections between it and the relays are bypassed with 0.01-μF capacitors.

The relay power supply is contained in a separate control unit. It is not shown on the schematic nor described here, as supply voltage is dependent upon the relays used. S1 through S4 are mounted on the front panel, along with a power on-off switch (not shown on the schematic). A five-lug terminal block is placed on the rear of the box and each terminal is bypassed with a 0.01-μF capacitor.

The assembled boxes are shown in Fig 7. Fig 8 shows how the power divider and switching box is installed. The complete array appears in Fig 9.

Testing

After assembly is completed and all wiring checked for errors, the input SWR should be measured. Terminate J8 and J9 with 50-Ω loads (each load should be capable of dissipating half of the maximum input power). Apply low power at J1 and measure the input SWR at the bottom, middle and top of the 160-, 80-, and 40-meter bands. In all cases, the SWR should be less than 1.5 to 1. Worst-case SWR measured on the unit described here was 1.20 at 7.3 MHz. An SWR of 3 or higher results if either J8 or J9 is not terminated in 50 Ω, or if an open circuit condition exists between J8, J9 and the hybrid. If all appears normal after the

Delay-line Design for Steering to 90 Degrees

Author Bob Alexander, W5AH, presents an excellent idea for a directional array that uses two vertical antennas. Bob chose to steer from 0° (broadside) to 79°. How would you go about modifying his design if you wished to steer to 90°?

The answer is to change the length of the delay lines so that their total electrical length is equal to the physical spacing between the antennas. To do this you will need three binary weighted lengths (as Alexander has pointed out). You could think of them as binary digits. The sum of the digits is seven (1 + 2 + 4 = 7). That means that the delay-line electrical lengths are 1/7, 2/7 and 4/7 of the physical spacing between elements. The delay/spacing ratio is the sine of the arrival (or steering) angle, θ. The following table shows the relationships for a system designed to steer to 90°.

Fraction	(Sine)	Angle, Degrees
0/7	0.0000	0.00 (Broadside)
1/7*	0.1429	8.21
2/7*	0.2857	16.60
3/7	0.4286	25.38
4/7*	0.5714	34.85
5/7	0.7143	45.58
6/7	0.8571	59.00
7/7	1.0000	90.00 (End fire)

* = delay-line lengths

To determine actual line lengths, multiply the value in either the fraction or sine column (they're the same) times the physical spacing between elements times the velocity factor of the coaxial cable used for the delay lines.—*Chuck Hutchinson, K8CH*

tial measurements, repeat them using increased power. There should be no change in SWR.

Next, connect the control unit to the divider and switch box with the 5-conductor cable to be used in the final installation. Apply power and visually check that the correct relays are energized in each of the switch positions.

Antenna Requirements

The two antenna elements must have the same SWR with respect to each other at all frequencies of operation. In other words, the SWR curves should be identical. The impedance of each antenna with the other antenna removed should be as close to 50 Ω at resonance as is practicable.

They should not be spaced more than 2 λ at the highest operating frequency nor closer than ¼ λ at the lowest frequency. For 160- through 40-meter operation, the elements should be spaced between ¼ and ½ λ on 160 meters (1 to 2 λ on 40 meters). For best performance, the antennas should not be near any surrounding metal objects.

Installation

Once the antenna spacing has been selected, the three delay lines are fabricated from low-loss 50-Ω cable using the formulas given earlier. Length is measured from tip to tip of the coax connector center conductors, and must be as near as possible to the calculated values.

The divider and switching box is placed midway between the antennas. Feed lines from the box to the antennas can be any length so long as they are of equal length. Hookup is as shown in Fig 3.

Operation

The SWR seen by the transmitter will vary somewhat as the steered angle is changed. Although the SWR seen at the transmitter may be quite low, operation on frequencies where the antenna SWR exceeds 2:1 should be avoided, as damage to the divider can result. The impedance seen at the transmitter and that seen by the divider is the same only for broadside operation.

On receive, the pattern peaks and nulls will be most noticeable on signals arriving at relatively low angles. With signals arriving at high angles, the steered angle selected for best reception may not accurately indicate the azimuth of the transmitting station.

Table 1 lists the switch positions and corresponding steered angles. The increasing rate of change in the steered angle is compensated for by the fact that the beamwidth increases as the steered angle nears the end-fire direction.

Summary Comments

1) Element spacing need not be exactly ¼ or ½ λ, or any of the values listed in Fig 2. An element spacing of, for example, 0.32 or 0.86 λ will work just as well.

2) All coaxial cables, especially those used for the delay lines, should be of the low-loss variety. Lossy cables will hurt performance, especially where null depths are concerned. A loss of 0.8 dB in the delay lines will limit nulls to 20 dB or less.

3) Never change beam headings while transmitting, as the relays may be damaged.

4) Since the divider and switch-box assembly is to be mounted outdoors, the box and all connectors should be weatherproofed.

5) The beam-steering method described here is not new. It has, however, seen little use. Hopefully, the material presented here will encourage the use of this older technology.

I wish to express my thanks to Dick Fenwick, K5RR, for his assistance in preparing this article.

Notes

[1]R. C. Fenwick and R. R. Schell, "Broadband, Steerable Phased Arrays," *QST*, Apr 1977.

[2][The theoretical patterns of Fig 2 are based on no losses and equal *currents* in the elements, rather than equal power. Because of mutual coupling, element feed-point impedances will not be the same except at a steering angle of 0°, and equal power will therefore yield unequal currents. Further, the phase relationship of the currents may not be as intended. As a result, the lobes may be skewed slightly, and the deep pattern nulls shown in Fig 2 may not be realized in practice. Detailed information on this phenomenon appears in Chap 8 of the 15th edition of *The ARRL Antenna Book* (1988).—Ed.]

[3]The Omega-T 2000C Beam Steering Combiner is a patented product of Electrospace Systems Inc, Richardson, TX.

A Steerable Array of Verticals

By S. L. Seaton, K4OR
460 Windmill Point Rd
Hampton, VA 23664

T he array described in this paper can be used to cover the 40-, 20- and 15-meter bands with 5 dB gain in four (switchable) cardinal directions. A self-supporting vertical antenna (see Fig 1) is fed by a buried RG-8 coaxial cable from a Transmatch, as shown in Fig 2. It provides a good match on the 10- through 40-meter bands. The last 18 inches of the coaxial cable is fanned out to make an impedance transformation (Fig 3). The added length also aids in connecting to the ground system and the base of the antenna. A horn gap at the base of each element provides lightning protection.

This particular antenna is near the beach, and the 8-foot stainless-steel ground rods go down into the salt-water table. In addition, buried radials assure a reliable ground plane.[1] In less favorable locations a more extensive ground-radial system is needed.

Fig 1—Four of the free-standing antenna elements. The driven element is in the center of the array, near the tree.

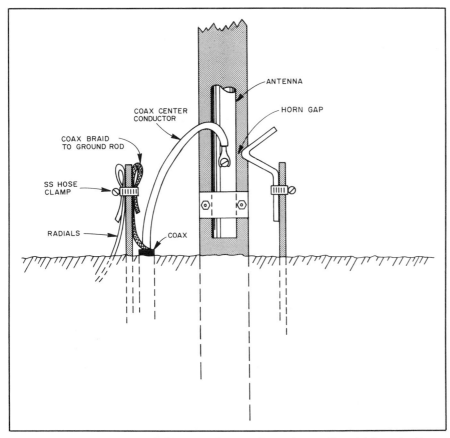

Fig 2—The driven element is fed by an underground run of coax. Use stainless-steel hose clamps but make sure the screw in the clamp is made of the same material—some are not. The horn gap is a lightning-protection device; see text.

Rotating the Beam

An orthogonal basis of a vector space is achieved in three bands by adding parasitic elements (see Fig 4) and a switching network at the base of each. A parasitic element acts as a reflector or a director, depending on its spacing from the driven element and upon its tuning. I used a convenient spacing of 16.33 feet and a tuning network at the base of each parasitic element.

[1]Stainless-steel radial wire is available from Defender Industries, 225 Main St, PO Box 820, New Rochelle, NY 10802-0820. To cut this wire, tape it about 2 inches then cut with a sharp cold chisel. Stainless steel is compatible with aluminum.

Fig 3—The base of the driven element. The wire projecting from the ground rod and almost touching the base of the antenna element is the horn (air) gap for lightning protection.

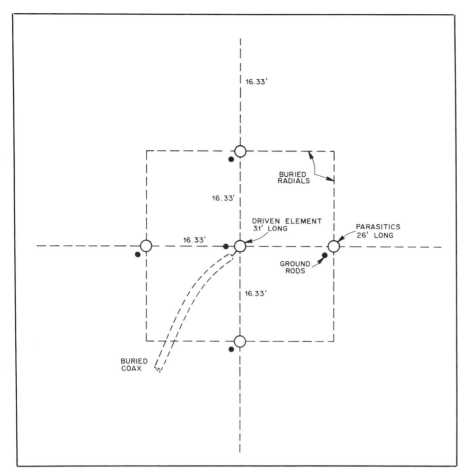

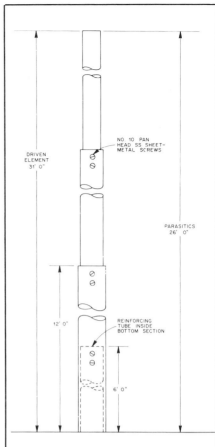

Fig 4—The vertical array and radial system. In this installation the radials are shown as dashed lines. Other situations without the availability of an excellent salt-water ground plane may require a more extensive radial system. The radials are 1/8-inch diameter stranded stainless-steel wire. The stainless-steel ground rods are 3/8-inch diameter × 8 feet long.

Fig 5—Element assembly. For a listing of aluminum tubing sizes see *The ARRL Antenna Book*, 15th edition, page 21-7, Table 5. Order all sections of 6061-T6 aluminum tubing with 0.058-inch wall thickness. You will need five 12-foot pieces for each of three outside diameters: 1-5/8 in., 1-3/4 in. and 1-7/8 in. Three more 12-foot pieces are needed with an outside diameter of 1-3/4 in. for reinforcements. Cut these into 6-foot lengths. Sheet-metal screws will keep the tubing in place and help maintain electrical contact. To fit the tubing together, first chill the inner tube and warm the outer tube.

The driven element is 31 feet long and the parasitic elements are each 26 feet long (Fig 5). The switching network is mounted on a Plexiglas® plate and consists of four banana sockets. One socket leads to ground, one to an open-ended 30-turn coil that is space wound one wire diameter on a 2-inch diameter form, another to ground through a similar coil of 9 turns, and the fourth is open (see Fig 6). A flexible lead connects the base of the antenna to a banana plug that is inserted into the selected socket (Fig 7). The method of mounting the antenna elements is shown in Fig 8.

For the 20-meter band the director is open ended and the reflectors (rear and sides) terminate in the open-ended 30-turn coils. On the 40- and 15-meter bands the director goes to ground and the reflectors (rear and sides) go through the 9-turn coils to ground. Rotation of the beam direction is done by re-plugging.

On the other bands this low-radiation-angle array gives the omnidirectional pattern of a vertical; however, it is possible to add elements to the base tuning network to accomplish directivity on the other bands if one wishes to do so, Fig 9.

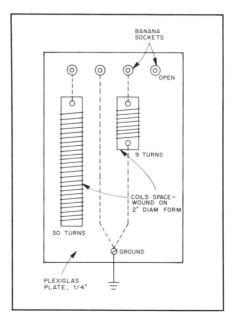

Fig 6—All tuning assemblies for the parasitic elements are mounted on a Plexiglas plate. The coils are space wound on 2-inch diameter forms such as plastic PVC pipe. If you're not sure about the dielectric properties of the coil-form material, you can test it in a microwave oven.

My antenna is exposed to high winds. In gales the elements look like loaded fishing poles, but have withstood 90-mi/h winds without taking a permanent set. Nevertheless, it is easy to lower the antennas in the light winds before a hurricane strikes.

While the parasitic elements lie on the locus of points describing a parabola, they are too far apart to act as a parabolic reflector, but tuning the side parasitic elements as reflectors greatly reduces the off-axis response, thus cutting down on interference (see Fig 9).

History

A three-element reversible array helped maintain long- and short-path schedules with western Australia and with New

Zealand for the better part of two decades on the 40- and 20-meter bands. Then ambition struck! The three-element array of verticals would not rotate without physically moving the parasitic elements. Now wait a minute—with two more post holes for a couple more verticals and some brain to replace the brawn, a 5-element array of verticals *will* rotate, electrically, that is.

Fig 7—To change the antenna pattern, just move the banana plug.

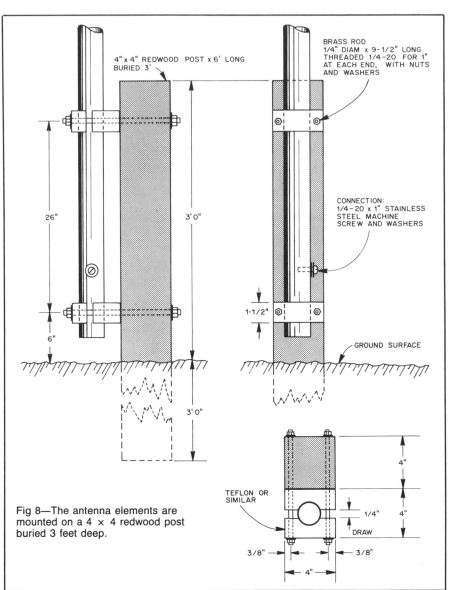

Fig 8—The antenna elements are mounted on a 4 × 4 redwood post buried 3 feet deep.

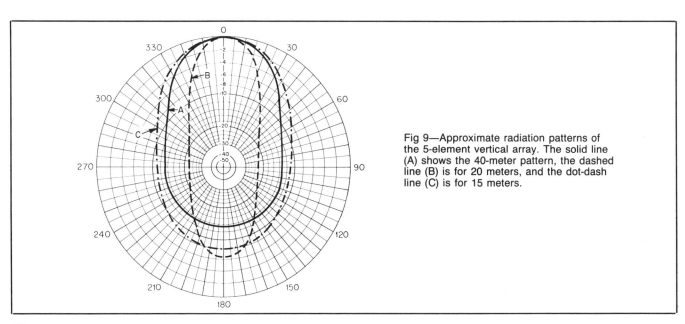

Fig 9—Approximate radiation patterns of the 5-element vertical array. The solid line (A) shows the 40-meter pattern, the dashed line (B) is for 20 meters, and the dot-dash line (C) is for 15 meters.

The Robert Tail—A DX Antenna

By Robert L. Brewster, W8HSK

2158 Westminster Rd
Cleveland Heights, OH 44118

During construction of the two towers at the new QTH of Tom Lee, K8AZ, the antenna configuration for each tower was decided upon for 10 through 40 meters. The north tower, closest to the shack, would be shunt fed on 160 meters. What about the 80-meter antenna?

Tom's towers are 120 feet high and 325 feet apart. The plane of the towers is broadside to Europe and the South Pacific. Several broadside 80-meter arrays were discussed and it was evident that whatever antenna was installed had to be confined within the plane of the towers because of the close proximity of the trees.

Development of Design

I gave Tom a revamped antenna design that I had used in W3 land many years ago. It was based on a design by Woody Smith, W6BCX.[1] I inverted a 40-meter bobtail, kept the phasing sections off the ground, and isolated the center vertical from the phasing lines.

Fig 1A illustrates only the current distribution. Fig 1B shows how the phase geometry of the Robert Tail is developed when a ¼-λ section from each end is rotated 90° with a coax-fed ¼-λ element added to the center in the same plane. The verticals are ½-λ apart.

The center of the total phasing line is a current loop (low impedance) and suitable for 50-Ω coax feed. The center conductor of the coax feed line is connected to the center vertical. The coax shield connects to the center of the phasing sections, which in turn connect to the two outside verticals.

The array beam pattern is a figure 8. Maximum response is perpendicular to the plane of the verticals, with 5 dB of gain over a comparative vertical groundplane antenna with similar foreground conditions. The displacement of the phasing lines from the verticals plus the current reversals at the center of each ½-λ phasing section result in very little radiation from the horizontal sections. This adds virtually nothing to the horizontal directivity of the array.

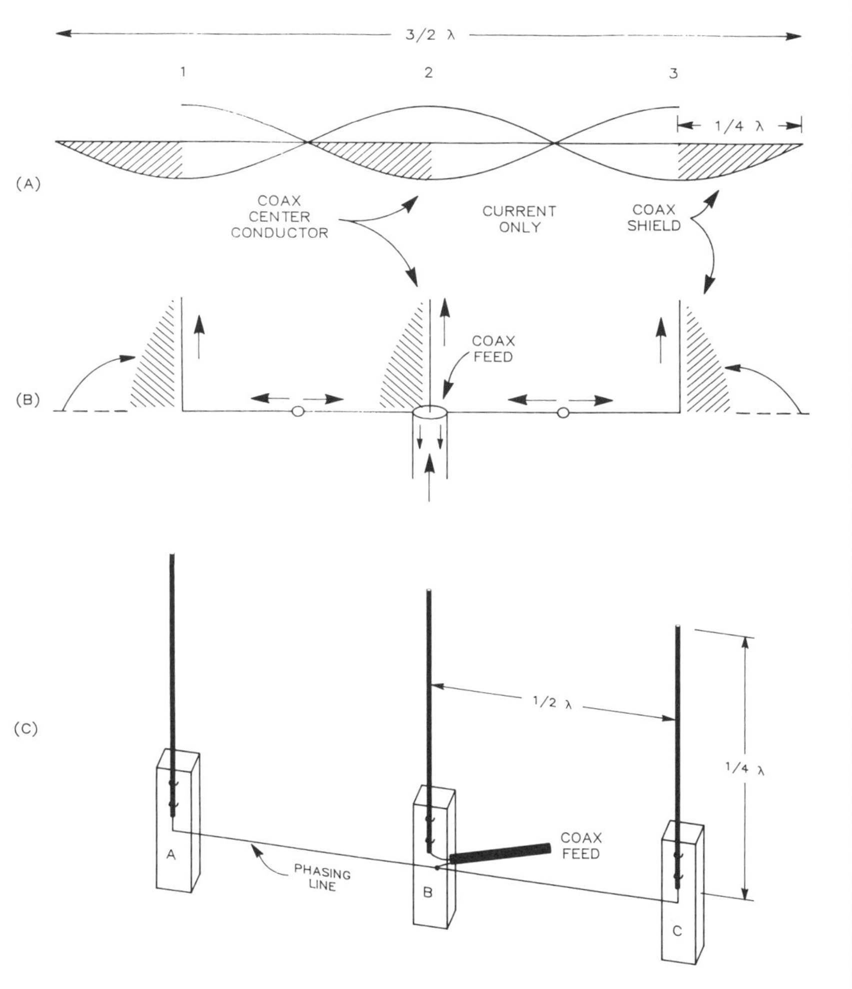

[1]Notes appear on p 24.

Fig 1—At A, current distribution on a 3/2-λ harmonic wire coincident with a coaxial-fed 1/4-λ vertical placed in the middle of the center current loop. At B, the phase relationship; currents in all three verticals are in phase and additive in the broadside direction. The beam pattern is perpendicular to the plane of the array. At C, the 40-meter version, the first Robert Tail array. It consisted of three self-supporting aluminum 1/4-λ verticals supported on ground-mounted vertical 10-foot 4 × 4-inch posts, 68 feet apart.

A previous 40-meter array, Fig 1C, consisted of three self-supporting aluminum ¼-λ verticals supported on 10-foot wooden 4 × 4 posts 68 feet apart. This design eliminated the tuner and the ground-radial system. It provided direct coax feed and proved to have gain in the broadside direction over a single ground-mounted ¼-λ vertical with 24 radials.

Tom and I proceeded to round up the necessary materials for an 80-meter wire array to be hung from a rope strung between towers. He found a roll of appliance wire, 1100 feet long and rated at

220 volts, at the local electronics surplus store. The tower pulleys were hung at the 120-foot level, and a ½-inch rope was strung on the pulleys between the towers. The total length of the rope is over 800 feet, providing enough slack to lower it to ground level for measurement and attachment of the verticals.

The array wire was laid out on the ground, measured and cut to the dimensions shown in Fig 2, with several inches added for insulator loops. To determine the total length of A_1 and A_2, use the harmonic formula based on 3/2 λ.

$$A_1 = \frac{492(3 - 0.05)}{3.52 \text{ MHz}} = \frac{1451.4}{3.52}$$
$$= 412\,'4'' \qquad \text{(Eq 1)}$$

$$A_2 = \frac{492(3 - 0.05)}{3.79 \text{ MHz}} = \frac{1451.4}{3.79}$$
$$= 383\,'0'' \qquad \text{(Eq 2)}$$

where

A_1 = total wire length (feet) for the 80-meter elements and phasing lines

A_2 = total wire length for the 75-meter elements and phasing lines

The center verticals are 5 inches apart and cut to the length calculated from

$$X = \frac{234}{3.79 \text{ MHz}} = 61\,'9'' \qquad \text{(Eq 3)}$$

$$Y = \frac{234}{3.52 \text{ MHz}} = 66\,'6'' \qquad \text{(Eq 4)}$$

where

X = center-element length for 75 meters
Y = center-element length for 80 meters

The centers and corners were marked with colored tape. The support rope was also measured and taped at the appropriate points. The corner insulators were strung on the wire before the end insulators and hanger ropes were installed. The hanger ropes had to be adjusted to allow for the slack in the ½-inch rope to keep the phasing sections parallel to the ground. The phasing sections are 30 feet above ground (Fig 3). A horizontal reference dipole was hung at 70 feet in the same plane as the array. We would have preferred an 80-meter ¼-wave vertical with ground radials as a reference antenna, but that was not feasible.

Fig 2 shows the present configuration of this array. The prototype design was hung in late 1985, and featured a common phasing line tying all four outside verticals together with the center verticals fed in common. It was our intent to see if two discrete resonances could be realized, one in the CW band DX segment and one in the phone band DX segment. Subsequent testing and SWR measurements ruled out the use of this prototype design. The array design was changed to separate three-element arrays. The physical dimensions were set to make the outside array resonant

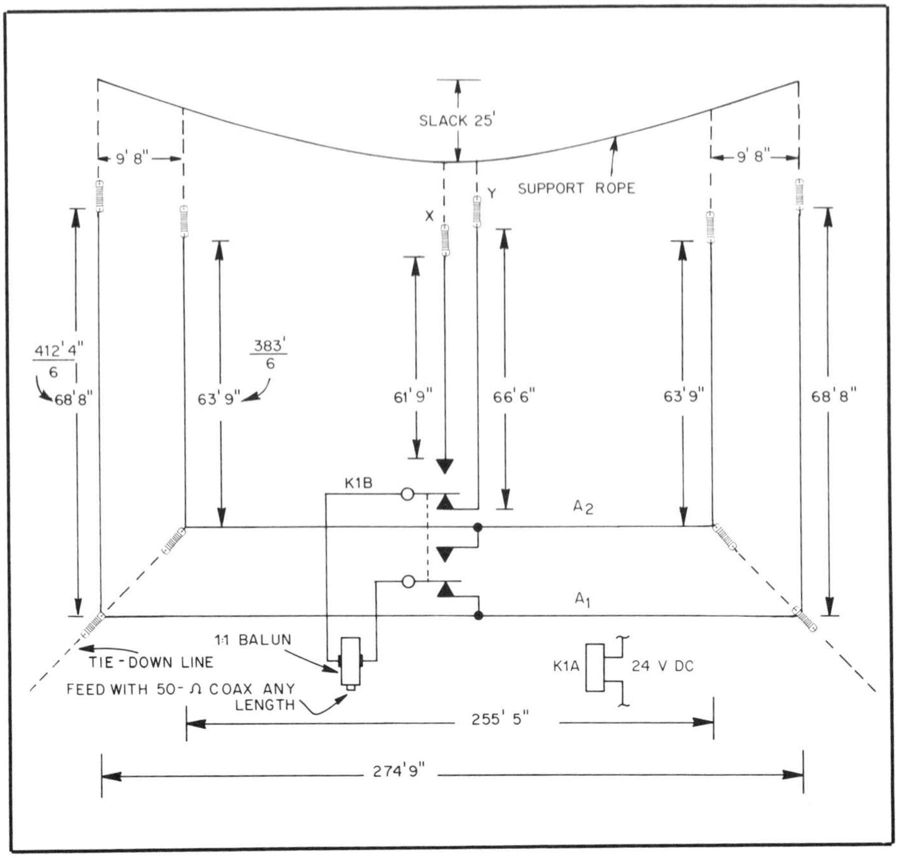

Fig 2—The 75- and 80-meter Robert Tail. Design equations given in the text can be used to determine array sizes for other bands.

K1—Shown in deenergized (CW) position. A DPDT relay can be used as shown here, or two SPDT relays can perform the same function (see text and Fig 6).

at the CW band DX segment and the inside array resonant in the phone band DX segment. The actual SWR curves for this configuration are shown in Fig 4.

Phone or CW Selection

Relay switching was added to the feed systems to permit instant selection of either the CW or phone array from the operating position. One relay selects the proper phasing line coincident with the other relay, which selects the corresponding center vertical. These relays and similar ones are available on the surplus market for under $5. The relays are operated at a low-voltage point and the insulation proves adequate at full legal power levels.

A 1:1 balun, together with the relays, are mounted on a Plexiglas® sheet (Fig 5), which is attached to and used as a spreader for the two center verticals and to support the center of the phasing sections. A new feed system has been built and is discussed later.

Design Considerations

Some considerations influenced the design of the Robert Tail, such as (1) how much is lost in efficiency, low-angle radiation and overall performance as the current loops are lowered ¼ λ when the phasing section is raised off the ground and no

Fig 3—The center verticals are hung from the center of the support rope and terminate at the feed panel. The insulated phasing lines are taped together for about 50 feet either side of the feed panel to reduce snagging on tree branches. There is no apparent adverse effect on the array performance.

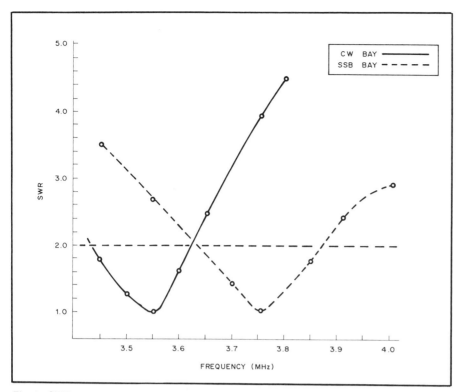

Fig 4—SWR curves based on actual readings. The operational bandwidth at a 2:1 SWR is about 200 kHz for both CW and SSB arrays. These results were recorded before the ferrite-bead choke balun was installed at the array input terminals.

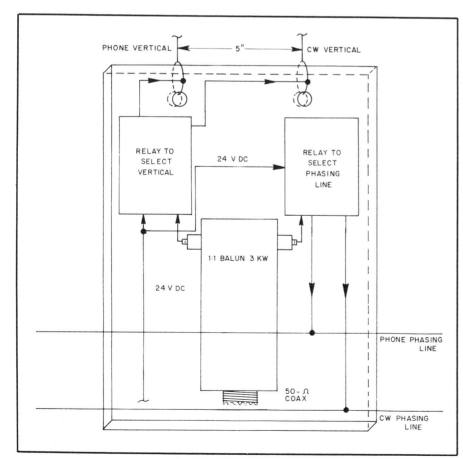

Fig 5—The feed-system panel is Plexiglas, 8 inches wide by 15 inches long and ¼ inch thick. The 24-V dc power supply line common to both relays is energized for 75 m and off for 80 m.

radial system is used, (2) elimination of the tuner at the base of the antenna is a distinct advantage, and (3) providing direct coaxial feed. Earlier I stated the array drive-point impedance would be suitable for 50-Ω coax feed. A single ¼-λ vertical, resonant and over a perfect ground system, has a self-impedance of 36 Ω. With the many configurations of ground-mounted vertical phased arrays, there is much concern about self-impedances matching the drive impedances within each segment of the arrays.

The mutual coupling between the three elements in the Robert Tail array and the resulting mutual impedances combined with ground losses make the determination of the drive-point impedance of this array quite complex. The summary effect of these array parameters results in raising the 36-Ω self-impedance to a higher value approaching 50 Ω. Characteristic of coax-fed vertical phased arrays is the worsening performance as the operating frequency is moved further from the array design frequency. This is because the drive-point impedance changes, with the resulting increasing mismatch. The vertical phase relationship deteriorates, resulting in a higher SWR and lower array gain. For these reasons, the 75-meter SSB array was placed inside the 80-meter CW array to maintain optimum SWR and gain figures in the respective DX segments of the band. The results have been outstanding.

An 80-meter dipole at 100 feet has its maximum vertical gain at 40° above the horizon. In the northern Ohio area the angles of most 80-meter incoming DX signals lie between 15 and 45 degrees. The difficulty of increasing dipole height for a lower angle of radiation makes the vertical phased array attractive from the construction standpoint.

Comments on Performance

The Robert Tail vertical phased array hears quieter than the reference dipole. There is no apparent noise pickup from the towers, in part due to the deep nulls off the ends of the array.

The Russian beacon (3.635 MHz, sends ..— with a long dash) can usually start to be heard about 2000Z (3:00 PM EST) on the array if the QRN is low. The reference dipole hears the beacon about 1½ hours later. Using a barefoot TS-830, comparative reports from DX stations showed a consistent 3- to 6-dB improvement using the array. In eastern Europe and Asiatic Russia there were more significant differences. When they could not hear signals from the reference dipole, signals from the array were often very readable.

The Robert Tail is not a Sweepstakes antenna and compares poorly with the reference dipole out to about 1500 miles. However, this is a big plus in the DX Contest. This array has considerable attenuation of high-angle signals, which has been demonstrated during the CQ WW and

ARRL DX Contests. It is not difficult to hear through the big guns on the East Coast, and it's a nice feeling during European sunrise to run Europeans simultaneously with ZLs and VKs.

Planned Changes

Installation of a ground radial system is under way. It will consist of several ground rods placed in a line directly under the Robert Tail and wired together. The radials will connect to the ground rods and be placed perpendicular to and on both sides of the plane of the array. The coax feed-line outer shield will be grounded to the radial system at ground level directly under the array feed box.

This ground system should reduce ground losses in the near field of this array but will have little effect on the low-angle pattern characteristics. The system will provide added lightning protection. In most areas, while driven ground rods are effective for lightning protection, their usefulness is often questionable at RF, especially in dry soil and/or at frequencies above 2 or 3 MHz. The radial wires, however, may offer measurable improvement at any location.—*Ed*.]

The phase relationship in the Robert Tail is critical to its performance. Our experience plus published problems with baluns have prompted a change. The 1:1 balun originally used in this feed system has been replaced by a ferrite-bead choke balun (Fig 6) using the construction data contained in an article by Walt Maxwell, W2DU.[2] The new feed-system circuit is identical to the circuit shown in Fig 5 except that the 1:1 balun is replaced by the ferrite-bead choke balun. The test results with the ferrite-bead choke balun installed confirms Walt's observation that the operational bandwidth is improved and no power loss is observed.

Referring to Fig 4, the SWR curves shown are actual readings taken before the ferrite beads were installed. With the beads installed the operational bandwidth at a 2:1 SWR increases an additional 40 to 50 kHz for each band segment.

A 24-V dc supply powers the relays from the operating position. The two wire nuts shown in the lower left corner of the feed box (Fig 6) will connect to the 24-V dc line that enters the box through the bottom lower left-hand corner.

When the power supply is off, both relays are in the CW mode. The right-hand relay connects the CW center vertical with the center conductor of the choke-balun coax. The left-hand relay connects the CW phasing line to the outside shield of the choke-balun coax. With the power supply on, both the relays switch to the phone mode.

The relays are attached to the back of the feed box with Velcro® strips which permit easy removal. Note that within each relay, all common terminals are shorted together to increase contact area. Spade-type female terminals afford easy connect/disconnect access to the relay terminal lugs. The SO-239 panel connector provides coaxial-cable input to the feed box. The feed-box lid, and all internal connections, are sealed with Dow-Corning RTV.

The performance of antennas using baluns in the feed line, and their effect on signal levels delivered to the receiver input, is rarely discussed. Simply stated, as the outer shield surface skin-effect current increases, the signal level delivered to the receiver input decreases. When properly installed, the choke balun will reduce this problem.

A parts list for the feed box is included in Table 1.

Use on Other Bands

The Q of the Robert Tail favors use at 40 and 20 meters and the WARC bands. A three-element array will provide operation over a large portion of each band with less than a 2:1 SWR. The use of two-, three- or five-element arrays in various configurations offers interesting design possibilities for around-the-compass coverage or super-gain antennas.

The following are sketches of array configurations that I have not modeled nor built but might create some interest. The values given for beamwidths and gain are estimated.

Fig 7 is a switchable four-element broadside array that consists of three sets of two-element arrays with one element common to all three arrays. Each selectable array is

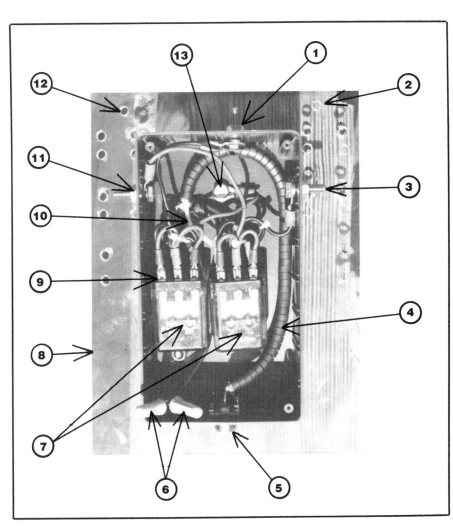

Fig 6—A close-up view of the new feed system with the choke balun. It is made from a 13-inch length of RG-141 with the outer jacket removed and 50 ferrite beads strung on the bare outer coax shield. The drilled holes in the Plexiglas panel are tie-down points for the phasing lines, coax cable and 24-V dc power line. Legend:

1—Connection to phone phasing line
2—Bottom tie point for phone vertical
3—Connection to phone vertical
4—Ferrite beads, no. 2673002401-0; 50 req'd.
5—SO-239 panel connector
6—Wire nuts on 24-V dc line
7—Relays
8—Plexiglas panel, 12 × 7 × ¼ in.
9—Common terminals shorted together
10—RG-141 coaxial cable
11—Connection to CW vertical
12—Bottom tie point for CW vertical
13—Connection to CW phasing line

Table 1
Feed-Box Parts List

Feed box, Radio Shack cat no. 270-224, plastic, 7-1/2 in. × 4-3/8 in. × 2-3/8 in.
Relays, Potter and Brumfeld model KUP11D15, 24-V dc, 10-A contacts, clear polycarbonate cover, solder-lug terminals.
Ferrite beads, no. 2673002401-0, Fair-Rite Product Corp.
Coaxial cable, RG-141, length 13 in.
Quick-slide female terminals, HWI no. 534005 3/16 in., quantity 18, hardware store item.
Coax connector, chassis mount SO-239.
Relay mounts, hoop and loop fasteners, Radio Shack cat no. 64-2345.
Weatherproof seal, Dow-Corning RTV, hardware store item.

Table 2
Array Dimensions

Design MHz	Outside Verticals	Center Vertical	Phasing Frequency Line Length
7.125	34'	32' 10'	135' 8'
10.100	23' 11'	23' 2'	95' 10'
14.125	17' 1'	16' 7'	68' 8'
18.125	13' 4'	12' 11'	53' 5'
21.200	11' 5'	11' 1'	45' 7'
24.925	9' 8'	9' 5'	38' 10'
28.400	8' 6'	8' 3'	34' 1'

The above dimensions are based on the formulas detailed in the text. The dimensions for verticals are for wire elements only. If aluminum tubing or electrical conduit is used for the vertical elements, these dimensions are to be shortened by 5 percent. The phasing-line lengths stay as indicated. For a two-element array (Fig 7) divide the phasing-line dimensions by 2.

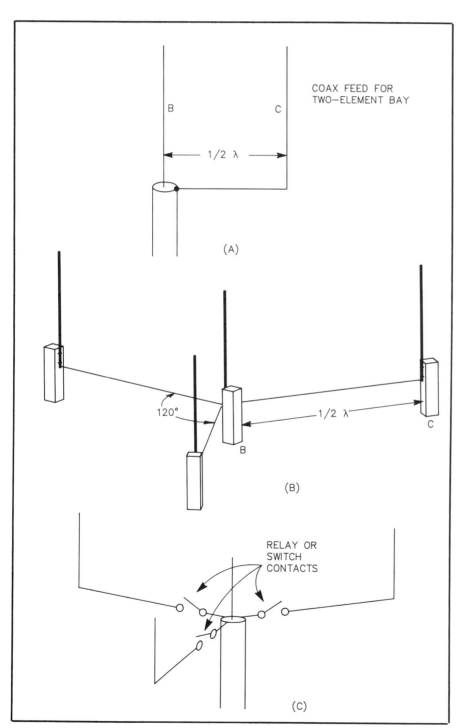

Fig 7—A switchable four-element broadside array which provides around-the-compass coverage.

set 120° apart around the common element and provides selectable azimuth coverage.

The center element is permanently connected to the center conductor of the coax feed line. The coax shield is fed through the switching relays mounted on the center vertical support. Each relay position connects one phasing line and shifts the beam pattern 120° while isolating the other two phasing lines. The two isolated verticals are 30° off the center line of the bidirectional beam of the driven array, which places them about 7 dB down from the beam centerline. Some slight skewing of the beam pattern may occur. The six nulls are switchable to possibly reduce offending QRM.

The array of Fig 8 develops a switchable unidirectional pattern which consists of two three-element arrays set ¼ λ apart and driven approximately 90° out of phase. The antenna relay introduces the delay line and provides the switching. [Ideally the radiation pattern would be a cardioid, but the phasing arrangement of Fig 8 likely does not yield element currents of equal amplitude that are exactly 90° out of phase. See the paper by Lewallen elsewhere in this chapter.—*Ed*.]

The beam pattern is perpendicular to the plane of the array. The azimuth beam is about 70° wide at the half-power points. The beam direction is toward the vertical with the lagging current (through the delay line).

Those who have the space and like their gain in double digits in one particular direction might consider using a five-element driven array with one five-element parasitic-reflector array and two five-element parasitic-director arrays. Table 2 dimensions apply only to the driven array. This configuration of parasitic elements is not shown.

Conclusion

I believe that broadside phased arrays mounted off the ground have a lot to offer as DX antennas, particularly for hams who need an alternative to the use of towers. For this type of array it is essential that mechanical and electrical array symmetry be established for good performance. These antennas and feed systems are not difficult to build or tune, and arrays for 40 meters and down can be serviced and maintained from step-ladder height.

My thanks go to Tom Lee, K8AZ, for providing the facility, material and assistance, and particularly for doing the climbing.

Notes

[1]W. W. Smith's "Bobtail" is described in W. I. Orr and S. D. Cowan, *All About Vertical Antennas* (Wilton, CT: Radio Publications, 1986), Chap 6, pp 145-148.

[2]W. Maxwell, "Some Aspects of the Balun Problem," *QST*, March 1983, pp 38-40.

[3]Ferrite beam materials are available from Fair-Rite Products Corp, 1 Commercial Row, Wallkill, NY 12589.

[4]Sleeve choke-balun kit, 2-30 MHz, available from Radiokit, PO Box 973, Pelham, NH 03076.

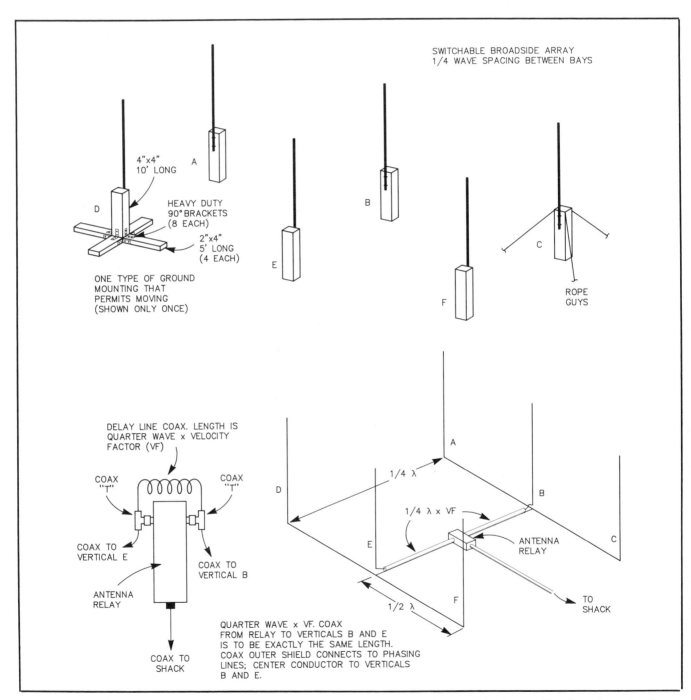

Fig 8—This array provides a switchable unidirectional pattern with about 7.5 dB gain over a single element. The dimensions of each three-element array are identical. The two coaxial cables connecting the arrays and relay box should be exactly the same length. (See text to calculate element and phasing-line lengths.)

The Simplest Phased Array Feed System...That Works

By Roy Lewallen, W7EL
5470 SW 152 Ave
Beaverton, OR 97007

Many amateurs having a phased-array antenna use the feed system shown in Fig 1, with the difference between the electrical lengths of the feed lines equaling the desired phase angle. The result is often disappointing. The reasons for poor results are twofold.

1) The phase shift through each feed line is not equal to its electrical length, and

2) The feed line changes the magnitude of the current from input to output.

This surprising combination of events occurs in nearly all amateur arrays because of the significant, and sometimes dramatic, change in element feed-point impedances by mutual coupling. The element feed-point impedances—the load impedances seen by the feed lines—affect the delay and transformation ratio of the cables. It isn't a small effect, either. Phasing errors of several tens of degrees and element-current ratios of 2:1 are not uncommon. Among the very few antennas which do work are arrays of only two elements fed completely in phase (0°) or out of phase (180°). This topic is covered in detail in *The ARRL Antenna Book*.[1]

It is possible, however, to use the system shown in Fig 1 and have the element currents come out the way we want. The trick is to use feed-line lengths which give the desired delay and transformation ratio *when looking into the actual element impedances*. More specifically, we choose the feed-line lengths to give the desired ratio of currents, with the correct relative phasing. This paper explains how to find the correct feed-line lengths, and includes a BASIC program to do the calculations. (See the Program 1 listing near the end of this paper.) Table 1 gives results for a 90°-fed, 90°-spaced, 2-element array.

Calculation of the feed-line lengths, with or without the program, requires knowledge of the element self- and mutual impedances. Most of us don't know these impedances for our arrays, so one of several approaches can be taken.

1) Measure the self- and mutual impedances using the techniques described in *The Antenna Book* (see note 1). If care-

[1]Notes appear on page 29.

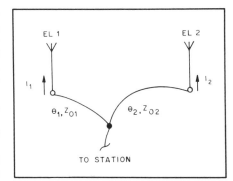

Fig 1—A typical feed system for two-element phased arrays. See text for a discussion of the feed-line lengths. Grounds and cable shields have been omitted for clarity.

fully done, this approach will lead to the best array performance.

2) Estimate the self- and mutual impedances. Methods and graphs are given in the *The Antenna Book*. This approach can lead to very good array performance if the array elements are straight and parallel, and with no loading elements or unusual features.

3) If the elements fit the above description, and in addition are self-resonant and close to ¼ λ high, you can use Table 1 instead of the program, if desired.

4) If you can't measure or estimate the self- and mutual impedances with reasonable accuracy, and your elements don't fit the description given in approach 3, you're likely to get poor results with this feed system. A better approach would be to use the L-network feed system described in *The Antenna Book*. It's quite simple and has the advantage of being adjustable. Adjustment methods also are given in *The Antenna Book*.

Using The Program

Program 1 was purposefully written in a very simple form of BASIC. It should run on nearly any computer without modification. If you encounter difficulty, the most likely cause is that the program was not copied exactly as printed.

The first prompts are for the self-R and X of the elements. These are the impedances which would be measured at the base of each element with the other element open-circuited at the base. The self-R includes any loss resistance. The remainder of the prompts are self-explanatory. Refer to Table 2, the sample run, for an example

---

Table 1

Phasing Line Lengths for a 90°-Fed, 90°-Spaced Two-Element Array

See Fig 1.

R_s*, Ohms	No. Radials per Element	Z_0, Ohms 1	Z_0, Ohms 2	Phasing Lines Elect Length (Deg) Line 1	Phasing Lines Elect Length (Deg) Line 2
65	4	50	50	No solution	
		75	75	No solution	
		75	50	30.36	104.96
				95.13	162.96
54	8	50	50	No solution	
		75	75	68.15	154.29
				132.60	184.95
45	16	50	50	No solution	
		75	75	58.69	153.48
				144.43	183.39
36	∞	50	50	80.56	154.53
				131.68	173.23
		75	75	51.61	155.40
				153.86	179.13

*Self-impedance, including losses.

Table 2

Sample Run of Program 1 for a 90°-Spaced, 90°-Fed Array

Calculations are for resonant elements, approximately ¼-λ high, with 8 ground radials per element.

```
RUN
SELF R, X OF LEADING ELEMENT (OHMS)
? 54,0
SELF R, X OF LAGGING ELEMENT (OHMS)
? 54,0
MUTUAL R, X (OHMS)
? 20,-15
EL.2:EL.1 CURRENT MAGNITUDE, PHASE (DEGREES)
—PHASE MUST BE ZERO OR NEGATIVE
1,-90
FEEDLINE 1, 2 IMPEDANCES (OHMS)
? 50,50
NO SOLUTION FOR THE SPECIFIED PARAMETERS.
  WOULD YOU LIKE TO TRY DIFFERENT
  FEEDLINE Z0'S (Y,N)? Y
FEEDLINE 1, 2 IMPEDANCES (OHMS)
? 75,75
```

	Z0 = 75 OHMS TO LEAD. EL. ELECT. L. (DEG.)	Z0 = 75 OHMS TO LAG. EL. ELECT. L. (DEG.)
FIRST SOLN.	132.6038	184.9522
SECOND SOLN.	68.1518	154.2918

```
Ok
```

of program operation.

Sometimes you might get the result, NO SOLUTION FOR THE SPECIFIED PARAMETERS. This doesn't mean there's a solution which the program couldn't find; it means that there really is no solution for the specified conditions. If this happens, try different feed-line impedances. I've found a combination of common feed-line impedances which will work with nearly every array I've wanted to feed, but there are some which can't be fed using this method.

Whenever there is a solution, there's also a second one. Both are computed by the program. It may be necessary to use the longer set of feed-line lengths in order to make the feed lines physically reach the elements. You can also add ½ λ of cable to *both* feed lines and maintain correct operation. For example, the array in the sample program run of Table 2 can be fed with two 75-ohm lines of the following lengths (given in electrical degrees).

68.15° and 154.29°
132.60° and 184.95°
248.15° and 334.29°
312.60° and 364.95° (or 4.95°)

The first two sets are the lengths given by the program. A half wavelength is added to both lines to make sets 3 and 4. Note that a *full* wavelength can be subtracted from the second line length in the last set.

Occasionally it's necessary to make the feed-line impedances different from each other. If you want to be able to switch the pattern direction but have unequal feed-line impedances, add ½ λ of line from each element to the phasing feed line. If both ½-λ

lines have the same impedance, directional switching will be possible while maintaining correct phasing.

Using the Table

Table 1 gives the feed-line lengths necessary to correctly feed a 90°-fed, 90°-spaced, 2-element array. The table is based on the following assumptions:

1) The elements are identical and parallel.
2) The ground systems of the elements have equal loss.
3) The elements are resonant when not coupled to other elements. A height of $237/f_{MHz}$ will be close to resonance for most vertical elements.
4) The elements are not loaded and do not have matching networks at their bases. Traps generally act like loading elements on the lower bands.
5) The elements are fairly "thin." HF antennas made from wire, tubing, or common tower sections fit this category.
6) Your ground isn't unusually dry or swampy. If it is, you may have more or less element self-resistance than shown for the number of radials. The resistance versus number of radials is based on measurements by Sevick.[2]

Since so many factors can affect ground losses and element self- and mutual impedances, the tables probably won't give exactly the best feed-line lengths for your array. But if the above assumptions apply, it's very likely that your array will work better using the recommended feed-line lengths. If the assumptions don't describe

your array, the table values won't be valid.

Two Four-Element Arrays

The Antenna Book (note 1) describes a feed system for two types of four-element arrays based on a combination of the "current forcing" method and an L network. Information on these arrays, the current forcing method, and practical advice on how to measure the various line sections can be found in Chapter 8 of *The Antenna Book*. The L network can be replaced by two feed lines, resulting in the feed systems shown in Figs 2 and 3. The principle is the same as for the two-element array, although the mathematics are a bit different due to the presence of the λ/4 or 3 λ/4 lines and the difficulty of including the mutual impedances between all elements. The mathematics are described in the next section.

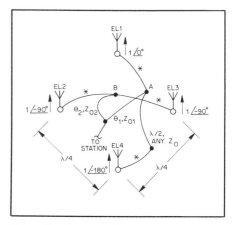

Fig 2—Feed system for the four-square array. Grounds and cable shields have been omitted for clarity. The lines marked "*" are all the same length, have the same Z_0, and are electrically either λ/4 or 3λ/4 long. The other lines are discussed in the text.

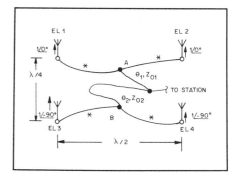

Fig 3—Feed system for a four-element rectangular array. Grounds and cable shields have been omitted for clarity. The lines marked "*" are all electrically 3λ/4 long and have the same Z_0. The other lines are discussed in the text.

Tables 3 and 4 give feed-line lengths for these two arrays. The same restrictions apply to the four-element tables as to the two-element table. They were calculated using modified versions of the BASIC program. These programs, which apply only to the four-square and rectangular arrays, are listed as Programs 2 and 3.

The Mathematics

For an array to work properly, the element currents need to have the correct relationship. So let's first look at the general problem of feeding two loads with a specific ratio of currents (Fig 4). The desired current ratio, I_2/I_1, is a complex number with two parts: magnitude M_{12} and angle ϕ_{12}. Both parts must be correct for the array to work as planned. Assuming for the moment that we know what the load impedance will be, we can write the following equation for feed-line no. 1.

$$V_{in} = I_1 Z_1 \cos \theta_1 + jI_1 Z_{01} \sin \theta_1 \quad \text{(Eq 1)}$$

where
V_{in} = voltage at the input end of the line
I_1 = current at the output end of the line
Z_1 = load impedance at the output end of the line
θ_1 = electrical length of the line in degrees or radians
Z_{01} = characteristic impedance of the line
V_{in}, I_1, and Z_1 are complex

This is the general equation which relates the output current to the input voltage for a lossless transmission line.[3] A similar equation can be written for the second feed line. Since the feed lines are connected together at their input ends, the input voltages are equal, and we can write

$$V_{in} = I_1 Z_1 \cos \theta_1 + jI_1 Z_{01} \sin \theta_1$$
$$V_{in} = I_2 Z_2 \cos \theta_2 + jI_2 Z_{02} \sin \theta_2$$

Rearranging to solve for the current ratio gives

$$\frac{I_2}{I_1} = \frac{(Z_1 \cos \theta_1 + jZ_{01} \sin \theta_1)}{(Z_2 \cos \theta_2 + jZ_{02} \sin \theta_2)} \quad \text{(Eq 2)}$$

This equation can be used to illustrate the problems of feeding unequal load impedances (present in the elements of most arrays). For example, if values that might be found in a 90°-spaced, 90°-fed array are

$Z_1 = 35 - j20 \ \Omega$
$Z_2 = 65 + j20 \ \Omega$
$Z_{01} = Z_{02} = 50 \ \Omega$
$\theta_1 = 90°$, and $\theta_2 = 180°$

then I_2/I_1 would be 0.735 at an angle of $-107°$, not 1 at an angle of $-90°$ as planned. In a real array, because of mutual coupling, the element feed-point impedances are modified by the currents

Table 3

Phasing Line Lengths for a Four Square Array

See Fig 2.

R_s,* Ohms	No. Radials per Element	λ/4 Line Z_0, Ohms	Z_0, Ohms A	Z_0, Ohms B	Phasing Lines Elect Length (Deg) Line A	Phasing Lines Elect Length (Deg) Line B
65	4	50	50	50	20.66	166.50
					147.94	204.91
			75	75	13.70	170.60
					158.00	197.70
		75	50	50	No solution	
			75	75	32.03	162.26
					133.53	212.18
54	8	50	50	50	25.82	166.32
					138.25	209.61
			75	75	16.80	170.01
					151.13	202.06
		75	50	50	No solution	
			75	75	45.11	167.22
					115.95	211.72
45	16	50	50	50	34.57	168.71
					123.73	212.98
			75	75	21.15	170.17
					141.45	207.23
		75	50	50	22.36	121.53
					134.77	261.34
			75	75	No solution	
36	∞	50	50	50	No solution	
			75	75	31.37	173.66
					121.79	213.18
		75	50	50	33.55	122.94
					120.01	263.50
			75	75	No solution	

*Self-impedance, including losses.

Table 4

Phasing Line Lengths for a Four Element Rectangular Array

See Fig 3.

R_s,* Ohms	No. Radials per Element	3λ/4 Line Z_0, Ohms	Z_0, Ohms A	Z_0, Ohms B	Phasing Lines Elect Length (Deg) Line A	Phasing Lines Elect Length (Deg) Line B
65	4	50	50	50	37.37	155.34
					132.87	179.10
			75	75	24.95	162.22
					150.41	181.47
		75	50	50	No solution	
			75	75	66.82	153.30
					87.34	161.46
54	8	50	50	50	59.02	151.97
					100.15	163.93
			75	75	35.79	157.33
					135.17	175.44
		75	50	50	No solution	
			75	75	No solution	
			75	50	29.44	112.55
					67.01	129.30
45	16	50	50	50	No solution	
			75	75	62.38	152.17
					98.08	159.50
		75	50	50	No solution	
			75	75	No solution	
			75	50	12.61	99.74
					64.34	114.66
36	∞	50	50	50	No solution	
			75	75	No solution	
		75	50	50	52.04	99.08
					176.88	274.80
			75	75	No solution	

*Self-impedance, including losses.

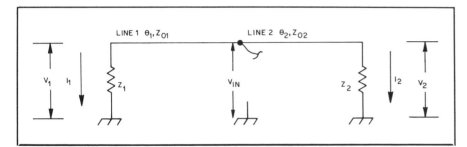

Fig 4—Feeding two load impedances with specific currents. This example assumes that Z1 and Z2 are not affected by mutual coupling.

flowing in the elements. But the element currents are a function of the element feed-point impedances, so Eq 2 can't be used directly to calculate currents in array elements. To write an equation which will do that, we need to modify Fig 4 to account for the effect of mutual coupling (Fig 5). From the diagram,

$$V_1 = I_1 Z_{11} + I_2 Z_{12} \qquad \text{(Eq 3)}$$

$$V_2 = I_2 Z_{22} + I_1 Z_{12} \qquad \text{(Eq 4)}$$

where

V_n = voltage at the feed point of element n

I_n = current at the feed point of element n

Z_{nn} = self-impedance of element n (the feed-point impedance when the element is totally isolated from all other elements)

Z_{12} = mutual impedance between the elements

All variables are complex

A slightly different form of Eq 1 is

$$V_{in} = V_1 \cos \theta_1 + jI_1 Z_{01} \sin \theta_1 \quad \text{(Eq 5)}$$

and for the second feed line

$$V_{in} = V_2 \cos \theta_2 + jI_2 Z_{02} \sin\theta_2 \quad \text{(Eq 6)}$$

V_1 and V_2 from Eqs 3 and 4 are substituted into Eqs 5 and 6. The right sides of Eqs 5 and 6 are set equal to each other, since V_{in} is the same for both feed lines. Finally, the resulting equation is rearranged to solve for I_2/I_1.

$$\frac{I_2}{I_1} = \frac{Z_{11} \cos\theta_1 - Z_{12} \cos\theta_2 + jZ_{01} \sin\theta_1}{Z_{22} \cos\theta_2 - Z_{12} \cos\theta_1 + jZ_{02} \sin\theta_2} \quad \text{(Eq 7)}$$

This is the same as Eq 2 except that an additional term containing mutual impedance Z_{12} appears in both the numerator and denominator. Given the element self- and mutual impedances and the lengths and impedances of the feed lines, Eq 7 can be used to find the resulting ratio of currents in the elements. The problem we're trying to solve, though, is the other way around: how to find the feed-line lengths, given the current ratio and other factors. Christman described an iterative method of using Eq 7 to solve the problem by beginning with an initial estimate of feed-line lengths, finding the resulting current ratio, correcting the estimate, and repeating until the answer converges on the correct answer.[4] This method gives accurate answers, and I used it for some time. The disadvantage of the approach is that convergence can be slow, and the iter-

ations can actually diverge for some arrays unless the program includes "damping."

Fortunately, an iterative approach isn't necessary, since Eq 7 can be solved directly for feed-line lengths. The method is straightforward, although tedious, and was done using several variable transformations to keep the equations manageable. The details won't be described here. The BASIC programs presented at the end of this paper use the direct solution method, and the validity of the results can be confirmed by substitution into Eq 7.

The feed system can be adapted to certain larger arrays by combining it with the current forcing method described in *The Antenna Book* (see Figs 2 and 3). The basic requirement is to make the *voltages* at points A and B have the proper ratio and phase angle. If this is accomplished, the elements will have correct *currents* because of the properties of the ¼-λ lines. *The Antenna Book* shows the use of an L network to obtain the voltage phase shift; the same thing can be accomplished by using two feed lines of the correct length.

To see how we can use the program to solve the problem, we'll rewrite Eq 2 to apply to the currents and impedances at points A and B:

$$\frac{I_B}{I_A} = \frac{Z_A \cos \theta_1 + jZ_{01} \sin \theta_1}{Z_B \cos \theta_2 + jZ_{02} \sin \theta_2} \quad \text{(Eq 8)}$$

Because $V_A = I_A Z_A$ and $V_B = I_B Z_B$, then

$$\frac{V_B}{V_A} = \frac{Z_B}{Z_A} \frac{I_B}{I_A} =$$
$$\frac{Z_B}{Z_A} \frac{Z_A \cos \theta_1 + jZ_{01} \sin \theta_1}{Z_B \cos \theta_2 + jZ_{02} \sin \theta_2} \quad \text{(Eq 9)}$$

Note the similarity to Eq 7, which is the equation the program solves for θ_1 and θ_2. We can use the program to solve Eq 9 if we

1) Enter Z_A when it prompts for the self-Z of element 1,

2) Enter Z_B when it prompts for the self-Z of element 2,

3) Enter 0,0 when it prompts for the mutual R, X, and

4) Enter $(V_B/V_A) (Z_A/Z_B)$ when it prompts for the desired current ratio. For the two four-element arrays,

$$V_B/V_A = 0 - j1 = 1 \underline{/-90°}.$$

The following steps are required to calculate Z_A/Z_B.

1) Measure or estimate the self- and mutual impedances of the elements.

2) Using the self- and mutual impedances and the current ratios, calculate the actual element feed-point impedances. The method is described in *The Antenna Book*.

3) Calculate the impedances looking into the λ/4 or 3λ/4 lines.

4) Where two of the lines are connected,

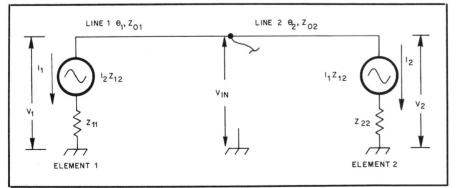

Fig 5—Feeding two antenna elements with specific currents. The voltage sources are added to account for mutual coupling.

calculate the parallel impedance. These will be Z_A and Z_B.

5) Calculate the ratio Z_A/Z_B.

Program 1 has been modified to do these calculations for you. The modified BASIC programs are listed as Programs 2 and 3. [All three programs are available on diskette for the IBM PC and compatibles; see information on an early page of this book.—Ed.] You must know the self-impedance of an element (all are assumed to be identical) and the mutual impedances between all elements in order to use the modified programs.

Closing Comments

I first solved Eq 7 for the feed-line lengths several years ago. However, I didn't try to publish the results because of the large amount of explanation which would have to go with it—why the common feed method frequently is disappointing, and explaining the current forcing and L-network feed systems, the role of mutual coupling in phased arrays, etc. I want to thank Jerry Hall, K1TD, for providing the opportunity to explain them in a forum which is readily available to amateurs —Chapter 8 of *The ARRL Antenna Book*.

Your array *will* work better if properly fed. This feed system isn't any more complicated than the one you've probably been using, but it's likely to give you much better results. Try it!

Notes

[1]G. L. Hall, Ed, *The ARRL Antenna Book*, 15th ed. (Newington: ARRL, 1988), Chap 8.

[2]J. Sevick, "The Ground-Image Vertical Antenna," *QST*, Jul 1971, pp 16-19, 22. Also "The W2FMI Ground-Mounted Short Vertical," *QST*," Mar 1973, pp 13-18, 41. (Summary information from these articles is presented graphically in Fig 23, p 8-23 of *The Antenna Book*—see note 1).

[3]*Reference Data for Radio Engineers*, 5th ed. (Howard W. Sams & Co, 1968).

[4]A. Christman, "Feeding Phased Arrays: An Alternative Method," *Ham Radio*, May 1985, p 58 and Jul 1985, p 74.

Program 1

Calculates Line Lengths for 2-Element Array

See Fig 1 and Table 2.

[The ARRL-supplied disk filename for this program is LEWALL1.BAS.—Ed.]

```
100 REM
110 REM   Simplified feed system for two-element arrays
120 REM   Roy Lewallen, W7EL
130 REM   7 Jan. 1985 — revised 15 May 1988
140 REM
150 REM   For an explanation of this program, see THE ARRL ANTENNA COMPENDIUM,
160 REM   VOLUME 2.
170 REM
180 REM   Angular calculations are done in radians;
190 REM   Inputs and outputs are in degrees.
200 REM
210 REM   Obtain the self and mutual impedances:
220 REM
230 PRINT
240 PRINT
250 PRINT "SELF R, X OF LEADING ELEMENT (OHMS)"
260 INPUT R1, X1
270 PRINT
280 PRINT "SELF R, X OF LAGGING ELEMENT (OHMS)"
290 INPUT R2, X2
300 PRINT
310 PRINT "MUTUAL R, X (OHMS)"
320 INPUT R3, X3
330 REM
340 REM   Obtain the desired current ratio
350 REM
360 PRINT
370 PRINT "EL.2:EL.1 CURRENT MAGNITUDE, PHASE (DEGREES)"
380 PRINT "— PHASE MUST BE ZERO OR NEGATIVE"
390 INPUT M,P
400 IF P<=0 THEN 490
410 PRINT
420 PRINT "PHASE MUST BE NON-POSITIVE, SINCE EL. 2 IS"
430 PRINT "DEFINED AS BEING THE LAGGING ELEMENT."
440 PRINT
450 GOTO 360
460 REM
470 REM   Obtain the feedline impedances
480 REM
490 PRINT
```

```
500 PRINT "FEEDLINE 1, 2 IMPEDANCES (OHMS)"
510 INPUT F1,F2
520 REM
530 REM   Calculate lengths of lines
540 REM
550 P1=3.14159
560 J=0
570 C1=COS(P*P1/180)/M
580 S1=SIN(P*P1/180)/M
590 H=R2+C1*R3+S1*X3
600 A=(R3+C1*R1+S1*X1)/H
610 B=S1*F1/H
620 C=((X3+C1*X1-S1*R1)*H-(X2+C1*X3-S1*R3)*(R3+C1*R1+S1*X1))/F2/H
630 D=F1*(C1*H-S1*(X2+C1*X3-S1*R3))/F2/H
640 Q=2-(A*A+B*B+C*C+D*D)
650 E=(A*A+C*C-B*B-D*D)/Q
660 F=2*(A*B+C*D)/Q
670 R=E
680 I=F
690 GOSUB 1030
700 G1=V
710 IF U<1 THEN 1150
720 X=ATN(SQR(U*U-1))
730 T1=(X+G1)/2
740 IF T1 >=0 THEN 760
750 T1=T1+2*P1
760 IF T1<P1 THEN 780
770 T1=T1-P1
780 I=C*COS(T1)+D*SIN(T1)
790 R=A*COS(T1)+B*SIN(T1)
800 GOSUB 1030
810 IF V>=0 THEN 830
820 V=V+2*P1
830 IF J=1 THEN 960
840 REM
850 REM   Print the results
860 REM
870 PRINT
880 PRINT " ","Z0=";F1;" OHMS",,"Z0=";F2;" OHMS"
890 PRINT " ","TO LEAD. EL.",,"TO LAG. EL."
900 PRINT " ","ELECT. L.(DEG.)","ELECT. L.(DEG.)"
910 PRINT
920 PRINT "FIRST SOLN.",T1*180/P1,,V*180/P1
930 J=1
940 X=2*P1-X
950 GOTO 730
960 PRINT "SECOND SOLN.",T1*180/P1,,V*180/P1
970 GOTO 1220
980 REM
990 REM   End of program
1000 REM
1010 REM   Rectangular-polar subroutine
1020 REM
1030 U=SQR(R*R+I*I)
1040 IF U=0 THEN 1080
1050 IF ABS(U+R)<.0000001 THEN 1100
1060 V=2*ATN(I/(U+R))
1070 RETURN
1080 V=0
1090 RETURN
1100 V=P1
1110 RETURN
1120 REM
1130 REM   No solution found
1140 REM
1150 PRINT
1160 PRINT "NO SOLUTION FOR THE SPECIFIED PARAMETERS."
1170 PRINT "  WOULD YOU LIKE TO TRY DIFFERENT"
1180 PRINT "  FEEDLINE Z0'S (Y,N)";
1190 INPUT A$
1200 IF A$="Y" THEN 490
1210 IF A$="y" THEN 490
1220 PRINT
1230 PRINT
1240 END
```

Program 2

Calculates Line Lengths for Four-Square Array

See Fig 2.

[The ARRL-supplied disk filename for this program is LEWALL2.BAS.—*Ed.*]

```
100 REM
110 REM   Simplified feed system for "four-square" array
120 REM   Roy Lewallen, W7EL
130 REM   15 May 1988
140 REM
150 REM   For an explanation of this program, see THE ARRL ANTENNA COMPENDIUM,
160 REM   VOLUME 2.
170 REM
180 REM   All elements are assumed to be identical and close to a quarter
190 REM   wavelength high.
200 REM
210 REM   Angular calculations are done in radians;
220 REM   Inputs and outputs are in degrees.
230 REM
240 REM   Obtain the self and mutual impedances:
250 REM
260 PRINT
270 PRINT
280 PRINT "SELF R, X OF ELEMENT (OHMS)"
290 INPUT R1, X1
300 PRINT
310 PRINT "MUTUAL R, X BETWEEN ADJACENT ELEMENTS (1-2, 1-3, 2-4, 3-4)"
320 INPUT R5, X5
330 PRINT
340 PRINT "MUTUAL R, X BETWEEN DIAGONAL ELEMENTS (1-4, 2-3)"
350 INPUT R6, X6
360 REM
370 REM   Calculate change in element impedances due to mutual coupling,
380 REM   assuming correct currents
390 REM
400 H1=-2*X5-R6
410 W1=2*R5-X6
420 H2=R6
430 W2=X6
440 H4=-1*R6+2*X5
450 W4=-1*X6-2*R5
460 REM
470 REM Obtain the line impedances
480 REM
490 PRINT
500 PRINT "Z0 OF 1/4 OR 3/4 WAVELENGTH LINES"
510 INPUT Z0
520 PRINT
530 PRINT "FEEDLINE 1, 2 IMPEDANCES (OHMS)"
540 INPUT F1,F2
550 P1=3.14159
560 REM
570 REM   Find Z looking into each 1/4 or 3/4 wavelength line
580 REM
590 Q=Z0*Z0/((R1+H1)^2+(X1+W1)^2)
600 T1=Q*(R1+H1)
610 U1=-1*Q*(X1+W1)
620 Q=Z0*Z0/((R1+H2)^2+(X1+W2)^2)
630 T2=Q*(R1+H2)
640 U2=-1*Q*(X1+W2)
650 Q=Z0*Z0/((R1+H4)^2+(X1+W4)^2)
660 T4=Q*(R1+H4)
670 U4=-1*Q*(X1+W4)
680 REM
690 REM   Find Z at points a and b
700 REM
710 Q=(T1+T4)^2+(U1+U4)^2
720 A1=((T1*T4-U1*U4)*(T1+T4)+(U1*T4+T1*U4)*(U1+U4))/Q
730 A9=((U1*T4+T1*U4)*(T1+T4)-(T1*T4-U1*U4)*(U1+U4))/Q
740 B1=T2/2
750 B9=U2/2
760 REM
770 REM   Calculate Za/Zb = (A1+jA9)/(B1+jB9)
780 REM
790 Q=B1*B1+B9*B9
```

```
800 M1=[A1*B1+B9*A9]/Q
810 M9=[A9*B1-A1*B9]/Q
820 REM
830 REM   Multiply Za/Zb by the desired voltage ratio of -j
840 REM   and convert to polar form
850 REM
860 R=M9
870 I=-1*M1
880 GOSUB 1410
890 M=U
900 P=V
910 REM
920 REM   Calculate lengths of lines 1 and 2
930 REM
940 J=0
950 C1=COS[P]/M
960 S1=SIN[P]/M
970 H=B1
980 A=[C1*A1+S1*A9]/H
990 B=S1*F1/H
1000 C=[[C1*A9-S1*A1]*H-[B9]*[C1*A1+S1*A9]]/F2/H
1010 D=F1*[C1*H-S1*[B9]]/F2/H
1020 Q=2-[A*A+B*B+C*C+D*D]
1030 E=[A*A+C*C-B*B-D*D]/Q
1040 F=2*[A*B+C*D]/Q
1050 R=E
1060 I=F
1070 GOSUB 1410
1080 G1=V
1090 IF U<1 THEN 1530
1100 X=ATN[SQR[U*U-1]]
1110 T=[X+G1]/2
1120 IF T >=0 THEN 1140
1130 T=T+2*P1
1140 IF T<P1 THEN 1160
1150 T=T-P1
1160 I=C*COS[T]+D*SIN[T]
1170 R=A*COS[T]+B*SIN[T]
1180 GOSUB 1410
1190 IF V>=0 THEN 1210
1200 V=V+2*P1
1210 IF J=1 THEN 1340
1220 REM
1230 REM   Print the results
1240 REM
1250 PRINT
1260 PRINT " ","Z0=";F1;" OHMS",,"Z0=";F2;" OHMS"
1270 PRINT " ","TO POINT A",,"TO POINT B"
1280 PRINT " ","ELECT. L.[DEG.]","ELECT. L.[DEG.]"
1290 PRINT
1300 PRINT "FIRST SOLN.",T*180/P1,,V*180/P1
1310 J=1
1320 X=2*P1-X
1330 GOTO 1110
1340 PRINT "SECOND SOLN.",T*180/P1,,V*180/P1
1350 GOTO 1600
1360 REM
1370 REM   End of program
1380 REM
1390 REM   Rectangular-polar subroutine
1400 REM
1410 U=SQR[R*R+I*I]
1420 IF U=0 THEN 1460
1430 IF ABS[U+R]<.0000001 THEN 1480
1440 V=2*ATN[I/[U+R]]
1450 RETURN
1460 V=0
1470 RETURN
1480 V=P1
1490 RETURN
1500 REM
1510 REM   No solution found
1520 REM
1530 PRINT
```

```
1540 PRINT "NO SOLUTION FOR THE SPECIFIED PARAMETERS."
1550 PRINT "  WOULD YOU LIKE TO TRY DIFFERENT"
1560 PRINT "   FEEDLINE Z0'S [Y,N]";
1570 INPUT A$
1580 IF A$="Y" THEN 520
1590 IF A$="y" THEN 520
1600 PRINT
1610 PRINT
1620 END
```

Program 3

Calculates Line Lengths for 4-Element Rectangular Array

See Fig 3.

[The ARRL-supplied disk filename for this program is LEWALL3.BAS.—*Ed.*]

```
100 REM
110 REM   Simplified feed system for rectangular array
120 REM   Roy Lewallen, W7EL
130 REM   15 May 1988
140 REM
150 REM   For an explanation of this program, see THE ARRL ANTENNA COMPENDIUM,
160 REM   VOLUME 2.
170 REM
180 REM   All elements are assumed to be identical and close to a quarter
190 REM   wavelength high.
200 REM
210 REM   Angular calculations are done in radians;
220 REM   Inputs and outputs are in degrees.
230 REM
240 REM   Obtain the self and mutual impedances:
250 REM
260 PRINT
270 PRINT
280 PRINT "SELF R, X OF ELEMENT [OHMS]"
290 INPUT R1, X1
300 PRINT
310 PRINT "MUTUAL R, X BETWEEN 90 DEG. SPACED ELEMENTS [1-3, 2-4]"
320 INPUT R5, X5
330 PRINT
340 PRINT "MUTUAL R, X BETWEEN 180 DEG. SPACED ELEMENTS [1-2, 3-4]"
350 INPUT R6, X6
360 PRINT
370 PRINT "MUTUAL R, X BETWEEN DIAGONAL ELEMENTS [1-4, 2-3]
380 INPUT R7, X7
390 REM
400 REM   Calculate change in element impedances due to mutual coupling,
410 REM   assuming correct currents
420 REM
430 H1=R6+X5+X7
440 W1=X6-R5-R7
450 H3=R6-X5-X7
460 W3=R5+R7+X6
470 REM
480 REM   Obtain the line impedances
490 REM
500 PRINT
510 PRINT "Z0 OF 3/4 WAVELENGTH LINES"
520 INPUT Z0
530 PRINT
540 PRINT "FEEDLINE 1, 2 IMPEDANCES [OHMS]"
550 INPUT F1,F2
560 P1=3.14159
570 REM
580 REM   Find Z looking into each 3/4 wavelength line
590 REM
600 Q=Z0*Z0/[[R1+H1]^2+[X1+W1]^2]
610 T1=Q*[R1+H1]
620 U1=-1*Q*[X1+W1]
630 Q=Z0*Z0/[[R1+H3]^2+[X1+W3]^2]
```

```
640  T3=Q*(R1+H3)
650  U3=-1*Q*(X1+W3)
660  REM
670  REM   Find Z at points a and b
680  REM
690  A1=T1/2
700  A9=U1/2
710  B1=T3/2
720  B9=U3/2
730  REM
740  REM   Calculate Za/Zb = (A1+jA9)/(B1+jB9)
750  REM
760  Q=B1*B1+B9*B9
770  M1=(A1*B1+B9*A9)/Q
780  M9=(A9*B1-A1*B9)/Q
790  REM
800  REM   Multiply Za/Zb by the desired voltage ratio of -j
810  REM   and convert to polar form
820  REM
830  R=M9
840  I=-1*M1
850  GOSUB 1380
860  M=U
870  P=V
880  REM
890  REM   Calculate lengths of lines 1 and 2
900  REM
910  J=0
920  C1=COS(P)/M
930  S1=SIN(P)/M
940  H=B1
950  A=(C1*A1+S1*A9)/H
960  B=S1*F1/H
970  C=((C1*A9-S1*A1)*H-(B9)*(C1*A1+S1*A9))/F2/H
980  D=F1*(C1*H-S1*(B9))/F2/H
990  Q=2-(A*A+B*B+C*C+D*D)
1000 E=(A*A+C*C-B*B-D*D)/Q
1010 F=2*(A*B+C*D)/Q
1020 R=E
1030 I=F
1040 GOSUB 1380
1050 G1=V
1060 IF U<1 THEN 1500
1070 X=ATN(SQR(U*U-1))
1080 T=(X+G1)/2
1090 IF T >=0 THEN 1110
1100 T=T+2*P1
1110 IF T<P1 THEN 1130
1120 T=T-P1
1130 I=C*COS(T)+D*SIN(T)
1140 R=A*COS(T)+B*SIN(T)
1150 GOSUB 1380
1160 IF V>=0 THEN 1180
1170 V=V+2*P1
1180 IF J=1 THEN 1310
1190 REM
1200 REM   Print the results
1210 REM
1220 PRINT
1230 PRINT " ","Z0=";F1;" OHMS",,"Z0=";F2;" OHMS"
1240 PRINT " ","TO POINT A",,"TO POINT B"
1250 PRINT " ","ELECT. L.(DEG.)","ELECT. L.(DEG.)"
1260 PRINT
1270 PRINT "FIRST SOLN.",T*180/P1,,V*180/P1
1280 J=1
1290 X=2*P1-X
1300 GOTO 1080
1310 PRINT "SECOND SOLN.",T*180/P1,,V*180/P1
1320 GOTO 1570
1330 REM
1340 REM   End of program
1350 REM
1360 REM   Rectangular-polar subroutine
1370 REM
1380 U=SQR(R*R+I*I)
1390 IF U=0 THEN 1430
```

```
1400 IF ABS(U+R)<.0000001 THEN 1450
1410 V=2*ATN(I/(U+R))
1420 RETURN
1430 V=0
1440 RETURN
1450 V=P1
1460 RETURN
1470 REM
1480 REM  No solution found
1490 REM
1500 PRINT
1510 PRINT "NO SOLUTION FOR THE SPECIFIED PARAMETERS."
1520 PRINT "  WOULD YOU LIKE TO TRY DIFFERENT"
1530 PRINT "  FEEDLINE Z0'S (Y,N)";
1540 INPUT A$
1550 IF A$="Y" THEN 530
1560 IF A$="y" THEN 530
1570 PRINT
1580 PRINT
1590 END
```

Unipole Antennas—Theory and Practical Applications

By Ron Nott, K5YNR
Nott Ltd
4001 LaPlata Hwy
Farmington, NM 87401

When properly designed, constructed and tuned, the folded unipole has the potential of having many advantages over more traditional antennas. It is often utilized as an "afterthought" antenna, sometimes installed on the tower or pole supporting a beam and/or VHF or UHF antennas. Compromises made to suit the particular situation may have negative effects on its ultimate performance, its advantages not being fully realized. While it is most familiar in the form of a vertical antenna for use on the HF bands, the same principles of tuning and performance apply regardless of polarization or frequency.

The folded unipole antenna has been said to be the same as a gamma-matched or perhaps a shunt-fed antenna. While there are similarities on the surface, a properly designed, constructed and tuned folded unipole is far superior to the other two mentioned in many respects.

In this paper, several properties are discussed and compared to the series-fed antenna as well as to the gamma- and shunt-feed methods. At the outset it should be said that a folded unipole is a truly tunable antenna. The dimensions for a standard series-fed antenna must be carefully calculated and then the ends sometimes pruned to get optimum performance from it. The dimensions on the unipole to be described are not critical, the input impedance being a function of how the stubs between the fold wires and the supporting structure are set.

Fig 1 is an illustration of the antenna. Compared to other antennas, it has broad bandwidth and low Q, and is inherently stable with changes of weather and season. Finally, it appears to be less ground-system dependent; that is, it still performs well even if the ground system is not ideal or real-estate boundaries limit the length of ground wires.

Geometry is often overlooked in the design of an antenna, sometimes for convenience and sometimes for economics. Length-to-diameter ratio has been known for many years to affect both the bandwidth and propagation velocity of an antenna.[1] This ratio is easily improved (decreased) by the application of good unipole design.

[1] Notes and references appear on p 38.

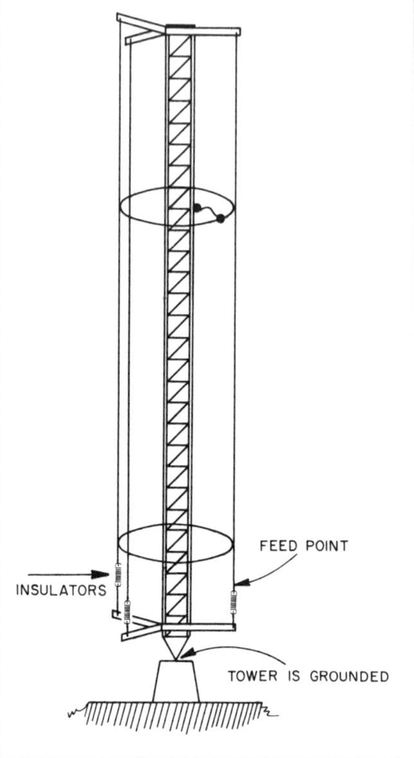

Fig 1—Used commercially for AM broadcasting, the folded unipole antenna can be adapted fairly easily for Amateur Radio use. Tuning may be accomplished by separate jumpers or one jumper and a commoning ring, as shown.

A good unipole uses several skirt wires rather than just one for both symmetry and effective diameter increase. A single wire may help establish a desired input impedance, but it does little to increase the effective diameter of an antenna. On the other hand, if three or more wires are equally spaced around a supporting structure, such as a tower, the effective antenna diameter approaches the diameter of the circle encompassing those wires (Fig 2). Additionally, each wire has an effect on the input impedance of the total antenna. A further effect is a reduction in the velocity of propagation within the antenna.

The folded unipole has been described as a "transmission line antenna." It might be viewed as a length of skeletonized coax cable, with the center conductor grounded while the outer conductor (wire skirt) becomes the feed point. Measurements

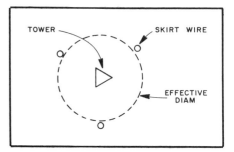

Fig 2—This top view shows placement of skirt wires that serve to increase the antenna's effective diameter.

made on model antennas indicate that the velocity may be on the order of 0.88c rather than the factor of 0.95c which is used in the rule-of-thumb formula for ¼ λ.

$$h = \frac{234}{f}$$

(approximately ¼ λ for a series-fed antenna; velocity is assumed to be 0.95c)

where

h = height, feet
f = frequency, MHz
c = the velocity of EM radiation (such as light)

Using the value of 0.88c, the formula becomes

$$h = \frac{216}{f}$$

(approximately ¼ λ for a unipole)

This makes a ¼-λ unipole about 7 to 8 percent shorter than a series-fed antenna, which may provide a slight advantage when building an antenna. However, we will see that there is nothing magic or sacred about a ¼-λ unipole antenna. With a series-fed antenna, the ¼-λ point is merely the first dimension at which the reactance goes through zero and the resistance is a manageable value. It makes input matching simple, but by a bit of juggling, we can make the folded unipole give us a good match without having to be concerned with precise height or length dimensions.

In the process of tuning, stubs are placed between the inner conductor (often a tower or pole) and the skirt wires. The location of the stubs can provide a wide variety of input impedances, depending on what is desired. Since most ham equipment has adopted a universal value of 50 Ω and zero reactance ($50 + j0$ Ω), that's the value that will be discussed here. Bear in mind, however, that almost any other value can be matched by proper design and tuning.

Theory: Superposition of Two Currents

During the latter years of the Vietnam War, our military wanted an antenna system that would quickly tune to any frequency between 2 and 30 MHz. The forces in the field were using a special version of the Collins KWM-2 which operated in this range. They wanted a quick QSY in the event of jamming, QRM or enemy listeners. Today's transceivers have optional auto-tuners to quickly match into any antenna input impedance (within reason), but they were not available back at that time. General Dynamics developed an antenna called the "Hairpin Monopole" which, with a special auto-tuner, attempted to accomplish this.[2] For an antenna only 14 feet in height, they were able to effect a good impedance match and reasonable coverage throughout the HF spectrum. What is important here is their analysis of the currents within the antenna.

Looking at the lower frequencies, it is obvious that a 14-foot antenna is electrically very short. *Superposition* is a math term that can lead to great complexity, but we will avoid equations and look at it graphically. What it means is that we must analyze the antenna as having two currents flowing in it simultaneously (Fig 3).

The first is a transmission line current, I_L. It enters the antenna terminal, flows upward to the top and then downward in the other conductor of the antenna toward ground. Note that its value is constantly changing to conform to transmission line theory, but if the trip from input terminal to ground terminal is much less than 180 degrees, it never reaches a value of zero or reverses direction.

The second current is the antenna current, I_A. It enters the antenna terminal in phase with I_L and remains so to the top of the antenna (or the tuning stub). However, it does not follow I_L when it turns and starts down the other conductor of the antenna. The other conductor, being in the very near field of the one connected to the input, has a current induced into it the way that a transformer primary winding induces a current into a secondary winding. Thus, in this "secondary" portion of the antenna, the two currents, I_L and I_A, are no longer in phase. But because their amplitudes and phases are not the same, they do not entirely cancel each other.

Superposition allows for the analysis of the resultant current. We may say that the current in the antenna "secondary" is less than in the "primary," but at no two points on the secondary will the resultant current be the same. It gets complicated, but

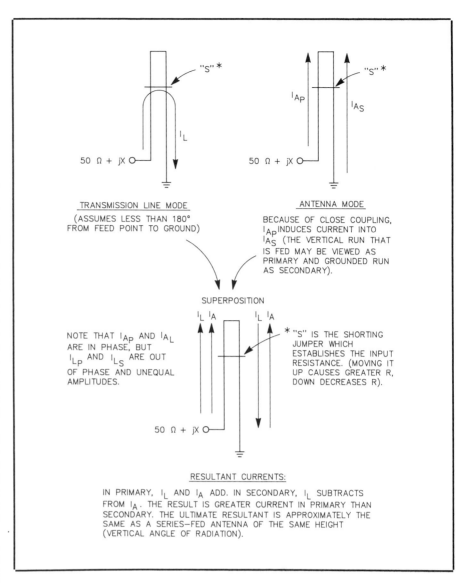

Fig 3—Current distribution on a folded unipole.

knowledge of such details is not necessary to construct a practical antenna. Most of the current will be in the skirt wires and a smaller portion in the supporting structure, such as a tower. The antenna will appear to have skin effect with most of the current being effectively in the form of a sheet on the effectively large outer diameter. The EM flux density will be much less than that surrounding a small-diameter conductor.

Tuning

I have built a two-capacitor tuner for my 48-foot tower, and it works perfectly (see Fig 4). I tested it on 1610 kHz (±28 degrees) and on several points in the 160-meter band (± 33 degrees), and in each case got the optimum match. The tune-up procedure I used may prove helpful.

To tune for $50 + j0$ Ω, connect a bridge from point A to ground. Set the R dial of the bridge to 50 Ω and leave it there. Adjust C_P and the reactance dial of the bridge until a null is found. Move the

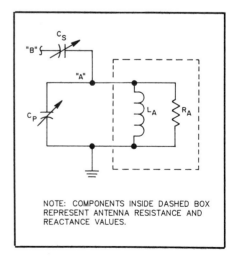

Fig 4—Tuner circuit for the author's "Hairpin Monopole" antenna. Components inside the dashed box represent antenna resistance and reactance values. See text for adjustment procedure.

bridge lead to point B. Set the bridge dials to $50 + j0 \ \Omega$. Adjust C_S until a null is found. Note that C_P and L_A are not resonated, but are deliberately detuned to a point down the slope of the impedance curve. The reactance is then resonated with C_S. Q is decreased by multiple conductors in the antenna, which decreases L_A.

Detuning

The unipole also has the capacity to function as a "detuning skirt." This occurs when it is deliberately tuned to parallel resonance, or nearly so, which then makes the tower or pole effectively disappear. Also, there may be times when a directional antenna system is to be omnidirectional (all the elements but one in the array are to effectively disappear). I have assisted in making two $\frac{1}{4}$-λ towers located only $\frac{1}{4}$ λ away from a tower broadcasting at 5 kW completely disappear from the effective field.

Another application of detuning is to negate the effects of a tower that is too tall—more than 5/8 λ for an AM broadcast antenna, for example. The skirt is sectionalized somewhere around the $\frac{1}{4}$-λ point with fiberglass insulators. The bottom portion becomes the antenna while the upper portion functions as a detuning skirt to make the upper part of the tower effectively disappear. *QST* has published excellent articles on the subject of conjugate tuning through lengths of coax.[3] A simple extension of this principle allows us to make use of "conjugate detuning" instead, which

eliminates the need for motorized components in weatherproof boxes up on the tower.

Practice

Nothing has been mentioned of how to make such an antenna for a typical ham application. In practice, a birdcage is built on a tower or pole with the top ends of the wires being either connected or insulated at their top ends. The bottom ends must be insulated, as they will be commoned together to become the feed point. For two reasons, the wires must be spaced out a substantial distance from the supporting structure: (1) to cause a large effective diameter, and (2) to minimize shunt capacitance between the wires and the structure. Skimping on the spacing defeats the advantages. How much is enough space? From 16 to 30 inches or more. This would provide an antenna of several degrees diameter even on 160 and 80 meters.

What are the stubs and how do you tune them? They are simply jumpers between the wires and the structure placed at a height that will present a reasonable value of resistance (45 to 60 Ω). On 40 meters, for example, they may be only about 12 to 15 feet up from the feed point. Will the antenna be resonant at this point? No, not even close. There will be a substantial value of inductive reactance, which means we don't need a loading coil but rather a loading capacitor. A variable capacitor is easier to tune than a coil, and a capacitor

is an inherently lower loss device than a coil.

In practical applications I have installed an air-variable capacitor in a weatherproof housing in series between the center conductor of the coax from the transceiver and the wire commoning the bottom ends of the skirt wires together. Conjugate matching might be done by calculating the length of the feed line to an appropriate length. However, I haven't gotten that far yet.

Conclusion

There is more to building a good antenna than just stringing up some wires of certain dimensions. Geometry is very important, and proper application of geometric principles can make the difference between a high-Q, narrow-band and sometimes unstable antenna and one of low Q, broad bandwidth and good stability in varying climatic conditions. The cage antennas of the early decades of ham radio very likely helped in settling down transmitting and receiving equipment that would have been wild had it been connected to antennas that were also difficult to deal with.

Notes and References

[1]See J. Devoldere, *Low Band DXing* (Newington: ARRL, 1987), p II-21.
[2]See J. A. Kuecken, *Antennas and Transmission Lines*, 1st ed. (Indianapolis: Howard W. Sams and Co, 1969).
[3]M. W. Maxwell "Another Look at Reflections," *QST,* Apr, Jun, Aug and Oct 1973, Apr and Dec 1974 and Aug 1976.
J. H. Mullaney, "The Folded Unipole Antenna," *Broadcast Engineering,* Jul 1986.

Magnetic Radiators–Low Profile Paired Verticals for HF

By Russell E. Prack, K5RP

2239 Creek Rd
Brookshire, TX 77423

The antenna described in this article consists of paired vertical radiators made from wire. The system does not require ground radials, nor does it require loading coils. The system provides ample bandwidth and it is fed directly with coaxial cable. It is an efficient radiating system which presents a very low profile.

Electric Versus Magnetic Radiators

Ordinary antennas such as dipoles and verticals are classified as *electric radiators*. These antennas generate a very high E component and a very low H component in the fields close in, near the antenna. Boyer shows that the E component creates ground losses and that the H component is virtually lossless.[1] Electric radiators exhibit a relatively high voltage at the antenna ends.

The magnetic radiator generates a very low E component and a very high H component close in to the antenna. Losses in the ground and in nearby objects are therefore minimized. Magnetic radiators exhibit a low voltage at the antenna ends and a relatively high current flowing within. See Fig 1. In the case of the electric radiator, Boyer shows that the radiation field, with a fixed E/H ratio of 377 Ω, predominates beyond ½ λ from the antenna. In the case of the magnetic radiator, the predominance of the radiation field occurs beyond 1½ λ from the antenna. Magnetic radiators can therefore be efficient and desirable in many applications.

Laport classifies the U antenna as a magnetic radiator.[2] In Fig 2, four U antennas are connected together and evolve into Dellinger's "coil aerial."[3] The coil is rectangular in shape and, although made from wire, is derived from Boyer's DDRR doublet.

The four U antennas forming the rectangular coil are referred to as a multiple U, or simply MU. The four vertical ends of the rectangle are the prime radiators. The horizontal wires produce two minor high-angle lobes. Note that there are no loading coils or radials used. The bottom horizontal runs of the antenna are located about 2 feet above ground.

The vertical ends of the antenna are 30 electrical degrees high. This is less than

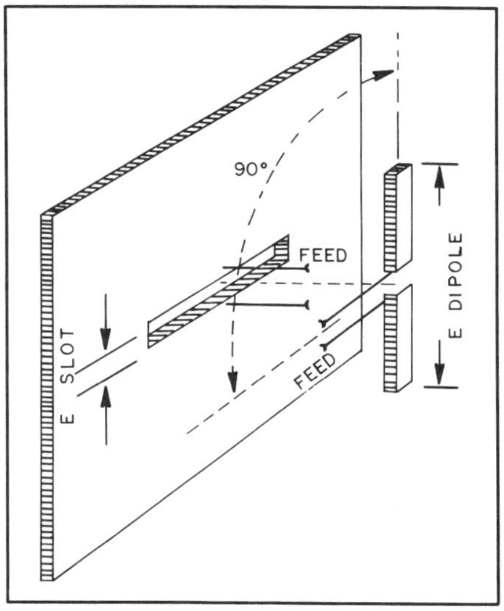

Fig 1—Insight into magnetic and electric radiators can be gained by considering the slot antenna. The slot is a magnetic radiator made by cutting a thin sliver from a large sheet. To the right, a complimentary electric dipole is made from the sliver. Near the antennas (near field), the H field component of the slot is proportional to the E component of the dipole. The H component of the dipole is proportional to the E component of the slot. Although the slot is horizontal, polarization is vertical. For comparison, the dipole is rotated 90° to a position of vertical polarization. (See notes 4, 5 and 6.)

0.1 λ, and indeed presents a low profile. This translates to about 11.5 feet for 40-m operation and about 21.5 feet for 80-m operation. (Although the wires at the far end cross at a small angle, these wires are referred to as vertical.)

The horizontal runs are approximately 150 electrical degrees long. This is about 58.2 feet for 40-m operation and 109.2 feet for 80-m operation. The total wire length is approximately 2 λ at the desired frequency. The configuration of the wire in the rectangular coil is such that the currents in the four vertical end wires are co-phased. Both the magnitude and phase of these currents are virtually constant throughout their 30° height.

Radiation Pattern

Fig 3 shows a plot of the radiation field of a model MU operating at 21.4 MHz. This plot, although made from crude measurements, resembles the theoretical free-space pattern for this type of antenna, also shown in Fig 3.

The measurements were made at sites ranging in distance from 0.8 to 1.5 miles and located at 45° intervals around the antenna, beginning in the suspected direction of the main lobe. All measurements were normalized to 1 mile. The deep null to the southwest is possibly the result of a nearby steel tower that is in this direction

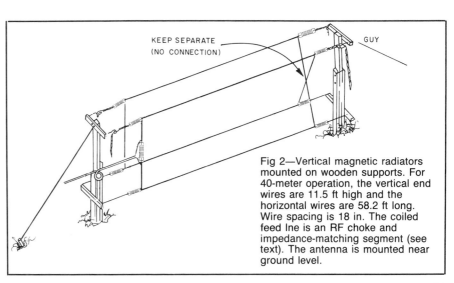

Fig 2—Vertical magnetic radiators mounted on wooden supports. For 40-meter operation, the vertical end wires are 11.5 ft high and the horizontal wires are 58.2 ft long. Wire spacing is 18 in. The coiled feed lne is an RF choke and impedance-matching segment (see text). The antenna is mounted near ground level.

[1]Notes appear on page 41.

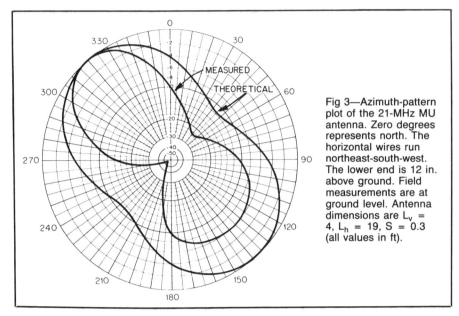

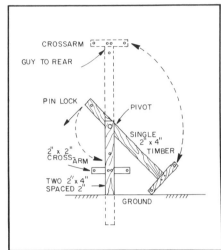

Fig 3—Azimuth-pattern plot of the 21-MHz MU antenna. Zero degrees represents north. The horizontal wires run northeast-south-west. The lower end is 12 in. above ground. Field measurements are at ground level. Antenna dimensions are L_v = 4, L_h = 19, S = 0.3 (all values in ft).

Fig 4—Typical wooden support. Setup and takedown can be accomplished by one person (use help if possible). Use treated lumber. To determine dimensions, see text.

and directly in line with the MU model. The smaller main lobe to the southeast is possibly the result of the measurement sites in this area being well below the antenna horizon. The plane of the model is oriented northeast-southwest. Maximum radiation is broadside, as it is from ordinary co-phased verticals.

Design and Construction

The following expressions were developed for use in designing the MU antenna.

L_t = 1988/f
L_v = 82/f
L_h = 415/f
S = 125/f

where

L_t = total wire length required, ft
L_v = height of each vertical wire, ft
L_h = length of each horizontal wire, ft
S = wire spacing, in.
f = center frequency of operation, MHz

For example, for a center frequency of 3.8 MHz, the following values are obtained from the expressions above.

L_t = 523.2 ft
L_v = 21.6 ft
L_h = 109.2 ft, and
S = 32.9 in.

The length of the horizontal runs, L_h, will usually give resonance at a slightly lower frequency such that the antenna can be pruned to resonance. The antenna should be pruned by clipping four short segments, equal in length, from the center of each horizontal run. Attempting to prune the antenna at the feed point will require the rerigging of the eight corner insulators (see Fig 2).

For efficient operation, the antenna should be made from no. 14 or larger wire.

Since the antenna is frequency sensitive to stretch in the horizontal runs, stranded wire is recommended. Good results have been reported when using no. 12 copper-clad steel.

Figs 4 and 5 show some of the construction details of the wooden supports used here for various MU antennas. On the 40-m model, it is recommended that the spacing between the cross arms on each support be made 14 feet. This will permit vertical tension to be applied to the vertical wires. For the same reason, 24-foot spacing is recommended for the 80-m model.

If the bottom two horizontal runs are placed within a few feet of ground level,

these runs should be protected or adequately marked to prevent someone from falling or becoming burned. If the 80-m model is mounted with the bottom near ground, two short supports bearing cross arms with insulators on the tips should be used to prevent the lower horizontal runs from sagging into the grass. These supports should be placed one-third of the distance between the end supports.

Feeding the MU

At resonance, the 40-m MU presents a measured input impedance of 115 + $j0$. (The impedance will change somewhat with different antenna heights above ground.)

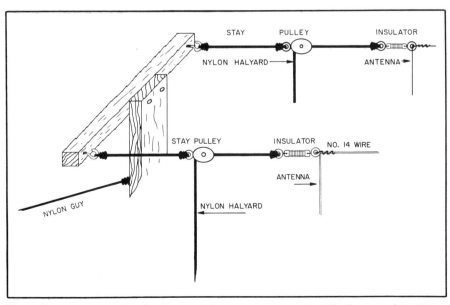

Fig 5—Upper cross-arm detail. Upper and lower cross arms are vertically spaced greater than L_v. (See text.)

This load is ideally matched to 50-Ω coaxial cable by ¼ λ of 75-Ω cable acting as a transformer. Since currents in the MU must be held equal, top and bottom, the 75-Ω cable should be made into an RF choke by winding it into a flat coil about 12 inches in diameter. This coil can then be taped or strapped together.

The 80-m MU presents a measured input impedance of $150 + j0$ at resonance (again, subject to minor change with antenna height). This load is matched by using a series segment of good 300-Ω ribbon. Feed the antenna with 50-Ω coaxial cable. Cut the cable at a point 16.9 feet from the antenna. At this point insert a 6.66-foot length of 300-Ω ribbon in series with the 50-Ω cable. This arrangement matches the 150-Ω load to any random length of 50-Ω feed line. (A velocity factor of 80% was assumed for both ribbon and cable. An operating frequency of 3.75 MHz was also assumed.) The 16.9-foot segment should be wound into a flat coil to form an RF choke, as was done with the 40-m MU. Additionally, it is well to form a second RF choke by winding another flat coil in a portion of the 50-Ω cable on the transmitter side of the ribbon, right at the ribbon.

With the feed arrangements just described, system bandwidth is about 4% of the center frequency. This is approximately 280 kHz on 40 m and 150 kHz on 80 m. The term bandwidth as used here refers to the frequencies between which the SWR rises to a ratio of 2:1 on the 50-Ω cable feeding the system.

A "20-Degree" MU

Joe Logan, WA4CPN, and Dr Tom McLees, WF5I, both aircraft pilots, became interested in the low-profile performance of the MU antenna. Hence a design for an 80-m "20°" MU was made with the vertical ends being but 14.5-feet high. Thus Joe, with Doc's assistance, built the first 14.5-foot MU for 80 m and erected it near Joe's flight strip. The advantage of low profile in this application is obvious.

On 3.75 MHz, Doc and Joe measured the input impedance of the 14.5-foot MU to be $110 + j0$. Hence, the 75 Ω transformer match was used. Shortly thereafter George McCulloch, WA3WIP, built the second 14.5-foot MU for 80-m operation.

For the lower profile 20° antenna, use the following.

$$L_v = 54.67/f$$
$$L_h = 442.33/f$$
$$S = 125/f$$

Operating Results

With the 14.5-foot MU on 80 m, Joe consistently receives good reports from Europe and Africa. George reports equally good results to Europe and to Australia via long path. On one test George reported that upon switching to a dipole antenna, neither the European nor the Australian via long path was able to hear him.

From this location, using the 21.5-foot verticals on 80 m, good reports are consistently received from South Africa and New Zealand, as well as from other continents. Using the 11.5-foot MU on 40 m, a daily schedule was maintained for a 4-month period with Paul Stein, G8NV/MM. Paul was aboard a cargo vessel operating in the South Atlantic, the Indian Ocean and the Far East. On 85% of the days, Paul reported that signals from the MU were S9 or more. Although the MU is not the best for local contacts, it has been found to be surprisingly effective as a long haul, low profile, radiating system.

Many thanks go to Dave Atkins, W6VX, Joe Boyer, W6UYH, Bob Lewis, W2EBL, and Paul Stein, G8NV. They generously donated their time and talents to the project.

Notes

[1]J. M. Boyer, "Surprising Miniature Low Band Antenna," Parts I and II, 73, Aug and Sep 1976.
[2]K. Henney, Ed., *Radio Engineering Handbook*, 4th ed. (New York: McGraw-Hill Book Co, 1950), p 660.
[3]J. H. Dellinger, "Principles of Radio Transmission and Reception with Antenna and Coil Aerials," *Bureau of Standards Sci Paper 354*, June 1919.
[4]H. Jasik, *Antenna Engineering Handbook*, 1st ed. (New York: McGraw-Hill Book Co, 1961), p 2.5 and Chap 8.
[5]*Reference Data for Radio Engineers*, 6th ed. (Indianapolis: Howard W. Sams & Co, subsidiary of ITT, 1975), pp 27-14, 27-15.
[6]J. D. Kraus, *Antennas*, 1st ed. (New York: McGraw-Hill Book Co, 1950), Chap 13.

A Multiband Loaded Counterpoise for Vertical Antennas

By H. L. Ley, Jr, N3CDR
PO Box 2047
Rockville, MD 20852

Did you ever wish you could have an efficient vertical antenna without installing "umpteen" radials? For most amateurs with limited available space for a radial system, the possibility of erecting a ground-mounted vertical antenna is frequently dismissed because ¼-λ radials will not fit in the usual residential city lot. Many turn to a roof-mounted vertical with resonant radials as an alternative. True, this is an efficient system with a number of advantages, but it removes the antenna to a point where it is difficult to reach for adjustments and experimentation. Another approach to the design of a ground-mounted vertical was presented by L. A. Moxon, G6XN.[1] Moxon points out that several amateurs, including himself, have found a tuned, loaded counterpoise to be an efficient substitute for the radial system with a conventional ground-mounted vertical radiator. Moxon makes a strong case that the counterpoise should be shorter than ¼ λ, as it normally would be for a roof-mounted vertical with insulated, resonant radials.[2] He sets as a lower limit of total physical counterpoise length about 1/8 λ, or approximately 18 feet for 7 MHz.[3] These concepts led me to explore the development of a tuned, loaded multiband counterpoise for a ground-mounted vertical antenna that has demonstrated on-the-air performance equivalent to a radial system of ten radials up to 37 feet long.

The counterpoise system is made from two 8-foot lengths of ¾-inch aluminum tubing and the associated loading inductance. These extend 8½ feet horizontally at either side of the base of the vertical antenna. The visual impact is acceptable, and the space requirements are so modest that the system will fit into most city residential lots.

Mechanical Details

I chose a Hustler 4-BTV antenna for the installation because it was available. Other multiband or single-band vertical antennas

should give comparable results. The antenna was ground-mounted according to the manufacturer's instructions with a 4-foot long, 2-inch diam steel pipe, sunk 2½ feet into the ground in the center of a 48 × 76 foot backyard area. The antenna base was insulated from the pipe by a section of split PVC tubing. The antenna tubing sections were adjusted, again following the manufacturer's directions, for a ground-mounted antenna with surface or buried radials.

In the counterpoise, the two 8-foot lengths of ¾-inch aluminum tubing are supported 8 inches above the ground by

ceramic standoff insulators on wooden stakes. The tubing sections are placed in a straight line, one on either side of the vertical, with the inner ends spaced 6 inches from the antenna. They are connected at their inner ends by a no. 12 wire jumper attached at the center to the loading inductance. Every effort was made to keep the installation symmetrical.

Operation of the counterpoise was compared with that of a radial system by switching between the two. The radial system consisted of 10 radial wires, symmetrically placed around the antenna. At the base of the antenna, each radial was

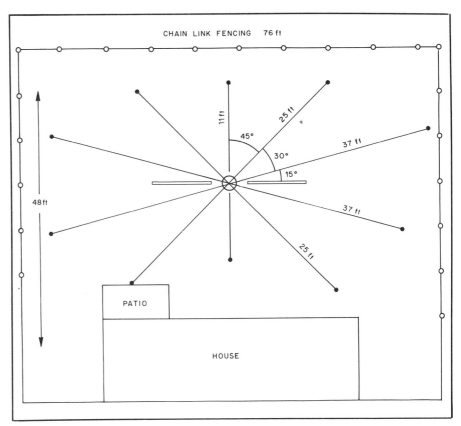

Fig 1—An overhead view of the antenna, radials, counterpoise, and surrounding area. The counterpoise is made of aluminum tubing and is perpendicular to the 11-foot-long radials.

[1]Notes appear on page 45.

connected with a solder lug to a stainless-steel disk, drilled and tapped for no. 8-32 machine screws. The radials were cut from no. 14 copper wire for the maximum length permitted by the size of the lot. Four of the radials were 37 feet, four were 25 feet, and two were 11 feet long. Each radial was pinned to the ground with 5-inch U-shaped "staples" cut and formed from coat hangers. The installation is outlined in Fig 1. In order to achieve symmetry in the radial installation, opposing radials were laid in a straight line, passing through the axis of the antenna. Originally, I gave thought to attaching the radials to the chain link fence surrounding the yard on three sides, but this was dismissed for reasons of loss of symmetry and of possible harmonic interference generated at rusty fence joints.

Electrical Details

For initial tests and evaluation of the vertical and radial system, an electrical ½ λ of 50-Ω RG-58 cable at 7.125 MHz was used with a Palomar noise bridge and appropriate low-reactance resistors installed in PL-259 plugs.[4] Initially, the bridge was set to zero on the reactance scale and balanced by adjustment of the station receiver frequency and the resistance potentiometer in the bridge. Each reading was repeated three times to permit averaging the result. The results are presented in Table 1. The measured antenna resistance varied between 45 and 70 ohms at average resonant frequencies of 7.10, 14.08, 21.40 and 28.27 MHz. These results were "distorted" by the fact that the feed line was not an exact multiple of ½-λ on all frequencies, although it was close in the 40-

and 15-meter bands. Nevertheless, the results indicated that the antenna and radial system would give a low SWR with 50-Ω cable on all bands, a fact that was confirmed in subsequent on-the-air testing.

The antenna tubing section lengths could have been adjusted on 14, 21 and 28 MHz to bring the resonant frequency closer to the center of the bands, but I decided at the beginning of this evaluation to use the lengths recommended by the manufacturer in all subsequent measurements and testing.

Next, a 2-inch diameter, 2½-inch long loading coil of 8 turns per inch was selected from the junk box (such as B&W no. 3900 or AirDux no. 1608T). Alternate turns of the coil were bent in toward the center on one side of the coil to permit tapping it at 1.5, 3.5, 5.5 turns and so on. The starting turn could be tapped easily from

the end of the coil. This inductance was connected from the coax braid connection on the insulated antenna mount through a relay to the counterpoise with small alligator clips. A 12-V 4-pole double-throw relay (Radio Shack No. 275-214) with all four similar contacts connected in parallel was wired to permit remote switching between the radial system and loaded counterpoise. A schematic of the installation is shown in Fig 2.

The station receiver, noise bridge and resistance standards were again put to work to determine the appropriate taps on the loading inductance for 40, 20, 15 and 10-meter operation with the counterpoise, using two separate sections of 50-Ω coax cut for ½ λ at 7.05 and 7.125 MHz. Results of these measurements are shown in Table 2. It is apparent that the loading of the

---

Table 1

Noise Bridge Data on Antenna and Radial System

Zero reactance was measured at all frequencies in this table.

Band	Resistance, Ohms	Freq, MHz
40 m	45	7.109
	50	7.110
	45	7.081
	Average 47	7.10
20 m	60	14.100
	50	14.019
	60	14.123
	Average 57	14.08
15 m	65	21.430
	65	21.378
	70	21.380
	Average 67	21.40
10 m	60	28.250
	65	28.276
	60	28.284
	Average 62	28.27

Measurements were made through a 7.125-MHz ½-λ section of RG-58 cable with Palomar noise bridge.

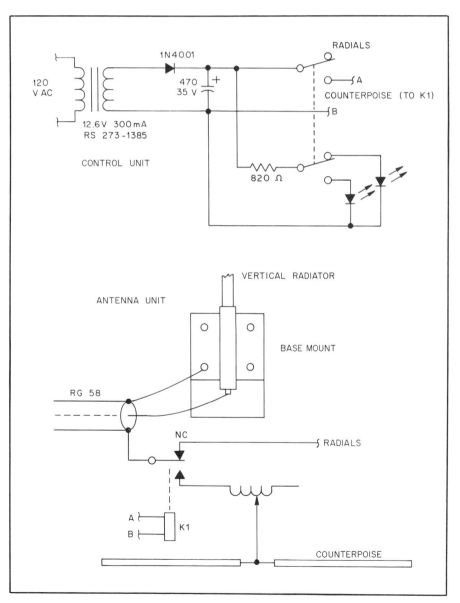

Fig 2—The remote switching arrangement used to switch between the counterpoise and ground radial system from the operating position.
K1—Radio Shack® four-pole double-throw relay, part no. RS275-214.

Table 2

Noise Bridge Data on Antenna and Counterpoise System

RG-58 cables were cut to ½ λ at 7.050 and 7.125 MHz.

Band	½λ Cable MHz	Measurement Freq, MHz	Loading Coil Tap, Turns	Resistance, Ohms	Reactance Dial
40 m	7.125	7.125	11.5	55	L 10
			9.5	40	C 2
			7.5	60	C 10
40 m	7.050	7.050	11.5	55	L 10
			9.5	45	0
			7.5	50	C 10
20 m	7.050	14.100	5.5	125	L 10
			3.5	50	0
			1.5	55	C 10
15 m	7.050	21.150	3.5	100	L 20
			1.5	75	0
			0.5	70	C 10
			1.5	70	0
10 m	7.050	28.000	None	60	C 10
		28.390	None	60	0

counterpoise could be adjusted to provide a nonreactive match in all four bands. Measured resistance values at a zero-reactance setting of the noise bridge varied between 45 and 75 ohms. At 28 MHz no added inductance was necessary to achieve system resonance at 28.390 MHz. This suggests that it may be desirable to extend the counterpoise, if the system is to be made resonant at a lower frequency on 10 m, by increasing the space between the inner ends of the tubing and increasing the length of the wire connecting them. With the collection of these results, it was clear that the loaded counterpoise provided an acceptable nonreactive antenna, ready for on-the-air testing.

On-the-Air Testing

Testing was conducted in three stages. First, SWR measurements were made in all four bands with the vertical working first against the radial system, and then against the counterpoise with the appropriate inductance connected in the circuit. Results of these tests are presented in Table 3. The radials and the loaded counterpoise gave comparable results in terms of SWR measurements.

The second stage of testing included checking the S-meter levels of received signals as the antenna was switched between the radial and the counterpoise system. No consistent difference was observed at levels above S-6. At levels of S-3 or lower, the radial system had a slight edge over the counterpoise, up to about one S unit. At no time was a signal heard on the system with radials that could not be heard with the counterpoise. This was true on all four bands.

The third stage of testing was use of the antenna in actual contacts, mainly on

Table 3

SWR Data

Band	Freq, MHz	With Radials	With Counterpoise
40 m	7.105	1.7:1	1.1:1
	7.125	1.7:1	1.1:1
	7.145	1.8:1	1.2:1
20 m	14.050	1.3:1	1.5:1
	14.150	1.7:1	1.8:1
	14.200	1.8:1	2.0:1
15 m	21.120	1.8:1	1.8:1
	21.150	2.0:1	1.7:1
	21.180	2.0:1	1.8:1
10 m	28.100	2.0:1	2.1:1
	28.300	2.1:1	2.0:1
	28.400	2.2:1	2.0:1

Measurements were made with a 25-watt-output transmitter and Heath HM-9 HF SWR/wattmeter.

40 m in the 7.100-7.150 MHz Novice/Technician segment. Contact would be initiated with the radials in use, and then, later in the contact, after RST reports had been exchanged, the system was switched to the counterpoise. I announced that I had "switched antennas" and asked if the other operator noted any change in signal strength. In every case, without a single exception, the response from the other operator was "no change," sometimes with a repetition of the original RST report.

Contacts with a 25-watt crystal-controlled transmitter covered a circle with a radius of approximately 400 miles. A quick review of the log and an atlas revealed no apparent preferred direction of communication of the counterpoise system. Further testing may reveal some favored direction for the counterpoise, but it is not apparent

at this point in evaluation of the system.

Discussion and Conclusions

I approached this project with considerable skepticism, despite the convincing arguments of Moxon that a vertical monopole with loaded counterpoise was electrically the near equivalent of a vertical dipole. The test results demonstrate that Moxon was correct, and, further, that the loaded counterpoise is equivalent to a modest system of ten radials in on-the-air testing.

Obviously, the Hustler 4-BTV antenna requires some "fine tuning" to adjust its element lengths for resonance in the desired portions of the bands covered. Do not be discouraged if the loaded counterpoise reduces the claimed bandwidth of your vertical antenna. I remind you that the loaded counterpoise will reduce the bandwidth of this or any other vertical antenna because of the inclusion of another tuned element in the system.

I would like to have tested the system in actual contacts on bands other than 40 m before preparing this paper, but because of time, license and propagation limitations was unable to do so. Performance on 40 m was considered the most critical test of the concept because the counterpoise length was approximately 1/8 λ at 7 MHz. Others can fill in the information for 20, 15 and 10 m, where the counterpoise is a greater fraction of a wavelength.

Some readers may question the use of only 10 radials for comparison with the counterpoise, pointing out that the efficiency of a vertical radiator is reduced if the ground resistance of the radial system is too high.[5] The radial system included four elements greater than 1/4 λ at 7 MHz, and another four at approximately 1/5 λ. Although this does not meet the commercial broadcast standard of 120 radials, it is probably a better radial system than used by most amateurs, and is certainly as good or as better than the usual rooftop insulated, resonant radial system. It is possible, as noted in *The ARRL Antenna Book*, that a system of twenty 1/8-λ radials might have been preferable to the system used here. But that would have doubled the number of wires that could become entangled in the lawn mower, an environmental hazard of some concern! Should you explore the use of a larger number of radials, I would be interested to learn the results.

There remains one unanswered question in my mind at this point in the project. To what degree are the favorable results observed with the counterpoise system attributable to the presence of an isolated radial ground screen below it? I have no answer to this question except to believe that the influence of the radials on the counterpoise is minimal because of another example given by Moxon in his book.[6] In his Fig 11.16, Moxon sketches a monopole with a short inductively loaded counterpoise for 20 m. In this instance the counterpoise loading is not lumped but rather distributed, with the 21-foot counterpoise bent roughly into the shape of a trombone at the base of the vertical ¼-λ monopole element. He points out that this antenna may be mounted at ground level, but that improvement in performance "of at least half an S-unit can be expected if the base height is raised to about 0.2 λ, or 14 feet at 14 MHz." Moxon's antenna had no radial system below it, and, except for the type of loading used, does not differ greatly from the design presented here.

Elevation of the base of the antenna is attractive for other reasons, namely reduced earth losses and environmental safety. On the other hand, it just is not always feasible to elevate the base of the vertical element by 0.2 λ, or 28 feet, at 40 m. In my installation, there are no young children in the household or immediate neighborhood, so the chance of accidental contact of persons with the counterpoise element is minimal in a fenced backyard. A potential hazard of RF burns exists with any ground-mounted vertical element, even when operated against grounded radials, so the counterpoise is not considered to pose a significantly greater hazard than a ground-mounted vertical. The advantages of having the antenna feed point easily accessible for adjustment of the counterpoise loading inductances is obvious, and even critical, if the taps are selected by an electrically or manually operated switch. If I operated high power, I would give thought to extending and guying the mounting pipe so the antenna and counterpoise might be mounted about 6 feet above the ground. This would significantly improve environmental safety, yet still permit access by ladder for adjustment or repair.

As yet, I have not found a mathematical method to calculate the inductance required to resonate the 17-foot center-fed counterpoise, as I did with a loaded short horizontal dipole.[7]

The data given here will provide a starting point if you are interested in installing a similar system. The values of inductance used for this ground-mounted vertical with loaded counterpoise are 4.3, 0.92 and 0.52 μH for 40, 20 and 15 m respectively. I will be pleased to hear from others installing a similar system, and will reply to comments or questions if a stamped return envelope is enclosed with the letter. So, good luck, down with radials, and up with the counterpoise!

Notes

[1]L. A. Moxon, *HF Antennas for All Locations* (RSGB: Potters Bar, Herts, 1982), pp 4, 154-157, 164-165.

[2]See Moxon (note 1), pp 144-145.

[3]See Moxon (note 1), p 155.

[4]B. S. Hale, Ed., *The ARRL Handbook for the Radio Amateur*, 66th ed. (Newington: ARRL, 1989), p 25-23.

[5]G. L. Hall, Ed., *The ARRL Antenna Book*, 15th ed. (Newington: ARRL, 1988), pp 2-33, 2-34.

[6]See Moxon (note 1), pp 164-165.

[7]H. L. Ley, Jr, "Short Loaded Half-Wave Dipole Design—The Easy Way," *The ARRL Antenna Compendium, Volume 1* (Newington: ARRL, 1985), pp 116-122.

A Multiband Groundplane for 80-10 Meters

By Richard C. Jaeger, K4IQJ
711 Jennifer Dr
Auburn, AL 36830

After several years of inactivity, I recently found time to get back on the air. I've always been a DX chaser and decided to pursue 5BDXCC. A review of my QSL collection found country totals close to 100 on 10, 15 and 20 meters, but only 43 on 40 and a paltry 7 on 80 meters. It was clear that my activity needed to concentrate on 40 and 80 meters.

Choosing the Antenna

I began to explore antennas for low-band operation. My past experience included a 40-meter full-wave loop that had performed reasonably well and low dipoles on 80 meters that had always left a lot to be desired. After studying a number of antenna articles and books, I became convinced that the low angle of radiation of a vertical antenna would offer the best opportunity for low-band DX operation. However, the same literature indicated that a large number of radials are needed for good performance, and I simply do not have the space to put an adequate radial system into the ground, particularly for 80 meters.

On the other hand, many European stations produce excellent signals on the high-frequency bands with groundplane antennas, and I set out to explore the possibility of using such an antenna on 40 and 80 meters. The book by Orr and Cowan provided some very useful information about the groundplane antenna.[1] Fig 1 gives the vertical radiation pattern of a groundplane with its base at a height of ½ λ, showing useful low-angle radiation below 10°.

Of even greater interest is Fig 2 which indicates that as few as 6 radials are needed if the base of the antenna is ¼ λ high, and only four are needed if the antenna is ½ λ high. If my 50-foot tower were used to support the groundplane, it would be almost a quarter wavelength high on 80 m and a half wavelength high on 40 m. At this point I opted to take down my beam and put up a groundplane. Not wanting to give up 10, 15 and 20 meters, I decided to make a multiband groundplane for 0 meters.

[1]Notes appear on page 49.

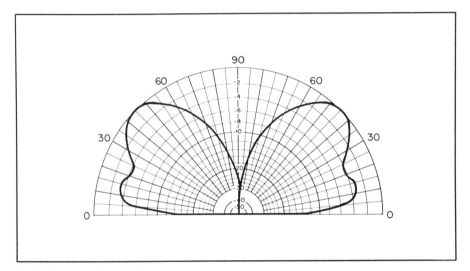

Fig 1—Vertical radiation pattern of a 7-MHz groundplane antenna with the base elevated ½ wavelength above ground. Calculated with MININEC, a method of moments procedure, for average soil conditions (dielectric constant = 13, conductivity = 5 mS/m).

The Multiband Design

Butternut verticals enjoy an excellent reputation, and the Butternut HF6V seemed to be a good choice for the vertical radiator. In addition, resonant radials are required for each band. Past and present editions of *The ARRL Antenna Book* have given several 80-10 meter dipole designs which require only a single set of 40-meter traps.[2] As an example, Fig 3 gives the computer-generated SWR curve of such an antenna showing the resonances near each of the five major bands. A pair of these trap dipoles at right angles to each other will provide a simple system of four radials for five bands. The inductance of the traps provides the additional advantage of reducing the overall radial length on 80 meters.

The first version of this antenna had four radials using 40-meter traps from SPI-RO Manufacturing[3] Details for home construction of similar traps can be found in *The ARRL Antenna Book*.[4] Another alternative would be to use the multiband dipoles available from Alpha Delta Communications.[5]

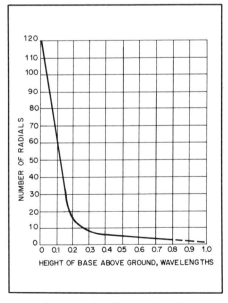

Fig 2—Number of radials required for a groundplane antenna versus height of the base above ground. Adapted from Orr and Cowan, *All About Vertical Antennas* (see note 1).

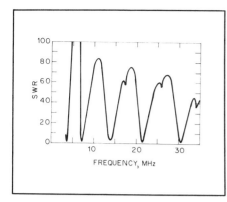

Fig 3—SWR versus frequency for a 5-band trap dipole obtained from computer simulation using the method of moments. The simulation assumes $\ell 1$ = 34 feet, $\ell 2$ = 23 feet and lossless traps. (See Fig 4.)

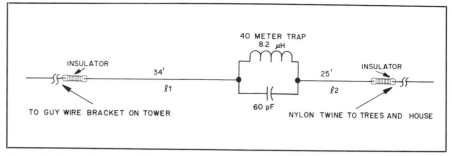

Fig 4—Multiband trap radial design with lengths chosen for operation at 3500 and 7050 kHz.

There is some disagreement in the literature on whether the radials should be exactly ¼ λ in length or approximately 2.5% longer than the standard formula for a ¼-λ wire antenna. I finally decided to start out with the longer length and trim the radials later if necessary. The inner section of each radial must be cut for 40-meter operation, and its length in feet is given by

$$\ell 1 = \frac{240}{f} \qquad \text{(Eq 1)}$$

where the frequency f is specified in MHz. In order to try to cover most of 40 meters, a design frequency of 7050 kHz was chosen, resulting in a length of 34 feet for the inner part of the trap radial.

Based upon the design information from SPI-RO and the other designs in *The Antenna Book,* the length of the radial beyond the trap was chosen to be 25 feet for an operating frequency of 3.5 MHz. To design for other frequencies, this length should be changed to

$$\ell 2 = 25 \times \frac{3.5}{f} \text{ feet} \qquad \text{(Eq 2)}$$

where f is the new frequency in MHz. The completed radial design is given in Fig 4. As it turned out, the length of the radials was never changed.

Concern is often expressed about trap losses. Measurement of a SPI-RO trap indicates resonance occurs at approximately 7120 kHz with a 10-μH inductor in parallel with a 50-pF capacitor. The equivalent series resistance of the traps was measured to be 2 ohms at a frequency of 3 MHz using an RF impedance bridge. Four radials provide an equivalent resistance of only ½ ohm. (It should be noted that this resistance is substantially higher than the dc resistance of each trap, which is less than 0.04 ohm.)

Fig 5 shows the calculated current on a trap dipole operating at resonance on 80 meters. The current at the trap position is approximately 70% of the current at the

feed point. Feeding 1000 watts into a 50-ohm antenna results in a current at the feed point of

$$I = \sqrt{\frac{P}{R}} = \sqrt{20} \text{ amperes} \qquad \text{(Eq 3)}$$

where I is the RMS value of the current. One quarter of this current flows in each radial, and the total loss in the four traps is

$$P = 4I_t^2 R_t = 4 \left(0.7 \times \frac{\sqrt{20}}{4} \right)^2 \quad (2)$$
$$= 4.9 \text{ watts} \qquad \text{(Eq 4)}$$

where I_t is the current flowing in the trap and R_t is the resistance of the trap at the operating frequency.

If the feed-point resistance of the antenna were only 10 ohms, as might occur in a short vertical, then the feed-point current would increase to 10 amperes, and the loss in the traps would be

$$P = 4 \left(0.7 \times \frac{10}{4} \right)^2 (2) = 25 \text{ watts}$$
$$\qquad \text{(Eq 5)}$$

This loss still represents only 2.5% of the total power being supplied to the antenna.

Antenna Construction

I removed my CL-33 beam from the tower and mounted the HF6V vertical on a piece of TV mast at the top of the tower, placing the base of the groundplane antenna approximately 55 feet in the air. The bottom of the HF6V was gently tapped into the top end of the TV mast. I drilled a hole through the mast and antenna base, and secured the two parts with a 2½-inch bolt.

The radials attach to a guy-wire bracket, obtained from Radio Shack, with two radials (or three for a six-radial system) attached to each of the bolts of the bracket. See Figs 6 through 8. The radials are constructed of no. 14 Copperweld® wire, and each radial is isolated from the guy-wire bracket by an insulator. The radials fan out from the tower with a 90° spacing for 4 radials and a 60° spacing for 6 radials. A wire ring connects one radial to the next at

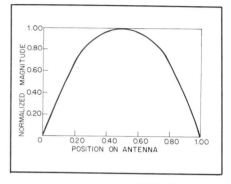

Fig 5—Simulated current distribution on a shorter trap dipole operating at resonance (3700 kHz) on 80 meters. Current at the trap position is approximately 70% of the feed-point current.

the base of the antenna. A single feed wire runs from one point on the ring up to the coax shield connection on the HF6V. The feed ring can be seen in Fig 8, which shows the completed antenna with its radials stretched out to various supports. The radials droop to help raise the feed-point impedance, and the ends are attached to various trees with 200-lb test nylon twine. The end of the lowest radial is approximately 30 feet above the ground.

The feed ring seems to work well, and there has not been any indication of an asymmetrical radiation pattern with the existing radial system. An alternative approach to feeding the radials would be to do away with the ring and run a bundle of feed wires from the radials up to the feed point. This alternative approach should help balance current flow in the radials.

The manufacturer's suggested procedure for adjusting the HF6V worked fine for tuning up the antenna. One perplexing problem was encountered. After making an adjustment, the SWR became high on 10 meters, as shown by curve A in Fig 9A. In addition, the SWR curve was independent of the 10-meter length adjustment. After pulling out some hair in frustration, I realized that I had replaced the coaxial cables between the transceiver, Transmatch and linear amplifier, changing the total length

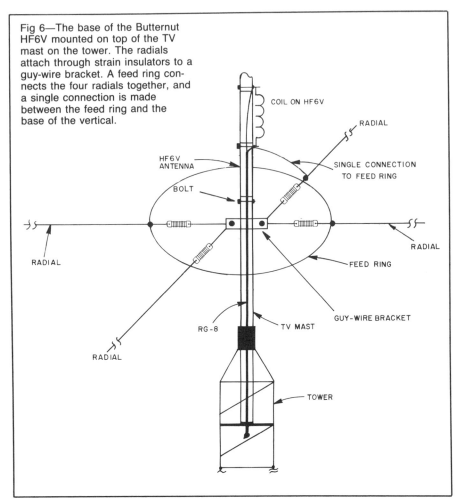

Fig 6—The base of the Butternut HF6V mounted on top of the TV mast on the tower. The radials attach through strain insulators to a guy-wire bracket. A feed ring connects the four radials together, and a single connection is made between the feed ring and the base of the vertical.

COIL ON HF6V

RADIAL

HF6V ANTENNA

SINGLE CONNECTION TO FEED RING

BOLT

RADIAL

RADIAL

FEED RING

RG-8

TV MAST

GUY-WIRE BRACKET

RADIAL

TOWER

of the coax feed line. Adjusting the length of the coax going to the transceiver eliminated the problem and resulted in curve B in Fig 9A. [The need for this change indicates the author probably has RF current flowing on the outside of the coax shield. The length (resonant frequency) of the radials will affect this current.—*Ed.*]

I have experimented and found that a similar problem can occur on 15 meters. It is also worth noting that the SWR circuit in my TS-940S usually gives a considerably more pessimistic estimate of the off-resonance SWR than does the bridge in my SB-200 amplifier.

Be prepared to raise and lower the antenna several times during the tuning process. I have a Hy-Gain gin pole which eases the job of dropping the tower and raising it again. The process of taking down the radials, lowering the tower, adjusting the antenna and putting the whole thing back up takes about an hour and a half.

It is easy to get the radials tangled after the tower is raised, particularly with six radials. After a couple of bad experiences trying to untangle the radials, I developed the following system. Each radial is taped to a leg of the tower before the tower is lowered, and the insulator at the end of each radial is marked with its compass direction using an indelible marking pen. Taping the radials helps keep them in their proper position, and marking the ends is an aid if they become entangled. When the tower is raised, be sure to untape and raise

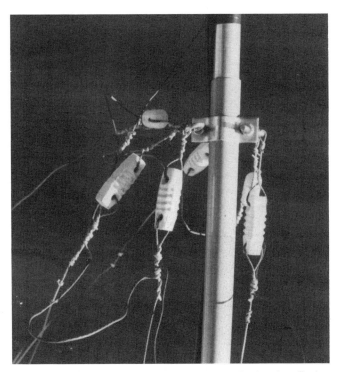

Fig 7—Shows the radials attached to the guy-wire bracket. Taping the radials to the tower can help eliminate tangled wires when erecting the antenna.

Fig 8—The 6-wire radial system with the groundplane in the air. Adding radials can compensate for inefficiency caused by lower antenna heights.

the radials in the reverse order that they were taped to the tower.

Results and Performance

The four-radial version of the antenna performed well, but the SWR curve was broader than expected on 80 meters, indicating that there still might be some substantial ground losses. So I ordered two more traps and went to a six-radial system. The two additional radials did indeed cause the SWR curve to become sharper. However, I have not really been able to notice any qualitative difference in on-the-air performance.

A photograph of the final antenna with six radials is given in Fig 8, with the SWR curves presented in Fig 9. The antenna easily covers the CW DX band on the low end of 80 meters and can be tuned to cover almost the full band on 40-10 meters with an SWR below 2:1.

Although the HF6V is also designed to work on 30 meters, the SWR of the groundplane is approximately 2.3:1 across the band. This is higher than desired, but it is easily within the range of most Transmatches. My noise bridge indicates that resonance is occurring at approximately 9.6 MHz, even with the antenna adjustment at the end of its range. Shorting out a portion of a turn on the 30-meter coil should help shift the resonance point into the 10-MHz band. An extra set of radials for this band would be another possibility. I have not had time to fully experiment with reducing the SWR on 30 meters.

This is by far the best low-band antenna I have ever had. Running approximately 800 watts output with this antenna produces a competitive signal on 40 meters and a solid signal on 80 meters. I often break through 40-meter pileups with only a few calls. I never knew so much DX activity existed on 40 and 80 meters, and I now routinely work Asia and the far Pacific on all bands. It is common to listen to 40 meters in the evenings and hear strong

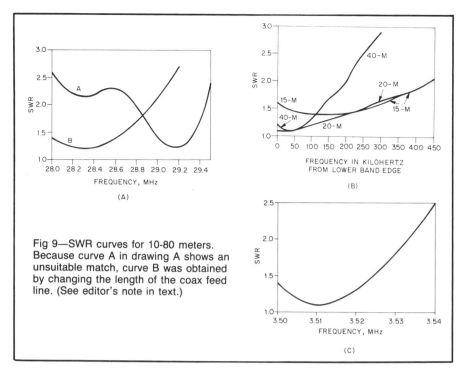

Fig 9—SWR curves for 10-80 meters. Because curve A in drawing A shows an unsuitable match, curve B was obtained by changing the length of the coax feed line. (See editor's note in text.)

signals from Europe, Africa, North America and South America, all coming in at once. In 18 months of casual but active operation I have worked 166 countries on 40 meters and 112 on 80 meters.

The low angle of radiation is effective on all bands. Although giving up 5 dB or more of antenna gain on 20 m is a disadvantage in big pileups, it has not been a problem for normal operation. Many times I have answered DX stations calling CQ on 20 meters where I seem to be one of the few stations answering the call. Fig 1 shows that there is substantial radiation from the antenna as low as 3 and 4 degrees from the horizon, and I think this provides an advantage on the bands. The multiband feature of the antenna is also so convenient

that I am considering keeping the groundplane as my permanent antenna.

I wish to thank Professor Lloyd S. Riggs and Scott Boothe of the Auburn University Electrical Engineering Department for providing the antenna simulations shown in Figs 3 and 5.

Notes
[1]W. I. Orr and S. D. Cowan, *All About Vertical Antennas* (Wilton, CT: Radio Publications, Inc, 1986).
[2]G. L. Hall, Ed. *The ARRL Antenna Book*, 15th ed. (Newington, CT: ARRL, 1988), p 7-10.
[3]SPI-RO Manufacturing Inc, PO Box 1538, Hendersonville, NC 28793; T-40 traps.
[4]See note 2.
[5]Alpha Delta Communications, Inc, P O Box 571, Centerville, OH 45459.

Tunable Vertical Antenna For Amateur Use

By Kenneth L. Heitner, WB4AKK/AFA2PB
2410 Garnett Ct
Vienna, VA 22180

For many years, I have used a simple random-wire antenna for the lower HF bands. It served my purposes well, since most of my operation is at one frequency, about 4.6 MHz. This frequency is used by the MARS program in which I participated. However, I also wanted to be able to operate on the 40- and 80-meter amateur bands, as well as the 7-MHz MARS frequencies. While my coupler allowed me to operate at these frequencies, performance of the random-wire antenna was less than optimal.

About two years ago, I began to work on a better approach. In addition to covering the frequencies of interest, I also wanted an antenna that would be relatively small and easy to construct from available materials. The antenna orientation that I chose was vertical. This is because I wanted to emphasize the low-angle radiation, especially on the 7-MHz frequencies used for longer distances. Quite frankly, this is not the best configuration for communication distances under 1000 miles, where high radiation angles are desirable. However, the vertical antenna is also easier to put up in many cases, and lends itself to situations where not much horizontal space is available.

Basic Design

The basic antenna is shown in Fig 1. Technically, it is a vertical folded monopole, fed with an omega match. The antenna itself is constructed from two pieces of aluminum tubing, 1 and 1¼ inches OD respectively. I joined them with about a 9-inch overlap by wrapping the smaller tube with thin aluminum sheet and sliding it into the larger tube. To assure that the final joint was rigid, I drilled small holes in the region of overlap, and used six no. 6 sheet-metal screws to fasten the two parts firmly together. I used plated screws to avoid corrosion.

The fold in the antenna is designed to have a characteristic impedance of about 600 ohms. This is another compromise. If the impedance is much lower, my experience is that the antenna tends to become too lossy. If the impedance is much higher, the spacing between the fold wire and the pole becomes too large. The fold was constructed of aluminum wire (about no.

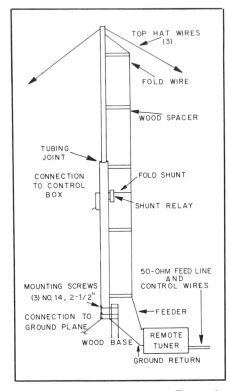

Fig 1—Tunable vertical antenna. The total height is approximately 15 feet. Radials are not shown.

12), and held in place by wood spacers with small Lexan® insulators at the end. Since the voltages involved are not that high, this arrangement works well at the 100-watt power level.

The fold runs the full height of the antenna. On the 7-MHz band, this provides more positive reactance than is needed to tune the antenna. Thus, the fold has a shunt placed across itself by a small relay. The relay is closed for 7-MHz operation. Almost any relay will do if it has reasonable contact spacing when open. My advice is not to operate the relay when transmitting. The relay I used is a Radio Shack No. 275-244 with the contacts wired in series.

The antenna is made to appear electrically longer by use of a top hat. In this case, a three-wire top hat is used. This approximately doubles the electrical height (length)

of the antenna. If more wires or a skirt are added, the upper frequency range of the antenna would be lowered.

A ground plane was formed from six radial wires, 20 to 30 feet long, running away from the base of the antenna. Since the antenna was roof mounted, the radials were terminated at the edge of the roof. Ground wires were also connected to the copper vent pipes on the roof, thus effectively tying in the water pipes as part of the ground system.

The tuner circuit is shown in Fig 2. The circuit is known as the omega match, and is described in several books on antennas. What is unique here is that the capacitors are made remotely tunable with two small permanent-magnet gear motors. The specific motors I used are Autotrol no. PX-200. Similar small dc gearhead motors are available from C & H Sales. (See Fig 2 caption.) In order to achieve very fine control over the capacitance value, a mechanical vernier can be put between the motor output shaft and the capacitor. I used a model similar to the Radio Kit no. S-50.[1] Typically, these verniers are designed for only 180-degree rotation of the output shaft. Thus, the vernier must be disassembled, and the stop tabs removed to allow for continuous rotation.

The tuner components were mounted on a small wood board. The entire tuner was placed in a large plastic refrigerator dish with a tight-sealing cover. The connections were made by mounting binding posts on the refrigerator-dish walls and wiring directly to the components of the tuner. Thus, the box could be easily connected at the roof, and it was essentially water proof.

Performance Characteristics

The antenna is tuned to the operating frequency by watching the reflected reading on an SWR meter. It is sometimes difficult to find the initial null because only one capacitor is varied at a time. Once the null is found, the two capacitors can usually be varied to produce a fairly deep null, indicating the antenna is resonant. Resonance should be achieved before full power is applied to the antenna.

[1]Radio Kit, PO Box 973, Pelham, NH 03076, tel 603-437-2722.

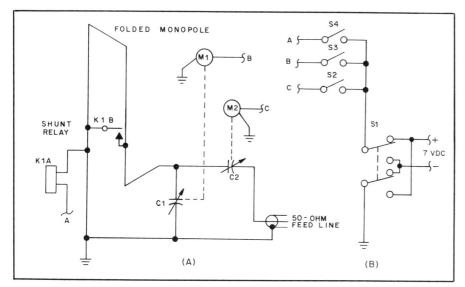

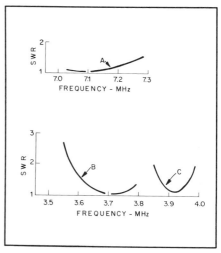

Fig 2—Remote controlled omega-match tuner (A) and control box (B).

C1—150-pF air variable.
C2—100-pF air variable.
K1—Shunt relay, DPDT Radio Shack no. 275-244.
M1, M2—Small dc gear motors. Available from C & H Sales, PO Box 5356, Pasadena, CA 91107-0356, tel 800-325-9465.
S1—DPDT switch (toggle or rotary).
S2, S3—SPST push-button switch.
S4—SPST switch.

Fig 3—SWR curves for tunable vertical antenna. These curves are with a top hat and partial skirt.
Curve A—Tuned to 7.1 MHz, shunt relay closed.
Curve B—Tuned to 3.7 MHz, shunt relay open.
Curve C—Tuned to 3.9 MHz, shunt relay open.

Once the antenna is tuned to a particular frequency, operation on adjacent frequencies does not require retuning. Fig 3 shows how the SWR varies with frequency. The bandwidth on 80 meters is at least equal to that of a trap vertical, with the advantage of retuning if desired. On 40 meters, the antenna is adequate across the entire band with one setting.

Based on a limited amount of operating experience at 4.6 MHz, the antenna can perform within a few decibels of a random wire. The random wire still has the advantage for close-in communications of producing more high-angle radiation. On 7 MHz, this antenna appears to outperform the random wire. Certainly, theoretical considerations favor the vertical antenna for longer HF circuits.

As with any do-it-yourself project, some care in the choice of materials and the methods of assembly will produce a better product. My experience with the aluminum tubing and plated hardware is that they withstood the elements fairly well. I think this particular antenna design will be useful to amateurs who wish to operate on HF, but have limited space in which to erect HF antennas.

A 5/8-Wave VHF Antenna

By Don Norman, AF8B
41991 Emerson Court
Elyria, OH 44035

This antenna grew out of a series of experiments with feed-line decoupling. Ralph Turner, W8HXC, and I began exploring feed-line decoupling after Ralph discovered considerable RF on the feed line of a popular commercial 2-meter vertical antenna. The writer built a 5/8-λ antenna and began a series of experiments with feed-line decoupling.

Since this was a homemade antenna, the radials were attached to the mast with a homemade ring clamp and could be repositioned very easily. A series of tests and measurements proved to my satisfaction that 1/4-λ radials do not belong near the matching network on a 5/8-λ antenna. Quarter-wave radials positioned 3/8 λ below the matching network worked quite nicely and yielded excellent feed-line decoupling.

I believe that a 5/8-λ antenna with 1/4-λ radials placed 3/8 λ below the matching point acts in the same manner as the venerable extended double Zepp antenna. In fact, the antenna works well when the matching network is removed and it is center fed with 300-Ω balanced line. The balanced line must be led away from the antenna at right angles for more than 1 λ.

When the decoupling experiments were finished, there were a large number of odds and ends on hand and a search was begun for a design of an antenna that the average amateur could build with ordinary hand tools and hardware store and Radio Shack items. A sketch of the antenna is presented in Fig 1.

The radiator is cut from ¾-inch OD aluminum tubing and the supporting mast is a 1¼-inch OD television antenna mast. The matching rod and radials are made from hard-drawn aluminum clothesline wire. The center insulator is made from a pipe coupling for the sort of semiflexible plastic water pipe that is joined with molded plastic fittings and stainless steel hose clamps. Fig 2 shows three views of the plastic pipe fitting.

Fig 2A is a sketch of the coupling before anything is done, 2B is a cutaway of the coupling inside the mast, and 2C is the radiator inside the insulator. The ¾-inch aluminum radiator is a loose fit inside a 1-inch pipe coupling, and the 1-inch pipe coupling is a loose fit inside the TV antenna mast. These loose fits are tightened with shims cut from aluminum beverage cans. The radiator is installed in the insulator by inserting the tubing halfway through the

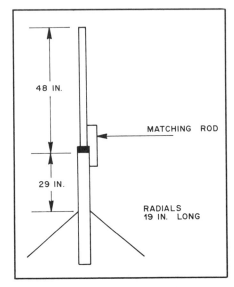

Fig 1—Diagram of the 147-MHz 5/8-λ antenna. It is designed to be built with readily available parts and ordinary hand tools.

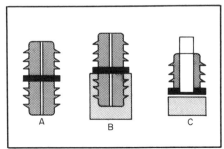

Fig 2—Three views of the plastic pipe fitting inside the center insulator.

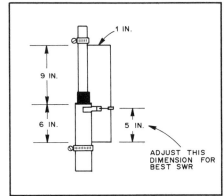

Fig 3—Detailed view of the matching network.

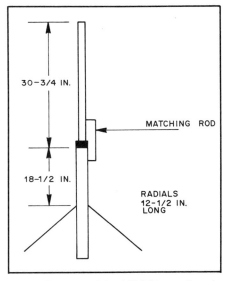

Fig 4—Diagram of the 220-MHz version of the antenna.

coupling and drilling a hole and installing a self-tapping sheet-metal screw. The radials are attached to the mast with self-tapping sheet-metal screws.

Fig 3 is a sketch of the matching network of the 5/8-λ antenna. The matching rod is bent up from a 19-inch length of hard-drawn aluminum wire. Measure 2 inches from one end and make a right-angle bend. Make another right-angle bend 1 inch from that one. Measure 2 inches from the other end and make a right-angle bend. Make another right-angle bend 1¼ inches from the end. You should now have a U-shaped piece of wire with the U 1 inch deep at one end and ¾ inch deep at the other.

Drill a 3/8-inch hole through the mast and insulator between two of the mast attachment screws. Fish the coax through this hole. Attach the matching rod to the radiator and mast with stainless steel hose clamps according to the dimensions in Fig 3. (Be sure to file any paint or anodizing off the mast and radiator.) Ground the coax shield to the mast under one of the mast attachment screws. Attach the coax center conductor to the matching rod with a homemade clamp. Adjust the coax tap position on the matching rod for best SWR.

The same design works well at 220 MHz. A 220 antenna was designed, constructed and tested in 1982, but only recently became popular locally. Fig 4 gives overall dimensions for the 220 antenna. Materials and construction of the 220-MHz version are the same as for the 147-MHz version; only the lengths are different.

Fig 5 is the matching detail for the 220-MHz antenna. The antenna may be built for center frequencies other than 147 and 220 MHz. Formulas are (all dimensions in inches, f = center frequency)

radiator—7056/f

radials—2793/f

radial attachment point below top of mast—4263/f

matching rod—2205/f

matching rod attachment point above

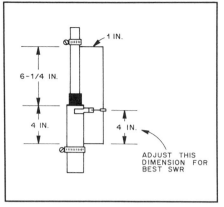

Fig 5—The matching network for the 220-MHz version.

top of mast—1323/f

The matching rod is spaced 1 inch from the radiator for both the 2-meter and 1¼-meter bands.

This antenna design works well, whichever frequency it is constructed for. We have found that matching is easier if the feed line is cut in multiples of a half wavelength at the most common operating frequency. A half wavelength in inches is determined by the equation

L (inches) = 5904/f × VF

where

f = frequency in MHz

VF = velocity factor of the particular cable

Some Experiments with HF and MF 5/8-Wave Antennas

By Doug DeMaw, W1FB

PO Box 250
Luther, MI 49656

Is a 5/8-wave radiator practical for HF and MF operation? This paper describes the results of some related tests. I have been asked many times if there is anything to be gained from using a 5/8-wave antenna at MF and HF. I have used the standard 5/8-wave vertical antenna for 2-meter mobile operation for a number of years, and I find it entirely acceptable for communications at VHF. But, what about lower-frequency use? I built two 5/8 antennas to explore the practicality of this type of radiator. My observations follow.

The Nature of the 5/8-Wave Antenna

A full-size 5/8-wave radiator presents a capacitive reactance of approximately 165 Ω. Therefore, it becomes beneficial to insert an inductive reactance of equal value at the feed point in order to make the antenna resistive. This accounts for the manner in which series-fed 5/8-wave mobile antennas are configured. The coil may be tapped (shunt feed) to provide a 50-Ω match to RG-8 or RG-58 feed line (illustrated in Fig 1). This approach is my preference for single-band use of a 5/8-wave antenna.

Comparisons—1/4 Wave to 5/8 Wave

I erected a 1/4-wave sloping wire (45° pitch) for 40 meters. It was worked against 20 on-ground radials and matched by means of a broadband transformer to a 50-Ω feed line. I then constructed a 5/8-wave sloping antenna with a matching/loading coil. It was used also with 20 on-ground radials. I used an antenna switch in the ham shack to allow quick comparisons in signal strength while evaluating the antenna performance. The antenna switch also controlled remote relays that selected inductors which detuned the unused antenna when the tests were made. This was done to minimize interaction between the antennas.

I observed local, medium-distance and DX signals with the two antennas. Stateside signals received at my Michigan QTH averaged 3-6 dB louder with the 5/8-wave antenna, but there were certain times when some signals were equal to or louder than those heard via the 5/8-wave, while using the 1/4-wave antenna. Antenna comparisons during transmit yielded similar overall results.

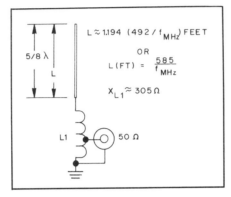

Fig 1—Example of a typical 5/8-wave VHF mobile antenna. L1 actually resonates the system by canceling capacitive reactance at the feed point. L may be trimmed to obtain an SWR of 1. The coil tap is selected to provide a match to 50-Ω feeder cable.

The story is somewhat different for DX operation. The 1/4-wave radiator had the edge 75% of the time. The difference in signal reports was on the order of one S unit, favoring the 1/4-wave antenna. However, there were times when the 5/8-wave radiator was approximately one S unit better than the shorter antenna. The time of day and propagation conditions at that time caused the change in antenna effectiveness. In any event, both antennas worked very well for communications to Europe and Asia when band conditions were favorable.

The 5/8-Wave Antenna on 160 Meters

I have used 1/4- and 3/8-wave inverted-L antennas on 1.8 MHz over the years. My curiosity prompted me to erect a 5/8-wave antenna in 1987. I replaced a full-wave, vertically erected, rectangular 160-meter loop with the 5/8-wave radiator. The loop had lower-corner feed to provide vertical polarization, and the loop was fed with 450-Ω tuned feeders to permit multiband operation.

I deployed a 5/8-wavelength wire (309 feet) for the new antenna. No loading/matching coil was used. Rather, I chose to feed the system with 450-Ω ladder line to allow multiband operation. Fig 2 shows the arrangement I am using. One side of the feeder is connected to the buried 160-m

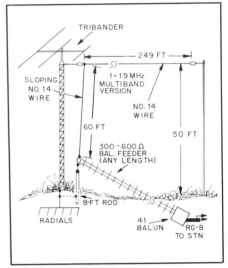

Fig 2—Details of the W1FB 160-meter, 5/8-wave inverted-L antenna. Balanced feed line is used to permit multiband use from 1.8 to 30 MHz. The far end of the wire may be bent toward ground if space is limited. In-ground or on-ground radials are required for proper antenna performance.

radial system (22 wires) and the other side of the line is attached to the end of the 5/8-wave wire. A 4:1 balun transformer is located just outside the ham-shack window. This converts the balanced feed to RG-8 coaxial line (8 feet of line to my Transmatch in the shack).

I cut the 160-m 5/8-wave wire for 1.9 MHz, the center of the band. With the arrangement in Fig 2 the SWR is 1.3:1 at 1.9 MHz without using the Transmatch. I need the Transmatch, however, for the low and high ends of 160 meters, and during harmonic operation on 80 through 10 meters.

Performance on 160 meters has been excellent. I receive better signal-strength reports than I did from the 1/4- and 3/8-wave inverted-L antennas. This analysis is based on reports from the same stations I worked regularly with the smaller antennas, out to 1000 miles from Luther, Michigan. I hear weak signals better than with the previous inverted-L antennas and the loop, including DX signals on top band. The 2:1 SWR bandwidth on 160 m is 100 kHz without the Transmatch.

Harmonic Operation of the 160-M 5/8-Wave Wire

Regular schedules are kept on 80 meters with my son, N8HLE, in Tolland, Connecticut—some 1000 miles from here. My signal improved between 5 and 10 dB over what it was with the 160-m loop antenna. In a reciprocal manner, his signal increased also. W8EEF (MI) also takes part in these weekly schedules. Norm uses comparable power to mine (AL-80A amplifiers), but uses an inverted-V dipole antenna for 80 meters (height is 60 feet). Previously, he matched or exceeded my signal strength in Connecticut. Now, my signal averages 5-10 dB stronger than his, which adds further proof that the inverted L is working well. On 80 meters it functions as a 1¼-wave radiator. My son has since replaced his inverted-V dipole with an 80-meter 5/8-wave inverted L, and he uses the same type of feed system that I do (Fig 2). He reports excellent performance from 80 through 10 meters. These results suggest the practicality of this antenna, should it be dimensioned for 40 meters.

I have the horizontal part of my 160-m 5/8-wave L running east and west. I discovered during some QSOs with Bill Orr, W6SAI, on 12 meters that it has gain and good directivity on the higher harmonics. Bill reported that my signal was sufficiently loud in Menlo Park, California to suggest that I am using a beam antenna and an amplifier. I operate 12 meters with a barefoot FT-102 (100 W output). There is obviously some strong east-west directivity

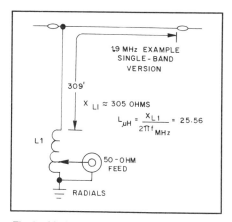

Fig 3—Method for feeding a 5/8-wave inverted L for single-band operation. A 1.9-MHz example is shown. L1 consists of 75 turns (close wound) of no. 14 enamel wire on a 1-inch OD low-loss coil form. Winding length is 4 inches. The tap on L1 is chosen experimentally to provide a match to 50-Ω coaxial line. Avoid shorted turns when placing the tap on T1.

from the wire at the higher frequencies (as with a long wire).

Single-Band Use

If you prefer to use coaxial-cable feed, the arrangement in Fig 3 may be followed. This will permit burying the feed line, which may appeal to city dwellers. L1 is adjusted to provide resonance in the portion of the band where you plan to operate. Next, the 50-Ω feeder is tapped experimentally on L1 (near the grounded end) to obtain an SWR of 1:1. The SWR meter may be used at the feed point during these adjustments.

Antenna resonance may be checked by way of a dip meter. After the feeder is matched to the antenna, recheck the antenna resonance and adjust L1 accordingly. There will be some interaction between the two adjustments. When checking antenna resonance, be sure to disconnect the feed line from L1, as it will cause misleading frequency measurements.

Summary Remarks

Try to make the vertical portion of your 5/8-wave inverted L as high as practicable. This will aid overall performance. Similarly, try to elevate the horizontal part of the antenna as high as you can. Also, the more 1/4-wave radials you install the better the antenna efficiency. Ideally, we should use 120 radials as the minimum. I seem to obtain good results with only 22 radials (130 feet long, each), but more wires will be laid next summer.

Dip-meter frequency tests may be made by coupling the dipper probe coil to the grounded end of L1 in Fig 3. Remove coil turns if the frequency is too low, and add turns if it is too high. You may avoid pruning the coil turns by adjusting the overall wire length while observing the SWR meter. If coil adjustment is contemplated, add three or four extra turns beyond the calculated amount. This will allow leeway in resonating the system.

Yagi-Type Beam Antennas

New Techniques for Rotary Beam Construction

By G. A. "Dick" Bird, G4ZU/F6IDC
111, Sur Le Four
Malves en Minervois
11600 Conques sur Orbiel, France

Equipped with lightweight, compact, modern transistorized equipment, many radio amateurs have the opportunity of taking an HF transceiver on holiday or business trips to exotic locations. Operation with a mobile whip or the odd piece of wire often produces rather disappointing results, and with a short-duration one-man DXpedition there is generally little possibility of shipping a normal commercial beam antenna. However, with a little ingenuity, and using somewhat unorthodox techniques, a weekend should suffice to construct a very effective beam antenna at negligible cost using locally available materials and simple tools.

The various designs to be presented are intended mainly for use on the 10- or 15-m bands; they could even be scaled up for 20 m, if so desired. Experience indicates that 10 and 15 m are probably the most rewarding bands for low-power DX work, particularly on north/south paths. With the higher MUFs to be expected, at least for the coming six to seven years, this will become even more evident.

Novices and others may well find some of the designs attractive for use at a permanent location. They are easy to construct, light in weight with low wind resistance, and the material costs are negligible. The performance, regarding gain and front-to-back ratio (F/B), will certainly not fall short of a commercial product.

Before proceeding further, I must stress that rotary beam elements do not *have* to be made of aluminum tubing, and they do *not necessarily have to be straight*. V-shaped elements do in fact offer certain electrical advantages relating to gain, bandwidth and F/B ratio. No. 165 copper wire will support power levels up to several kilowatts with negligible loss!

The first design to be described, the "Jungle Job," was developed some years ago for use in Africa where the only materials available were bamboo rods and ordinary household electric-light wire. The subsequent designs are more sophisticated developments exploiting the advantages of rigid, lightweight plastic materials. They offer somewhat higher gain and perhaps better visual impact for the neighbors.

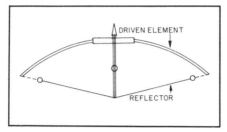

Fig 1—The Jungle Job two-element beam is a simple and lightweight structure.

The Jungle Job

As will be observed from Fig 1, the Jungle Job looks rather like a bow and arrow! The "arrow" is the supporting boom. It can be of 1-inch-square timber or stout bamboo of similar section about 5 to 6 feet long (for the 10-m band). The arrowhead on the end to indicate direction of fire is an optional extra!

The "bow" is fabricated with two 9-foot bamboo canes, end to end, tightly lashed with string to a supporting member of ½-inch × 1-inch timber approximately 3 feet long. This supporting member was fixed to the boom with a ¼-inch iron bolt and wing nut. No drill being available with the original model, the holes were made with a red-hot nail! As no coax fittings were obtainable, a 5-ampere household plug and socket were mounted at the center of the

bamboo-cane structure, to form the termination of a dipole radiator wound in a very open spiral (about one turn per foot) along the length of the bamboo canes.

The classic formula for a wire dipole, 468/f, would suggest 16 feet 6 inches tip to tip or 8 feet 3 inches each side. This is probably just about correct for no. 16 or no. 18 enameled copper wire. In my particular case, the only wire available was plastic-covered flexible wire. Because of the loading effect of the plastic covering, the length needs to be shortened by 3 to 5%. The reflector, however, needs to be about 3 to 5% longer than the radiator. The simplest solution seems to be to cut 16 feet 6 inches of twin lead, split it down the center, use one length for the reflector, and the other cut in the center for the radiator. The tips of the radiator could be subsequently trimmed by an inch or so for best SWR.

Obviously the reflector forms the "string" of the bow. It is tensioned by two pieces of nylon cord to the tips of the bow, producing the form shown in Fig 1.

The feed impedance is around 35 to 40 ohms, according to height, and the SWR at band center will be better than 1.3:1 in most cases. The F/B will be around 25 dB, which is much better than a normal two-element beam. The reason for this superior performance is explained in Appendix 1.

For a permanent installation, a more sophisticated appearance (at some addi-

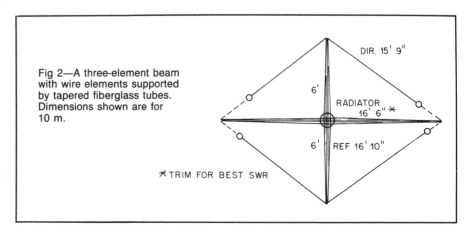

Fig 2—A three-element beam with wire elements supported by tapered fiberglass tubes. Dimensions shown are for 10 m.

DIR. 15' 9"

RADIATOR 16' 6" ✶

REF 16' 10"

✶ TRIM FOR BEST SWR

tional cost) could be realized by replacing the bamboo canes with tapered fiberglass fishing-rod blanks, and the use of a thin-wall aluminum tube for the boom. The total weight should be well under five pounds and the wind resistance is very low, so TV mast sections will provide adequate support if attached to the side of the house or a convenient chimney stack.

I doubt whether the complexity of a balun is justified, but if desired, a ¼-λ sleeve removed from larger diameter coax can be slipped over the lightweight 50-ohm coax feeder, or the feeder can be formed into a four- or five-turn line choke close to the feed point.

Structural and Size Considerations

The beam in Fig 2 is a three-element array that represents an attempt to overcome many of the limitations and shortcomings of commercial monoband beams.

Most experts would agree that the gain and bandwidth of a three-element beam are proportional to the length of the boom, and that with close spacing between elements, it is not possible to maintain peak performance over the whole width of an amateur band. (A boom length of 0.4 to 0.5 λ is optimum with 3 elements).

Unfortunately, transportation and distribution services normally impose a limit on the length of packages they are required to handle. Beam manufacturers are therefore obliged to cut both the boom and radiating elements in two or more pieces to facilitate transit. A split boom produces obvious mechanical weaknesses, which is aggravated by the end loading resulting from the almost universal use of tubular elements for the director and reflector. Mechanical failure makes very bad publicity and manufacturers therefore tend to favor boom lengths much shorter than optimum.

Many prospective users may be concerned about the size of the turning circle as they wish to avoid overhanging neighboring property. It is true that a shorter boom will reduce the turning circle, and this may in some cases be a deciding factor in purchasing brand X as against brand Y.

In this respect, readers may be interested to note that because of the use of V-shaped parasitic elements, the beam of Fig 2 has a smaller turning circle than even a very close spaced beam of conventional design. It therefore provides more gain and bandwidth in less space (and almost certainly at much lower cost). See Fig 3.

Construction

Because the central wire-dipole radiator and the V-shaped parasitic elements have a weight of around 3 ounces each, and because of relatively low wind resistance, the boom and support for the radiator can be fabricated with light, but reasonably rigid material. Selected bamboo canes, with the knots smoothed down with a sanding

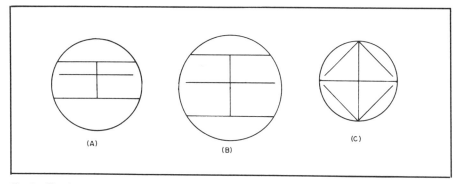

Fig 3—Turning circles (to same scale). At A, three elements spaced 0.1 to 0.15 λ have a good F/B but forward gain is no more than 7 or 8 dBi. B shows three elements spaced 0.2 λ, having 10 to 10½ dBi gain with only a moderate F/B. With a spacing of 0.2 λ at C, 10 dBi with a good F/B is possible.

disk and a couple of coats of silver paint, can look quite reasonable.

A more elegant solution is to use tapered fiberglass rods, such as fishing-rod blanks or spreaders sold for quad antennas. These would be fixed solidly in a short length of aluminum or plastic tube at the center to support the compression force of the mounting brackets, which can be fabricated with sheet metal and U bolts. Because of the wide spacing, no critical adjustments are necessary.

The feed impedance is about 40 ohms (as compared with only 15 or 20 ohms for a close-spaced beam). As with the previous antenna, a line choke or ¼-λ balun can be used if so desired. Wide-spaced beams have a reputation for poor F/B. With this particular antenna the F/B is excellent (around 25 to 30 dB) because of the V-shaped reflector.

Summary

The majority of HF operators settle for a 3-element Yagi or a loop type of antenna such as the cubical quad. Adding further elements will produce some extra gain but this unfortunately follows the rule of diminishing returns. Assuming a ½-λ boom, it would be necessary to increase the boom length *eight* times to 4 λ for an improvement of 6 dB or one S point! A better solution is *vertical stacking,* which, with appropriate spacing, can provide up to *5 dB* of additional gain. Users of stacked arrays have an enviable reputation for a very potent DX signal, but mechanical and structural problems (plus cost!) have restricted the use of stacked arrays to the "lucky few."

A gain of around 14 dBi can be realized without critical adjustments. Material cost is moderate and weight and wind resistance are suprisingly low.

Appendix 1—Gain and F/B

When tuned for maximum gain, a Yagi type of beam generally shows a rather mediocre F/B. The F/B can be improved

by retuning the elements but this results in a loss of gain. In theory, an *infinite* F/B is possible if the currents in the radiator and parasitic elements are equal and suitably phased (phase shift equal to physical spacing in fractions of a wavelength).

One method of equalizing currents is to couple the elements tightly together by using close spacing (with straight tubular elements). This unfortunately results in a narrow bandwidth and very low radiation resistance, which introduces matching problems and unacceptable resistive losses.

A better method is to use V-shaped reflectors. The capacitance between the tips of the radiator and reflector provide the critical coupling required for a high F/B, while the points of maximum current at the center of the elements can remain widely spaced to ensure maximum gain, wide bandwidth and reasonably high radiation resistance. The optimum spacing between the tips of the radiator and reflector will generally be between 8 and 14 inches on 10 m, and proportionally more for 15 or 20 m.

The spacing between the tips of the *director* and radiator should be at least double these figures, preferably 2 feet to 2 feet 6 inches. Readers who find attractive the low cost and simplicity of this form of construction may well ask, "Would four elements give even more gain, and is it possible to construct a two-band or three-band antenna using similar techniques?"

A second director will not provide more than about 1 dB of additional gain, which is barely noticeable. Taking advantage of the light weight and low wind resistance, a better approach would be to mount the antenna somewhat higher than could be contemplated with a more massive system. On long-haul DX contacts, this can often provide an improvement of 6 dB or more.

Appendix 2—Multiband Systems

In this appendix, I present several advanced systems offering the possibility of multiband coverage. A simple and effective way of covering more than one band is to (1) scale up the dimensions of the Jungle Job (Fig 1) for, say, the 15-m band and then (2) add a second 10-m driven element in parallel at the feed point, plus a 10-m reflector appropriately spaced as in Fig 4. When two elements for different frequencies are close together there is always some interaction. For best SWR it will generally be necessary to slightly *shorten* the longer element and *lengthen* the shorter element. Folding them back a couple of inches will generally suffice, but start with elements a little on the long side. It is *possible* to cover 10, 15 and 20 m using such techniques, but the array becomes rather cumbersome, and there are probably better solutions, as shown in Fig 5.

Most readers are familiar with the center-fed "Zeppelin" type of radiator, to use the common misnomer. When fed with 300- to 600-ohm balanced lines via a Transmatch, it can provide a resistive load to the transmitter over a range of at least 1½ octaves, and near the high end of the range can (without directors or reflectors) provide a gain of up to 5½ dBi (see extended double Zepp in any antenna book). Using the Zepp as a driven element, the forward part of the boom can remain 6 feet long because it has to support only the 10-m director. The beam is thus 3 elements on 10 m and 2 elements on 15 and 20 m. For those who *insist* upon 50-ohm coax feed and are willing to accept the additional cost, the radiator could be replaced by a commercial three-band trap dipole, which

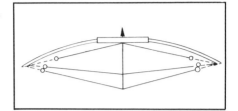

Fig 4—Another band could be added by placing an additional director and radiator inside the "bow" used for the lower frequencies.

is usually of similar length.

Because of its simplicity, this center-fed antenna is used extensively for 160, 75 and 80 m operation. Admittedly, a major change of operating frequency will require some adjustment of the Transmatch, but even this chore becomes less significant now that more and more stations are equipped with *automatic* Transmatch units.

I must stress that unlike a ½-dipole, the flat-top section does not require trimming to any critical length, providing it is not shorter than about 0.4 λ at the lowest frequency to be covered. The feeder can be any convenient length as dictated by local circumstances, although reference to a handbook will indicate that certain lengths may simplify the tuning and loading procedures.

For a 10/15/20-m radiator (which will also cover 24 MHz and the 27-MHz CB

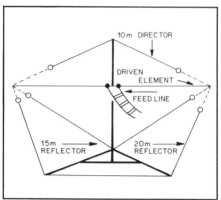

Fig 5—Three bands may be accommodated by feeding the driven element with a tuned open-wire line and adding elements as shown. The 20-m reflector is strung around the back of the array in the shape of a U.

band), a suitable tip-to-tip length would be 26 feet, or in other words, 13 feet each side of center. A feeder length of around 55 to 60 feet will provide voltage feed on all bands.

With a fair-sized tuning capacitor, a single tuned circuit in the Transmatch will in most cases cover the entire range of frequencies. A more sophisticated alternative might be switched, pretuned circuits for each band. This may seem laborious, but it is probably no worse than adjusting a gamma-match system at the top of a 50-ft tower!

The Attic Tri-Bander Antenna

By Kirk Kleinschmidt, NTØZ
Assistant Technical Editor
ARRL HQ Staff

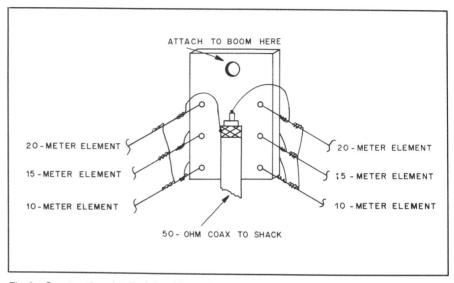

Fig 2—Construction detail of the driven element. The center supporting insulator is made from a 2- × 3-inch piece of plastic or acrylic sheet, such as Plexiglas®. Bare copper wires tie the driven element halves for each band together.

When I moved to Connecticut from Minnesota, I left behind plenty of room for my antenna farm. To make matters worse, I was forced by time constraints and economics to move into an apartment where I would have little opportunity for outside antennas.

As it turns out, I shouldn't have been worried. I live on the third floor of a three-family, wood-frame house, with a large walkup attic in which to erect my antennas. The first antenna I put up was a full-wavelength 40-meter loop. It is run around the perimeter of the attic floor in a horizontal configuration. The loop worked so well that, after working a lot of DX on 40 meters, my thoughts turned to an indoor *gain* antenna for 20 through 10 meters. The result is the 3-element, attic-mounted wire tri-bander described in this paper.

The tri-bander boom (made from ¼-inch nylon rope) runs down the attic main roof beam. See Fig 1. The wire elements are fastened to the rafters for support, effectively making the antenna an inverted-V wire beam.

The wire elements (constructed of 18-gauge zip cord) are fastened to the rope boom through plastic insulators (see Fig 2), and are fanned out through the use of plastic spreaders, shown in Figs 3 and 4.

The driven elements for all bands are tied together at the feed point and fed by a single run of 50-ohm coax (see Fig 2). Driven-element dimensions are derived from the standard dipole element-length formula (468/f, where f = frequency in MHz), and from information contained in *The ARRL Handbook*.

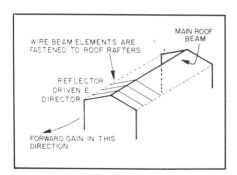

Fig 1—Bird's-eye view of the attic tri-bander, showing the outline of the house and roof lines.

The beam (3 elements on 15 and 10 meters, 2 elements on 20 meters) uses an element-to-element spacing of 6 feet. That's consistent with optimum forward gain on 20 and 15 meters and acceptable forward gain on 10 meters. Front-to-back and front-to-side ratios were not considered in this design, and indeed, they're not that good. Forward gain ranges from approximately 5.5 dBd on 20 meters to 7.5 dBd on 15 and 10 meters. For a $10 wire beam, that's okay by me!

Building the Wire Beam

The first step is to cut the wire elements to length. See Table 1 for the lengths I used. If you'd like to tailor your beam for different parts of the band, the driven elements are cut to standard wire dipole lengths (468/f); the director elements are 5% shorter; the reflector elements are 6% longer.

I used 18-gauge zip cord which, after unzipping the two wires as element halves, helped keep each side of the dipole elements the same length. After cutting the elements, the next step is to fabricate the insulators and spreaders. I used plastic strips cut from an old Tupperware™ cheese container. You'll need three center insulators, four "long" spreaders and six "short"

spreaders. (See Figs 2 and 4 for details.)

After all of the parts have been fabricated, the antenna is ready to be assembled. In my installation, one end of the rope boom is fastened to the attic wall, a few inches below the roof center span. Next, the three center insulators are attached to the boom (they're spaced 6 feet apart).

Starting with the driven element, the wire elements are "laced" onto the center insulator and, for the moment, left hanging loose. The 20-meter elements are fastened to the top hole, the 15-meter elements are fastened to the middle hole, and the 10-meter elements are fastened to the bottom hole. When all wires are in place, the elements on each side of the center insulator are tied together with a bare copper wire. A single piece of 50-ohm coaxial cable is soldered to the driven elements, with the coax shield going to one side of the center insulator, the center conductor to the other. A single feed line isn't the best way to feed the antenna, but it is the simplest. See Fig 2.

The wire elements making up the director are laced onto their center insulator as well. These elements are soldered together (to make a parasitic element) at the center insulator, but are not connected in parallel. See Fig 5 for details. The reflector elements are attached in the same way. Remember,

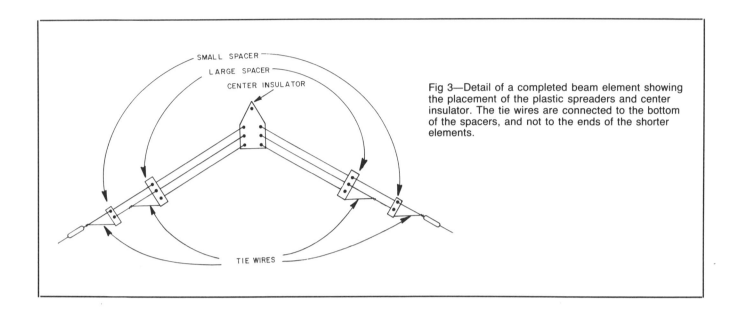

Fig 3—Detail of a completed beam element showing the placement of the plastic spreaders and center insulator. The tie wires are connected to the bottom of the spacers, and not to the ends of the shorter elements.

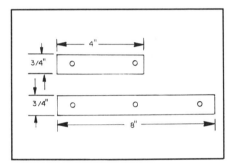

Fig 4—Dimensions for the plastic spacers (spreaders). Six small and four large spacers are needed. The spacers are constructed of plastic or acrylic strips.

Table 1

Element Lengths for the Tri-Band Wire Yagi

Lengths shown are in feet.

Element	20 Meters	15 Meters	10 Meters
Director	31.4	20.9	15.7
Driven	33.0	22.0	16.5
Reflector	—	23.2	17.5

Dir length = 0.95 × 468/f
DE length = 468/f
Refl length = 1.06 × 468/f

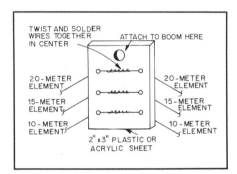

Fig 5—Construction detail of the Yagi director element.

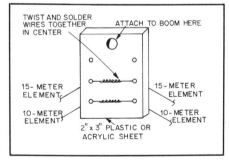

Fig 6—Construction detail of the Yagi reflector element. Note there are only two reflector elements—one for 15 meters, and one for 10 meters. The 20-meter portion of the Yagi has only a driven element and a director.

the reflector has only two elements because the 20-meter portion of the beam has only two elements—a driven element and a director. Fig 6 shows details.

When all the elements have been connected to their respective center insulators, the antenna is raised into position. One end of the rope boom is fastened to the wall with an eye-bolt; the other end is tied around the chimney, conveniently located in the middle of the attic. The center roof beam is about 13 feet above the floor, putting my rope boom at about 12 feet above the floor. Because of this, the 20-meter elements run down the length of the rafters and touch the floor. (Actually, about a foot or so of the 20-meter elements touch the floor.)

Starting with the driven element, spread the wires out as shown in Fig 3, sliding the appropriate spreaders on the wires before fastening the wire elements to the wooden rafters. If your attic doesn't have a lot of inaccessible nooks and crannies, your wire beam should go up in no time. Unfortunately, mine didn't! When everything's in place, your antenna should look like that shown in Fig 1.

After the antenna's in place and all of the wire elements have been adjusted for proper tension and symmetry, run the coax away from the beam to your shack. Preferably, the coax should be run away from the antenna at a right angle.

Tune-Up and On-the-Air Performance

Tuning the wire beam is not difficult, but may be a bit time consuming. Apply a low-power signal to the antenna, starting with 20 meters and proceeding to 10 meters. Note the SWR values at several points on each band. If the lowest-SWR frequency on any one band is too high, a short piece of wire (a few inches long) must be added to each leg of the driven element for that band. This will lower the frequency of lowest SWR. Before proceeding, measure the SWR at several points on each band again. There is some interaction because of the common feed point. If the lowest-SWR frequency on any one band is too low, trim an inch or so off each leg of the driven element for that band. This will increase the frequency of lowest SWR. Before proceeding, measure the SWR at several points on all bands. Once again, there is some interaction. Repeat the process until the SWR is as low as possible on all bands.

If this sounds like a bunch of running back and forth, it is! The fruits of your labor are well worth the effort, however. I was able to get the SWR down to 2:1 or less over the entire 20- and 15-meter bands, and a large portion of the 10-meter band.

After I finished installing my antenna, I started listening on 15 meters, just after supper time. As I had hoped, signals from the west (boresight for my antenna) were considerably stronger than those received on the 40-meter loop. (The loop was fed through a Transmatch that was adjusted for a 1:1 SWR—my former antenna for the high bands.)

To make a long story short, signals are 2 to 5 S units stronger from the wire beam than those received on the loop. The exact difference depends on distance, direction, and a bunch of other factors that I don't care to consider. The bottom line is the antenna works great! I've added a lot of new countries to my 5-band DXCC and QRP DXCC totals. It works so well, in fact, that I can hardly tell the difference between the wire tri-bander (at approximately 45 feet above the ground) and a conventional tri-bander located nearby at approximately the same height.

The 15-meter horizontal radiation pattern for the wire tri-bander is shown in Fig 7. Since the installation of the original wire beam I've added another. It's aimed due east. The two beams are mounted

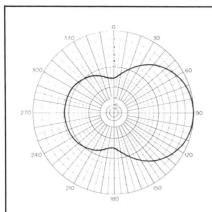

Fig 7—Calculated horizontal radiation pattern for the wire beam on 15 meters. The beam exhibits more than 7 dBd forward gain, although the front-to-back and front-to-side ratios are not as good as a completely horizontal Yagi. The pattern for 10 meters is nearly identical to the 15-meter pattern. The pattern on 20 meters (only two elements) shows less forward gain and a lower front-to-back ratio.

back-to-back (separated by about 8 feet) along the main roof beam of the attic. The performance of this antenna matches the original. All in all, it's the best-performing electrically switchable $25 array I've ever owned!

If your ham-radio spirit is dampened by the lack of outdoor antenna space, and if you're lucky enough to have a large attic, give this antenna a try. You may be pleasantly surprised by its performance. And if you have the room, there's nothing to keep you from installing the wire beam outside.

Yagi Beam Pattern-Design Factors

By Paul D. Frelich, W1ECO
72 County St
Dover, MA 02030

Several years back while studying the good work of Forrest Gehrke, K2BT, on vertical phased arrays,[1] I wondered why precautions were made that certain spacings and phasings must be followed carefully. So I dug out the old textbooks,[2,3] and here is what I relearned: (1) A 2-element array, driven with equal currents, has a pattern null (when it exists) that is controlled by the element spacing and the phase of the drive currents. The null can be placed at any desired angle, measured from the array axis. (2) The pattern-product theorem is a very useful design tool, ie, for an array of antennas all having the same pattern, the overall pattern is the product of the array pattern and the individual antenna patterns, which, being a common factor, can be factored out and multiplied by the array pattern. The same applies to an array of subarrays all having a common pattern. The useful feature here is that the nulls of the subarrays all show up in the overall pattern since "nothing times something equals nothing." Perhaps this feature could be a useful tool for controlling sidelobe levels, for intuitively one would not expect a strong sidelobe level to exist between two closely spaced pattern nulls.

I then began to design some phased arrays using these two powerful concepts. I proved to myself, for example, that the diagonally directed pattern of a square 4-element vertical array, such as one might use on 40 or 80 meters, is nothing but the product of two cardioid patterns at right angles to each other, and that, within reason, similar patterns could be realized regardless of element spacing. I designed some interesting arrays. One I especially liked was an 8-element array in a circle, slightly greater than ½ λ in diameter. Four elements were left open-circuited and the remaining four were to be driven as a 2-element broadside array with another broadside array immediately behind, phased to minimize radiation to the rear. The overall horizontal pattern is the product of the patterns of a 2-element

broadside array and a 2-element phased array. Result: By switching the phasing network to the appropriate four elements, eight beams are available, overlapping at 2 dB down, with equal sidelobes down 25 dB.

It also occurred to me that very useful patterns could be developed for linear endfire arrays. The scheme is to build up the array in steps. The first step is to go from a single dipole to a 2-element array of two dipoles. The next step is to treat this array as a subarray and take two of these subarrays and form another 2-element array of the subarrays. Now, if this is done on the same axis and with the same spacing, we will end up with a 3-element array, as two of the dipoles will be occupying the same space and so can be replaced by a single dipole carrying the vector sum of their currents. It is obvious that by iteration we can build up the array to any desired size, and at each step a new 2-element array pattern is available for controlling null positions in the overall pattern. If we stop at three elements (dipoles), then the overall three-dimensional pattern would be the product of the dipole pattern and the two 2-element array patterns. I call these 2-element array patterns PDFs for "pattern design factors." In this paper I outline some of the problems that arise, and use the PDF approach to design Yagi antennas

Design Procedure

I want to apologize now to the poor reader who may be perplexed by the peculiar intermix of BASIC programming artifices and normal mathematical expressions used in any equations or statements that follow. I only hope that the reader understands. In outline the design procedure is as follows.

1) Select N, the number of elements.
2) Select null angles for the N − 1 PDFs.
3) Select the common element spacing.
4) Compute the PDF phase angles.
5) Compute the array pattern as the product of the PDFs. Save R0, the response on the nose of the main lobe.
6) Compute the overall dipole and array pattern and the directivity.

7) Compute the element currents from the array buildup and PDFs.
8) Read mutual impedance v spacing data, and set up a mutual impedance table Zmut.
9) Interpolate the Zmut table amd set up an array mutual impedance matrix.
10) Set up the linear equation set.
11) Compute the back voltages in each element caused by other currents.
12) Set drive voltages to zero, as if the elements are parasitically driven.
13) Compute Z_{self} from $Z_{self} + V_{back}/I = 0$ for each element.
14) Compute the element length, knowing the resistive part of Z_{self}.
15) Compute the element reactance, knowing the element length and radius.
16) Compute the element tuning, knowing the element and Z_{self} reactances.

This completes the design of the parasitic elements. Now, on to consider each element as the driven element. But please note, the power radiated by the array is known since the currents, the pattern response, and the directivity are known.

17) Set up the relationship, power into the array equals power radiated by the array.
18) Compute the resistance of Z_{drive}, knowing the current and radiated power.
19) Compute the resistance of Zd_{self} from $Zd_{self} + V_{back}/I = Z_{drive}$, reals.
20) Compute the element length, knowing the resistive part of Zd_{self}.
21) Compute the element reactance, knowing the element length and radius.
22) Compute the reactance of Z_{drive} from $Zd_{self} + V_{back}/I = Z_{drive}$, imaginaries.
23) Compute, along the way, percent shifts of the element lengths from λ/2.
24) Print out, along the way or afterwards, all variables of interest.
25) Plot the array pattern responses, using the TAB command.
26) Plot "cumulative power radiated" and pattern together, using TAB.

This completes the outline of the design procedure. I will now review and augment the steps, put in some theory and a few equations, and eventually show the results of some designs I have worked out.

[1]Notes appear on page 70.

Review of Procedure

First select the number of elements, N, and design the pattern of an N-element array as the product of (N − 1) PDFs. Each PDF is the pattern of a 2-element phased array, the basic building block. Any null in a PDF pattern, since we are dealing with products, shows up as a null in the overall pattern, so normally there would be (N − 1) nulls there. How the Yagi array is built up from the building blocks and the equation for the PDFs will be discussed shortly.

Next select locations for the (N − 1) null angles. The array pattern is a pattern of revolution about the array axis, so we have to deal only with angles from 0 to 180°. Divide this range into N equal sectors, and put a null in each sector except the first, where the main lobe is located. I use a design variable fraction called FRC to move all the nulls en masse within their sectors. For example, if N = 6 and FRC = 0.6, there will be 5 nulls located at 48, 78, 108, 138, and 168°.

Select the common uniform element spacing, and compute the corresponding set of PDF phase angles. Next, compute the overall array pattern from the product of the PDFs, and check the sidelobe levels. Save R0, the response on the nose of the main lobe. Also compute the directivity as the ratio of the intensity on the main-lobe axis to the average intensity. Compute the average intensity as the product of the array pattern and a dipole pattern, squared and summed over all elemental areas of a whole sphere divided by the area of the whole sphere. If the array is just a single dipole with a current of 1 ampere, use the "magic 2-kilometer sphere."

The practical units for R0 will be 60 millivolts/meter. The intensity will be 9.5491 microwatts/square meter ($0.06^2/377$), and the average intensity will be 5.8211 μW/m^2. The total area is 4 × PI × 10^6 square meters so that the total power radiated is 73.12 watts, and that is why the dipole has a radiation resistance of 73.12 ohms. The dipole and the array patterns have different reference axes. The product of the patterns for many particular directions are used to compute the directivity. These particular directions point from the center of the sphere to the center of 2880 little elemental areas 2.5° by 9° on the surface of the sphere.[4] Precomputed factors involving the dipole pattern, strips 2.5° by 90° and symmetry, allow the directivity to be calculated as the summation of 72 product terms instead of 720 complicated computations that involve spherical trigonometry and inverse trigonometric functions. Taking these interrelationships into account in earlier programs led to long running times when computing the directivity. I have cut the running time down considerably by using the precomputed factors for the interrelationships, which are stored in data statements.

Next compute the individual element currents using the PDF phase angles. Assume 1 A at 0° for the front element at the origin and let the array build, one element at a time and to the rear, for each new PDF. Essentially the build-up is this routine: (1) Clone the previous array and shift it one space to the rear; (2) Rotate the currents of the clone by the phase angle of the PDF in use; (3) Vectorially add the currents for all pairs of elements that can merge into one; and (4) Merge the two arrays. Repeat until all the PDFs are used.

Until now, the element currents, except for the first, were not needed. As an optional check, compute the overall pattern using the element currents and the spacing. Compare this against the pattern computed using the PDFs.

Read the MX and MY mutual impedance values versus element spacing from the data statements and set up the mutual impedance table. I have adapted and modified Table 1.3 found in Lawson, W2PV,[5] using more precise values for the MX and MY entries at S = 0 and correcting the obvious mistake at S = 1.85. I found that the discontinuity at S = 2 between the tabular values and the values produced by the approximations beyond S = 2 by Lawson's Eqs 1.15 and 1.16 were giving anomalous results in some of my design solutions. This was particularly so in larger arrays when an element would move from one side of S = 2 to the other as small changes in array spacing were made. I revised these equations to include near-field inverse square and inverse cube terms.[6] The discontinuity is much less and the array solutions behave much better.

Next, using multiples of the array element spacing, interpolate the mutual impedance table by a parabolic fit to three adjacent data points and set up the array mutual impedance matrix. Let ZMX(J,K) and ZMY(J,K), and temporarily ZMX(J = K) = MX(0) and ZMY(J = K) = MY(0), ie, Z(1,1), Z(2,2),..., equal the approximate 73.12 + j42.46 ohm impedance of a half-wave dipole. We will soon compute new values for these self-impedances.

Next, before proceeding, let us examine, for example, the set of linear equations that relate the array currents, impedances and voltages of a small 4-element array.[7]

$$V(1) = I(1)*Z(1,1) + I(2)*Z(1,2) + I(3)*Z(1,3) + I(4)*Z(1,4) \quad \text{(Eq 1A)}$$

$$V(2) = I(1)*Z(2,1) + I(2)*Z(2,2) + I(3)*Z(2,3) + I(4)*Z(2,4) \quad \text{(Eq 1B)}$$

$$V(3) = I(1)*Z(3,1) + I(2)*Z(3,2) + I(3)*Z(3,3) + I(4)*Z(3,4) \quad \text{(Eq 1C)}$$

$$V(4) = I(1)*Z(4,1) + I(2)*Z(4,2) + I(3)*Z(4,3) + I(4)*Z(4,4) \quad \text{(Eq 1D)}$$

Everything on the right-hand side of these equations is known. The currents were just computed from the PDF phase angles and the spacing, the mutual impedances were just computed from the mutual impedance table and stored in the array mutual impedance matrix, and the self-impedances were set to the impedance of a dipole. If that be true, then the voltages on the left-hand side are known by a straightforward computation of the terms on the right-hand side. *Voila!* Supply these drive voltages and the result is a phased array with all elements driven, one in which we are not particularly interested. We shall return to the parasitically driven Yagi array shortly. If these "voltage" equations are divided through, each in turn by I(1), I(2), I(3) and I(4), then the drive impedances are also known, and can be expressed alternatively as

$$ZDRIVE(1) = V(1)/I(1) \quad \text{(Eq 2)}$$

$$ZDRIVE(1) = Z(1,1) + (I(2)*Z(1,2) + I(3)*Z(1,3) + I(4)*Z(1,4))/I(1) \quad \text{(Eq 3)}$$

$$ZDRIVE(1) = Z(1,1) + (VBACK(1))/I(1) \quad \text{(Eq 4)}$$

where VBACK(1) is the sum of the back EMFs induced into element 1 by the currents flowing in the other elements, and so on for the others.

Return to Eqs 1A through 1D. Consider each element as a candidate for being driven parasitically. Set all the drive generator voltages V(J) to zero. Divide each equation through by I(K). Remember, the currents I(K) and the mutual impedances Z(J,K) are known, and hence the equations can now be rearranged to solve for the self-impedances Z(J = K) which, for the assumed currents, would make the drive voltages zero, both parts, real and imaginary. These solutions, using Eqs 2 and 4, are given by

$$ZSELF(J) = Z(J = K) = -VBACK(J)/I(K) \quad \text{(Eq 5)}$$

and except for sign were designated earlier as part of ZDRIVE(J) in Eq 4.

Next, compute the half-length KL(J) in radians for each element (K = 2PI/ Lambda), given ZSELFA(J), the real part of ZSELF(J), and the equation

$$R(KL) = 104.093(KL)^2 - 190.970(KL) + 116.256 \quad \text{(Eq 6)}$$

Then compute the corresponding XKL = X(KL) from the equation

$$X(KL) = 29.263(KL)^2 - 26.862(KL) + 12.452 \quad \text{(Eq 7)}$$

Next, knowing X(KL), KL the element length, and ELRAD the element radius, compute the element reactance from the equation

$$ELX = XKL - 120*(LOG(2*KL/ ELRAD) - 1)/TAN(KL) \quad \text{(Eq 8)}$$

The three equations above were adapted from a table and equation found in Jasik concerning the input impedance of a center-driven antenna.[8] The coefficients for the first two equations were computed as a second-order fit to the last three data entries of Jasik's table.

Next, series tune each element such that ZSELFB(J), the imaginary part of ZSELF(J), will be realized. Compute XTUN(J), given by

$$XTUN = ZSELFB - ELX \qquad (Eq\ 9)$$

Now, with the real and imaginary parts canceled, each element's drive generator voltage is zero for the assumed array currents, and so the pruned and tuned element can be shorted and not driven in lieu of this.

Check KL the parasitic element half-lengths. If all the element half-lengths fall within plus or minus 10% of PI/2, then all could be used as parasitically driven elements, and the design is almost complete except for selecting the driven element.

Every element has the potential of being the driven element. Consider this: The lengths we have just computed made the real part of the drive impedance zero, so for any one element, if its length is slightly increased, the real part will also increase. So we increase the length and resistance until the power into the array equals the power out of the "magic 2-kilometer sphere," which is already known for the assumed currents, from R0 and the directivity. The power in is I(J)²*ZDRIVEA(J) watts, and the power out of the sphere, for the assumed 1 ampere in the first element is 120*R0²/directivity watts. Anything else violates the laws of physics. By equating these powers and solving for the resistive part of the drive impedance for each element as the driven element, we have

$$ZDRIVEA(J) = 120*(R0/I(J))^2/ \\ DIRECTIVITY \qquad (Eq\ 10)$$

and also

$$ZDSELFA(J) = ZDRIVEA(J) \\ + ZSELFA(J) \qquad (Eq\ 11)$$

Values in this equation are reals. From the result, lengths can be computed as before and then, from the imaginaries, the associated reactances determined for each element as a driven element. Thus, each element can be a parasitic element or a driven element.

Hence, in this design method of parasitic arrays there are two critical lengths for the elements: (1) if parasitically driven, that length which reduces the loop resistance to zero, ie, real part of ZDRIVE to zero, and (2) if the driven element, that length which yields a real part of ZDRIVE such that the array gain equals the directivity, as prescribed by the assumed current distribution. In effect, all parasitic elements after pruning and tuning act as if their drive generators are shorted; only the driven element accepts real power, which, barring losses, is radiated.

My BASIC programs are listed at the end of this paper. [The programs are available from the ARRL on diskette for the IBM PC and compatibles; see information on an early page of this book.—*Ed.*] A typical run was conducted for N = 5, a 5-element PDF Yagi, spacing = 54° (0.15 λ) and FRC = 0.30. The rear element is driven. There is no reflector. The elements in the front are long, and elements are electrically tuned with either coils and capacitors or transmission-line stubs. The gain is just over 12 dBd.

The printouts from the programs are long, multipage presentations in a format that cannot be suitably reproduced in this paper. For the 5-element Yagi, however, they reveal the four null angle locations of the PDFs and the element currents, rectangular and polar. The mutual impedance table, adapted from Lawson, is printed out, and so is the array matrix in a peculiar form. The program can interpolate the Zmut table, but it so happens the matrix values correspond to the tabular values because of the 0.15-λ spacing. Also printed out are the element lengths and the percent deviation from PI/2. The tuning reactances for the parasitic elements are printed out. Values for R0, dBd, dBi, directivity, and half-power half-beamwidth are displayed also.

There are two array pattern plots, the same as E-plane plots for the ranges 0 to −70 dB and 0 to −50 dB. Combined on the second plot is the cumulative radiated power plot, on a normalized scale of 0 to 1. Starting at the nose of the beam, this is how the radiated power builds up in the 72 summation rings, each 2.5° wide, when computing the directivity. It depicts the effectiveness of a directive beam. If it hits 0.8 at the half-power beamwidth, I would say that you have a good beam on your hands. This particular design hits about 0.6 at the −3 dB beamwidth.

Sidelobe Levels and Control

One very important factor in the design of multi-element end-fire antenna arrays, whether the elements are actually driven or parasitically driven, is the average sidelobe level with respect to the unidirectional main-lobe level. In a highly directional array, ie, one with a very narrow main lobe, the sidelobe levels must be relatively low to realize a high directivity factor. Consider, for example, an end-fire array in free space having a main-lobe beamwidth of 0.2 radian (about 11.5° wide), centered within a 1-kilometer-radius sphere having a total area of 4 PI square kilometers. Most of the energy radiated outward by this beam would pass through an area on the sphere of 0.01 PI square kilometer. The maximum directivity factor could be 400, that is if no power were radiated through the remaining area which is 399 times as large.

In order to achieve 80% of the maximum directivity, only 20% of the total power radiated must pass through the remaining

area. It works out that the average sidelobe level, including the power radiated by the main lobe between the −3-dB points and the first null, must be 32 dB below the main-lobe level, and the directivity would drop from 26 dB to 25 dB. In a good design most of that 20% would be due to the remaining part of the main lobe. If the average sidelobe level is down 26 dB, then half of the transmitted power is wasted in the sidelobes and the directivity drops to 23 dB. Of course, the situation becomes worse with narrower beamwidths; even lower average sidelobe levels are in order. Not to worry. I gather from recent literature that earth-moon-earth communication is perfectly possible with directivities of 23 to 25 dB.

The real design problem is to devise a small array that can realize these numbers and have small sidelobes to match—in other words, to design an array that really has an effective narrow beam. A program run for an 18-element array showed the nature of this problem, where it was expected to realize a gain of 24 dBi. The cumulative radiated power curve produced by the program gives a new insight into the effectiveness of a narrow beam with high sidelobe levels. The directivity is low; as a transmitter it sprays energy everywhere, and as a receiver it listens to disturbing noises from everywhere, overpowering what we want to hear in the main lobe. And yet it has a very narrow beamwidth in the main lobe, with the same target-tracking problems of a good array that narrow beamwidth. There is some control over the sidelobe levels by moving the nulls backwards, but we can get almost the same directivity with a smaller array. It is apparent that, for large values of N, the present scheme is not optimum for locating the pattern null angles. So we conclude that low sidelobe levels become more and more important as the directivity goes up and the beamwidth goes down.

Designing the patterns first, using the PDF nulls to control the sidelobes, has proved to be interesting and very effective. I have only begun to explore the possibilities.

Pattern Design Factors

The array pattern of an N-element end-fire array can be designed, having N−1 selected null angles, by using the PDFs and the design procedures outlined earlier. Each PDF controls one null. The response, with the 2s factored out separately, is the product of all the 2-element PDF pattern responses, and is given by

$$RESP = PDF(1)*PDF(2)*\ldots \\ PDF(N-1)*2^{N-1} \qquad (Eq\ 12)$$

The question remains, what is an optimum way to select these null angles? I reasoned as follows. Put the nulls everywhere we expect sidelobes, but never in the unidirectional main lobe region. So I tried a very simple scheme. Divide the interval 0 to

180° into N equal sectors and put a null angle in the center of each sector except the first, which is where the main lobe is. The computed patterns were remarkable—low sidelobe levels, nice sharp main lobe, great front-to-back ratio, and a directivity approximately equal to 20 (log N) − 1 dB. I reasoned why not. After all, an end-fire array keeps energy from spreading in two dimensions simultaneously, by virtue of having an array pattern which is a pattern of revolution about the array axis.

A few words are in order about the standard pattern response equation for the pattern design factors.

$$R = 2*\cos(PHI - S*\cos(B)) \quad \text{(Eq 13)}$$

This is found in Jasik,[9] but I have rewritten the equation in BASIC notation for the variables defined here. PHI is half the phase difference of the currents, S is half the spacing of the two elements, and B is the angle to a far-field observation point. BASIC computes trigonometric functions properly only if angles and phase angles are expressed in radians. Hence, PHI must be in radians, S must be converted to radians by multiplying the half-spacing by 2*PI/Lambda, and B must be converted to radians by multiplying it by PI/180 before computing its cosine.

Actually this equation represents twice the real part, ie, twice the cosine of the phasor or vector observed at the far field point due to the rear element of a 2-element array, placed at S and −S on the X axis. See Fig 1 for the array geometry and Fig 2 for the phasor. Fig 2 is a great help in trying to picture and estimate beforehand the pattern that Eq 13 represents. The interactions of phase, spacing and bearing (PHI, S and B) are readily apparent, and estimates of the pattern lobes and null angle can be made very quickly. "What if" questions can be answered immediately, too.

The phasor swings back and forth about the drive-current phase angle for one complete revolution of the observation point around the horizon, somewhat akin to a modulation process. A null is produced whenever the phasor swings through PI/2, ie, 90°. The same conclusion could have been reached for Eq 13 by realizing that the cosine of PI/2 is zero, producing a null. The null condition is PHI − S*COS(B) = PI/2. Hence, the phase necessary to produce a null at angle B, given the spacing S, is

$$PHI = PI/2 + S*\cos(B) \quad \text{(Eq 14)}$$

Given the angle at which we want a null, put it into Eq 14 and compute the phase angle. Conversely, given the phase angle, we could solve for the null angle as an arc-cosine, which is tricky in BASIC. But as can be seen from Fig 2, there may not even be a null. Eq 14 says that there will always be a null if PHI is within the limits PI/2 + S and PI/2 − S, because of the COS

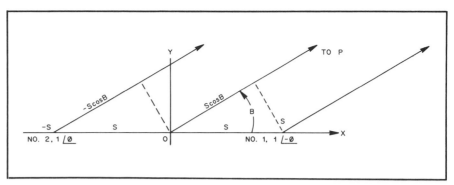

Fig 1—Two-element array geometry and phasing.

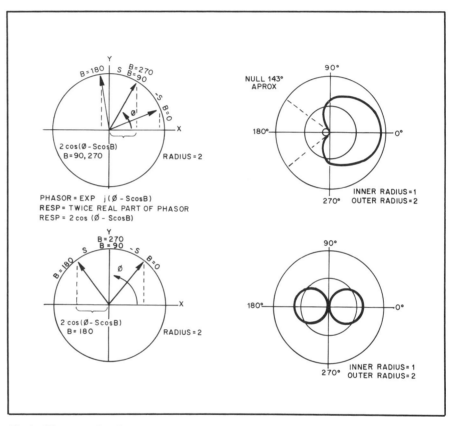

Fig 2—Phasor and pattern response.

limits +1 and −1. Also, to preclude multiple nulls, the spacing (2*S) must be less than PI (aka 180°, λ/2, or a half wavelength.)

Design Characteristics

I have made nine design runs of a 5-element PDF designed Yagi. Tables 1 through 3 summarize their characteristics for three values of FRC, and three values of spacing. BW/2 is half of the half-power beamwidth; % DELTAL(N) is the percentage deviation of the length of the parasitic elements from a half wavelength, as likewise % DELTADL(N) is for the driven ele-

ment. X TUN(N) is the required series tuning reactance for the parasitic elements. R0 is the relative response of the main lobe (at the peak), a numeric; R0 = 1 for a dipole. FRC locates the pattern nulls within their sectors. There are nine designs summarized in the tables, but only the design FRC = 0.75 and SPACING = 63° has an extra-long first element, being 16.23% longer than a half wavelength. So one should make either FRC or the spacing smaller, or both.

My work so far has uncovered the following general characteristics.

1) Pattern: Sharp main lobe, low side-

Table 1
Array Parameters v Spacing for FRC = 0.25

N = 5

Spacing	45.00	54.00	63.00	degrees
Boom length	0.50	0.60	0.70	λ
DBI	14.34	14.26	14.16	dB/isotropic
DBD	12.19	12.11	12.01	dB/dipole
BW/2	19.50	19.70	19.90	degrees
R0	0.22	0.42	0.71	RSP NUM
R DRIVE	0.21	0.79	2.32	ohms
X DRIVE	−2.55	−0.78	6.26	ohms
% DELTAL1	0.47	1.44	3.56	DEV PI/2
% DELTAL2	1.09	1.58	2.18	DEV PI/2
% DELTAL3	−0.14	−0.24	−0.37	DEV PI/2
% DELTAL4	−1.03	−1.43	−1.88	DEV PI/2
% DELTADL5	−0.43	−0.47	0.01	DEV PI/2
X TUN1	−4.63	−15.19	−38.61	ohms
X TUN2	−36.51	−47.86	−60.81	ohms
X TUN3	−30.26	−35.96	−41.37	ohms
X TUN4	−13.95	−15.30	−16.30	ohms

Table 2
Array Parameters v Spacing for FRC = 0.50

N = 5

Spacing	45.00	54.00	63.00	degrees
Boom length	0.50	0.60	0.70	λ
DBI	13.61	13.53	13.43	dB/isotropic
DBD	11.46	11.38	11.28	dB/dipole
BW/2	22.30	22.60	22.90	degrees
R0	0.40	0.65	1.26	RSP NUM
R DRIVE	0.83	3.06	8.71	ohms
X DRIVE	−3.79	3.64	25.78	ohms
% DELTAL1	1.38	4.02	8.71	DEV PI/2
% DELTAL2	2.26	3.22	4.22	DEV PI/2
% DELTAL3	−0.31	−0.57	−0.98	DEV PI/2
% DELTAL4	−2.01	−2.83	−3.79	DEV PI/2
% DELTADL5	−0.52	0.01	1.96	DEV PI/2
X TUN1	−10.11	−36.25	−87.76	ohms
X TUN2	−49.45	−66.54	−85.33	ohms
X TUN3	−29.24	−33.54	−36.51	ohms
X TUN4	−3.51	−0.41	4.17	ohms

Table 3
Array Parameters v Spacing for FRC = 0.75

N = 5

Spacing	45.00	54.00	63.00	degrees
Boom length	0.50	0.60	0.70	λ
DBI	12.83	12.73	12.61	dB/isotropic
DBD	10.68	10.58	10.46	dB/dipole
BW/2	24.80	25.20	25.70	degrees
R0	0.65	1.22	2.01	RSP NUM
R DRIVE	2.67	9.60	26.67	ohms
X DRIVE	1.58	29.61	98.02	ohms
% DELTAL1	2.94	8.24	16.23	DEV PI/2
% DELTAL2	3.51	4.87	6.08	DEV PI/2
% DELTAL3	−0.59	−1.14	−2.01	DEV PI/2
% DELTAL4	−2.97	−4.18	−5.64	DEV PI/2
% DELTADL5	0.07	2.59	8.82	DEV PI/2
X TUN1	−20.19	−73.02	−166.71	ohms
X TUN2	−63.67	−86.61	−109.98	ohms
X TUN3	−27.27	−29.09	−27.86	ohms
X TUN4	6.71	14.19	24.21	ohms

lobes, high front-to-back ratio.

2) Directivity: 20 (log N) for small N to 20 (log N) − 2 for large N.

3) Element currents: Related to binomial coefficients, peak in center diminishes as the spacing increases, uniform progressive phase shift.

4) Element self-impedances: Directly computed from linear equation set.

5) Element lengths: Directly computed from element self-impedances.

6) Element tuning: Directly computed from element self-impedance, length and radius.

7) Array gain: Driven-element length pruned so that gain over an isotropic source equals the directivity.

8) Drive impedance: As spacing increases, the main lobe response increases, which allows the drive impedance to increase. However, at the same time, the directivity slowly decreases, so there is a trade-off.

9) Spacing versus element-length tolerance: Not all spacings yield element lengths that deviate only slightly from the nominal half wavelength.

10) Design parameter fraction: This parameter moves all the pattern nulls en masse within their angular sectors. As the nulls are moved forward, for small arrays, directivity first increases as the sidelobes rise and then decreases at FRC = 0, where the array becomes bidirectional. As FRC moves the nulls the other way, the sidelobes diminish greatly but the main lobe widens and the directivity lessens. Meanwhile, the main lobe response increases. If N is large, moving the nulls backwards can improve the directivity, but for N greater than 12 the equal spacing of the nulls does not seem to be optimum. Further work is needed here. Perhaps a Dolph-Chebyshev approach is called for. There is a trade-off between directivity and the drive impedance as a function of the element spacing.

11) In general, the element lengths are longer at the front of the array than at the rear, just the opposite of the old-style Yagis. This is so because the elements are pruned in length to realize a desired real part and then tuned to negate the excess reactance. How this affects the bandwidth deserves future study. Sacrificing bandwidth for high gain in a small array may indeed be a fine economic trade-off. After all, by definition, all our amateur bands are bandwidth limited, or they wouldn't be called bands.

Conclusions

The PDF design procedure by straightforward computations is capable of yielding some remarkable high-gain Yagis, as evidenced by Fig 3 and Tables 4 and 5. At

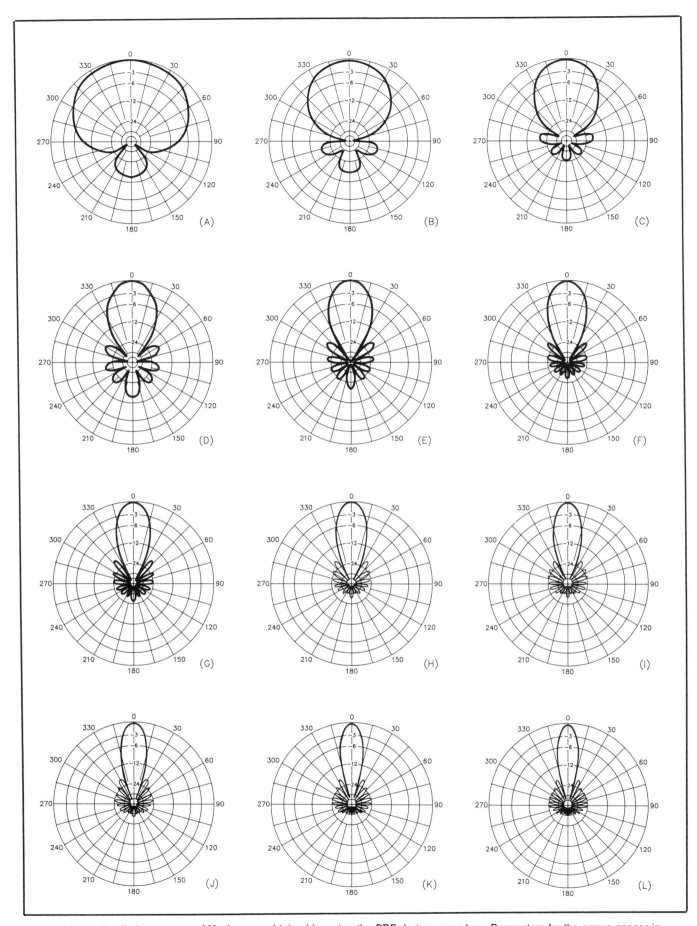

Fig 3—Azimuthal radiation patterns of Yagi arrays obtained by using the PDF design procedure. Parameters for the arrays appear in Tables 4 and 5.

no time is it necessary to invert a huge matrix to solve for the element currents, given the element lengths, and then compute the pattern and hope for the best. Instead we start with a good pattern, and compute the element lengths to realize that pattern. Further studies must be made to optimize the null angle locations for large arrays, beginning at N = 13 or so.

Notes

[1]F. Gehrke, "Vertical Phased Arrays," *Ham Radio*, May, June, July, October and December, 1983, and May, 1984.
[2]Terman et al, *Electronic and Radio Engineering*, 4th ed. (New York: McGraw-Hill, 1955).
[3]H. Jasik, Ed., *Antenna Engineering Handbook*, 1st Ed. (New York: McGraw-Hill, 1961).
[4]M. E. Shanks and R. Gambill, *Calculus, Analytic Geometry Elementary Functions* (New York: Holt, Rinehart and Winston, Inc, 1973).
[5]J. L. Lawson, *Yagi-Antenna Design* (Newington, CT: ARRL, 1986). Table 1.3 is given in Chapter 1, page 10.
[6]See note 3.
[7]See notes 2, 3 and 5.
[8]See Chap 3, p 2 of Jasik (note 4).
[9]See Chap 2, p 16 of Jasik (note 4).

Table 4

Parameters Producing the Radiation Patterns of Fig 3

Element lengths are given in Table 5. The rear-most element is driven, ie, there is no reflector. All drive-point reactances are inductive.

Fig 3 Pattern	Number of Elements	Element Spacing, Degrees	Boom Length, λ	FRC	Gain, dBd	BW/2, degrees	Drive Point R, Ohms	Drive Point X, Ohms
A	2	30.0	0.0833	0.44	5.07	59.90	15.73	25.14
B	3	45.0	0.25	0.40	7.87	37.02	9.63	18.89
C	4	54.0	0.45	0.40	9.95	27.28	6.20	13.54
D	5	63.0	0.70	0.25	12.00	19.86	2.32	6.26
E	6	66.0	0.9167	0.30	13.36	16.93	1.48	0.62
F	7	72.0	1.20	0.40	14.28	15.35	2.25	9.60
G	8	81.0	1.575	0.37	15.35	23.25	2.53	11.70
H	9	81.5	1.811	0.40	16.26	11.90	1.55	10.10
I	10	90.0	2.25	0.40	16.98	10.76	2.40	13.98
J	11	89.0	2.472	0.40	17.81	9.72	1.08	20.88
K	12	93.0	2.8417	0.40	18.44	8.89	2.01	26.51
L	13	95.0	3.1667	0.40	19.03	8.19	0.73	36.94

Table 5

Element Lengths Producing the Radiation Patterns of Fig 3

Also see Table 4. Element no. 1 is the forward element. The highest numbered element is driven, at the rear of the array. Lengths are indicated in percent change from a resonant ½ λ.

Fig 3 Pattern	No. 1	No. 2	No. 3	No. 4	No. 5	No. 6	No. 7	No. 8	No. 9	No. 10	No. 11	No. 12	No. 13
A	− 1.717	1.675	—	—	—	—	—	—	—	—	—	—	—
B	9.466	− 3.545	1.122	—	—	—	—	—	—	—	—	—	—
C	9.243	1.452	− 3.294	0.689	—	—	—	—	—	—	—	—	—
D	3.560	2.176	− 0.373	− 1.878	0.013	—	—	—	—	—	—	—	—
E	0.854	2.309	0.711	− 0.925	− 1.833	− 0.394	—	—	—	—	—	—	—
F	− 0.591	2.726	1.582	− 0.122	− 1.541	− 2.140	0.405	—	—	—	—	—	—
G	− 1.433	2.333	1.820	0.561	− 0.726	− 1.725	− 2.108	0.431	—	—	—	—	—
H	− 2.101	1.651	1.735	0.949	− 0.056	− 1.005	− 1.706	− 1.887	0.255	—	—	—	—
I	− 0.474	1.865	1.955	1.315	0.428	− 0.487	− 1.298	− 1.872	− 1.917	0.773	—	—	—
J	1.808	1.634	1.658	1.272	0.667	− 0.016	− 0.680	− 1.240	− 1.568	− 1.330	1.596	—	—
K	3.861	1.682	1.605	1.344	0.882	0.318	− 0.275	− 0.843	− 1.336	− 1.662	− 1.456	2.895	—
L	4.245	1.650	1.527	1.343	0.993	0.542	0.046	− 0.455	− 0.929	− 1.340	− 1.604	− 1.284	4.054

Program 1
BASIC Listing of PATNULL1 Program

[The ARRL-supplied disk filename for this program is FRELICH1.BAS—*Ed.*]

```
10 REM PATNULL1 PROGRAM  PAUL D. FRELICH  W1ECO  DATE 12-08-1987  REV 6-10-88
20 REM NULL ANGLES, PATTERN RESPONSES, AND ELEMENT CURRENTS
30 PI=3.141592654#
40 DIM ANG(32),PHI(32),RESP(181),AX(32),AY(32),BX(32),BY(32),MAG(32),PHASE(32),HF(181),ACC(181)
50 INPUT "ENTER NUMBER OF ELEMENTS IN END-FIRE ARRAY ",ELNUM
60 INPUT "ENTER UNIFORM ELEMENT SPACING < OR = 90 DEG. ",SPACING
70 INPUT "ENTER FRACTIONAL NULL PLACEMENT ",FRC
80 PRINT "NUMBER OF ELEMENTS="ELNUM,"SPACING="SPACING,"FRC="FRC
90 PRINT "I","NULL ANGLE","PHASE, PHI"
100 S=SPACING*PI/360
110 FOR I = 1 TO (ELNUM-1)
120 ANG(I)=(PI/ELNUM)*(I+FRC)
130 PHI(I)=PI/2+S*COS(ANG(I))
140 PRINT I,ANG(I)*(180/PI),PHI(I)*(180/PI)
150 NEXT I
160 ALPHA=0
170 GOSUB 840
180 R0=R
190 PRINT
200 PRINT "R0= "R0
210 PRINT
220 PRINT "N","BEARING","RESPONSE-DB","R"
230 DALPHA=PI/72
240 ACCUM=0
250 FOR N=0 TO 72
260 READ HF(N)
270 ALPHA=N*DALPHA
280 GOSUB 840
290 GOSUB 420
300 GOSUB 440
310 ALPHA=(N-.5)*DALPHA
320 GOSUB 840
330 ACCUM=ACCUM+HF(N)*R^2
340 ACC(N)=ACCUM
350 NEXT N
360 ACCUM=ACCUM*DALPHA/20
370 DI=4.343*(LOG(R0^2/ACCUM))
380 PRINT
390 PRINT "DI= "DI"DB"
400 PRINT
410 GOTO 460
420 RESP(N)=8.686*LOG(ABS(R/R0)+10^-30)
430 RETURN
440 PRINT N,ALPHA*180/PI,RESP(N),R
450 RETURN
460 AX(1)=1
470 FOR M=1 TO (ELNUM-1)
480 C=COS(2*PHI(M))
490 S=SIN(2*PHI(M))
500 FOR N=1 TO M
510 BX(N)=AX(N)*C-AY(N)*S
520 BY(N)=AX(N)*S+AY(N)*C
530 NEXT N
540 FOR N=1 TO M
550 AX(N+1)=AX(N+1)+BX(N)
560 AY(N+1)=AY(N+1)+BY(N)
570 NEXT N
580 NEXT M
590 PRINT "N","REAL","IMAG","MAGNITUDE","PHASE-RAD"
600 FOR N=1 TO ELNUM
610 ON SGN(AX(N))+2 GOTO 710, 640, 620
620 PHASE=ATN(AY(N)/AX(N))
630 GOTO 770
640 ON SGN(AY(N))+2 GOTO 690, 650, 670
650 PHASE=0
660 GOTO 770
670 PHASE=PI/2
680 GOTO 770
```

```
690 PHASE=-PI/2
700 GOTO 770
710 ON SGN(AY(N))+2 GOTO 760, 740, 720
720 PHASE=ATN(AY(N)/AX(N))+PI
730 GOTO 770
740 PHASE=PI
750 GOTO 770
760 PHASE=ATN(AY(N)/AX(N))-PI
770 ON SGN(PHASE)+2 GOTO 780, 790, 790
780 PHASE=PHASE+(2*PI)
790 PHASE(N)=PHASE
800 MAG(N)=(AX(N)^2+AY(N)^2)^.5
810 PRINT N,AX(N),AY(N),MAG(N),PHASE(N)
820 NEXT N
830 GOTO 900
840 R=2^(ELNUM-1)
850 A=S*COS(ALPHA)
860 FOR I = 1 TO (ELNUM-1)
870 R=R*COS(PHI(I)-A)
880 NEXT I
890 RETURN
900 FOR N= 0 TO 72
910 PRINT ACC(N)/ACC(72) SPC(0)
920 NEXT N
930 PRINT
940 FOR N=0 TO 72
950 PRINT HF(N) SPC(0)
960 NEXT N
970 PRINT
980 PRINT TAB(20)"CUMULATIVE POWER RADIATED, FRACTIONAL, O"
990 PRINT
1000 PRINT TAB(2)"0"TAB(9)".1"TAB(16)".2"TAB(23)".3"TAB(30)".4"TAB(37)".5"TAB(44)".6"TAB(51)".7"
     TAB(58)".8"TAB(65)".9"TAB(72)"1"TAB(74)"ANGLE"
1010 FOR M=0 TO 72
1020 J=2.5*M
1030 DB=RESP(M)
1040 ON SGN(50+DB)+2 GOTO 1050, 1050, 1060
1050 DB=-50
1060 A=INT(70*DB/50+72.5)
1070 B=2
1080 C=INT(70*ACC(M)/ACC(72)+2.5)
1090 FOR N=2 TO 73 STEP 1
1100 IF N=A THEN 1160
1110 IF N=C THEN 1190
1120 IF N=B THEN 1130 ELSE 1210
1130 PRINT TAB(B) "+" ;
1140 B=B+7
1150 GOTO 1210
1160 PRINT TAB(N) "*" ;
1170 IF N=B THEN 1140
1180 GOTO 1210
1190 PRINT TAB(N)"O";
1200 IF N=B THEN 1140
1210 IF N=73 THEN 1220 ELSE 1240
1220 IF J-INT(J)=0 THEN 1230 ELSE 1240
1230 PRINT TAB(73)J;
1240 NEXT N
1250 NEXT M
1260 PRINT TAB(1)"-50"TAB(15)"-40"TAB(29)"-30"TAB(43)"-20"TAB(57)"-10"TAB(72)"0"
1270 PRINT
1280 PRINT TAB(20)"PATTERN RESPONSE, DECIBELS  *"
1290 PRINT
1300 DB3=-3.010299956#
1310 FOR N=0 TO 72
1320 ON SGN(DB3-RESP(N))+2 GOTO 1400, 1380, 1330
1330 D1=RESP(N-1)-RESP(N)
1340 D2=RESP(N+1)-RESP(N)
1350 D3=DB3-RESP(N)
1360 BW=DALPHA*(D3*(D3*(D1+D2)-D1^2-D2^2)/(D1*D2^2-D2*D1^2)+N)*180/PI
1370 GOTO 1410
1380 BW=DALPHA*N*180/PI
1390 GOTO 1410
1400 NEXT N
1410 PRINT "R0"R0" ELNUM"ELNUM "SP"SPACING "FRC"FRC "BOOMLENGTH"(ELNUM-1)*SPACING/360 "DI"DI
     "HALF-BEAMWIDTH"BW
```

```
1420 DATA 0, .2180728, .6519813, 1.079234, 1.495576, 1.897017, 2.279923, 2.641095
1430 DATA 2.977833, 3.287975, 3.569931, 3.822683, 4.045778, 4.239305, 4.403849, 4.540447
1440 DATA 4.650522, 4.735827, 4.798367, 4.840344, 4.864085, 4.871982, 4.866445, 4.84984
1450 DATA 4.824465, 4.792504, 4.756006, 4.716868, 4.676813, 4.637392, 4.599973, 4.565743
1460 DATA 4.535709, 4.510699, 4.491374, 4.478221, 4.471564, 4.471564, 4.478221, 4.491374
1470 DATA 4.5107, 4.535708, 4.565743, 4.599974, 4.637392, 4.676813, 4.716869, 4.756007
1480 DATA 4.792504, 4.824465, 4.84984, 4.866445, 4.871983, 4.864084, 4.840344, 4.798368
1490 DATA 4.735827, 4.650522, 4.540447, 4.403848, 4.239304, 4.045777, 3.822682, 3.569931
1500 DATA 3.287974, 2.977833, 2.641093, 2.279923, 1.897015, 1.495574, 1.079232, .6519791
1510 DATA .2180707
1520 END
```

Program 2
BASIC Listing of PATNUL1P Program

[The ARRL-supplied disk filename for this program is FRELICH2.BAS—*Ed.*]

```
10 REM PATNUL1P PROGRAM   PAUL D. FRELICH   W1ECO   DATE 12-08-1987   REV 6-10-88
20 REM NULL ANGLES, PATTERN RESPONSES, AND ELEMENT CURRENTS
30 PI=3.141592654#
40 DIM ANG(32),PHI(32),RESP(181),AX(32),AY(32),BX(32),BY(32),MAG(32),PHASE(32),HF(181),ACC(181)
50 INPUT "ENTER NUMBER OF ELEMENTS IN END-FIRE ARRAY ",ELNUM
60 INPUT "ENTER UNIFORM ELEMENT SPACING < OR = 90 DEG. ",SPACING
70 INPUT "ENTER FRACTIONAL NULL PLACEMENT ",FRC
80 PRINT "NUMBER OF ELEMENTS="ELNUM,"SPACING="SPACING,"FRC="FRC
85 LPRINT "NUMBER OF ELEMENTS="ELNUM,"SPACING="SPACING,"FRC="FRC
90 PRINT "I","NULL ANGLE","PHASE, PHI"
95 LPRINT "I","NULL ANGLE","PHASE, PHI"
100 S=SPACING*PI/360
110 FOR I = 1 TO (ELNUM-1)
120 ANG(I)=(PI/ELNUM)*(I+FRC)
130 PHI(I)=PI/2+S*COS(ANG(I))
140 PRINT I,ANG(I)*(180/PI),PHI(I)*(180/PI)
145 LPRINT I,ANG(I)*(180/PI),PHI(I)*(180/PI)
150 NEXT I
160 ALPHA=0
170 GOSUB 840
180 R0=R
190 PRINT
200 PRINT "R0= "R0
210 PRINT
220 PRINT "N","BEARING","RESPONSE-DB","R"
230 DALPHA=PI/72
240 ACCUM=0
250 FOR N=0 TO 72
260 READ HF(N)
270 ALPHA=N*DALPHA
280 GOSUB 840
290 GOSUB 420
300 GOSUB 440
310 ALPHA=(N-.5)*DALPHA
320 GOSUB 840
330 ACCUM=ACCUM+HF(N)*R^2
340 ACC(N)=ACCUM
350 NEXT N
360 ACCUM=ACCUM*DALPHA/20
370 DI=4.343*(LOG(R0^2/ACCUM))
380 PRINT
390 PRINT "DI= "DI"DB"
400 PRINT
410 GOTO 460
420 RESP(N)=8.686*LOG(ABS(R/R0)+10^-30)
430 RETURN
440 PRINT N,ALPHA*180/PI,RESP(N),R
450 RETURN
460 AX(1)=1
470 FOR M=1 TO (ELNUM-1)
480 C=COS(2*PHI(M))
490 S=SIN(2*PHI(M))
500 FOR N=1 TO M
510 BX(N)=AX(N)*C-AY(N)*S
```

```
520 BY(N)=AX(N)*S+AY(N)*C
530 NEXT N
540 FOR N=1 TO M
550 AX(N+1)=AX(N+1)+BX(N)
560 AY(N+1)=AY(N+1)+BY(N)
570 NEXT N
580 NEXT M
590 PRINT "N","REAL","IMAG","MAGNITUDE","PHASE-RAD"
595 LPRINT "N","REAL","IMAG","MAGNITUDE","PHASE-RAD"
600 FOR N=1 TO ELNUM
610 ON SGN(AX(N))+2 GOTO 710, 640, 620
620 PHASE=ATN(AY(N)/AX(N))
630 GOTO 770
640 ON SGN(AY(N))+2 GOTO 690, 650, 670
650 PHASE=0
660 GOTO 770
670 PHASE=PI/2
680 GOTO 770
690 PHASE=-PI/2
700 GOTO 770
710 ON SGN(AY(N))+2 GOTO 760, 740, 720
720 PHASE=ATN(AY(N)/AX(N))+PI
730 GOTO 770
740 PHASE=PI
750 GOTO 770
760 PHASE=ATN(AY(N)/AX(N))-PI
770 ON SGN(PHASE)+2 GOTO 780, 790, 790
780 PHASE=PHASE+(2*PI)
790 PHASE(N)=PHASE
800 MAG(N)=(AX(N)^2+AY(N)^2)^.5
810 PRINT N,AX(N),AY(N),MAG(N),PHASE(N)
815 LPRINT N,AX(N),AY(N),MAG(N),PHASE(N)
820 NEXT N
830 GOTO 900
840 R=2^(ELNUM-1)
850 A=S*COS(ALPHA)
860 FOR I = 1 TO (ELNUM-1)
870 R=R*COS(PHI(I)-A)
880 NEXT I
890 RETURN
900 FOR N= 0 TO 72
910 PRINT ACC(N)/ACC(72) SPC(0)
915 LPRINT ACC(N)/ACC(72) SPC(0)
920 NEXT N
930 PRINT
940 FOR N=0 TO 72
950 PRINT HF(N) SPC(0)
955 LPRINT HF(N) SPC(0)
960 NEXT N
970 PRINT
975 LPRINT
980 PRINT TAB(20)"CUMULATIVE POWER RADIATED, FRACTIONAL, 0"
985 LPRINT TAB(20)"CUMULATIVE POWER RADIATED, FRACTIONAL, 0"
990 PRINT
995 LPRINT
1000 PRINT TAB(2)"0"TAB(9)".1"TAB(16)".2"TAB(23)".3"TAB(30)".4"TAB(37)".5"TAB(44)".6"TAB(51)".7"
     TAB(58)".8"TAB(65)".9"TAB(72)"1"TAB(74)"ANGLE"
1005 LPRINT TAB(2)"0"TAB(9)".1"TAB(16)".2"TAB(23)".3"TAB(30)".4"TAB(37)".5"TAB(44)".6"TAB(51)".7"
     TAB(58)".8"TAB(65)".9"TAB(72)"1"TAB(74)"ANGLE"
1010 FOR M=0 TO 72
1020 J=2.5*M
1030 DB=RESP(M)
1040 ON SGN(50+DB)+2 GOTO 1050, 1050, 1060
1050 DB=-50
1060 A=INT(70*DB/50+72.5)
1070 B=2
1080 C=INT(70*ACC(M)/ACC(72)+2.5)
1090 FOR N=2 TO 73 STEP 1
1100 IF N=A THEN 1160
1110 IF N=C THEN 1190
1120 IF N=B THEN 1130 ELSE 1210
1130 PRINT TAB(B) "+" ;
1135 LPRINT TAB(B) "+" ;
1140 B=B+7
1150 GOTO 1210
1160 PRINT TAB(N) "*" ;
```

```
1165 LPRINT TAB(N) "*" ;
1170 IF N=B THEN 1140
1180 GOTO 1210
1190 PRINT TAB(N)"O";
1195 LPRINT TAB(N)"O";
1200 IF N=B THEN 1140
1210 IF N=73 THEN 1220 ELSE 1240
1220 IF J-INT(J)=0 THEN 1230 ELSE 1240
1230 PRINT TAB(73)J;
1235 LPRINT TAB(73)J;
1240 NEXT N
1250 NEXT M
1260 PRINT TAB(1)"-50"TAB(15)"-40"TAB(29)"-30"TAB(43)"-20"TAB(57)"-10"TAB(72)"0"
1265 LPRINT TAB(1)"-50"TAB(15)"-40"TAB(29)"-30"TAB(43)"-20"TAB(57)"-10"TAB(72)"0"
1270 PRINT
1275 LPRINT
1280 PRINT TAB(20)"PATTERN RESPONSE, DECIBELS  *"
1285 LPRINT TAB(20)"PATTERN RESPONSE, DECIBELS  *"
1290 PRINT
1295 LPRINT
1300 DB3=-3.010299956#
1310 FOR N=0 TO 72
1320 ON SGN(DB3-RESP(N))+2 GOTO 1400, 1380, 1330
1330 D1=RESP(N-1)-RESP(N)
1340 D2=RESP(N+1)-RESP(N)
1350 D3=DB3-RESP(N)
1360 BW=DALPHA*(D3*(D3*(D1+D2)-D1^2-D2^2)/(D1*D2^2-D2*D1^2)+N)*180/PI
1370 GOTO 1410
1380 BW=DALPHA*N*180/PI
1390 GOTO 1410
1400 NEXT N
1410 PRINT "R0"R0" ELNUM"ELNUM "SP"SPACING "FRC"FRC "BOOMLENGTH"(ELNUM-1)*SPACING/360 "DI"DI
     "HALF-BEAMWIDTH"BW
1415 LPRINT "R0"R0" ELNUM"ELNUM "SP"SPACING "FRC"FRC "BOOMLENGTH"(ELNUM-1)*SPACING/360 "DI"DI
     "HALF-BEAMWIDTH"BW
1420 DATA 0, .2180728, .6519813, 1.079234, 1.495576, 1.897017, 2.279923, 2.641095
1430 DATA 2.977833, 3.287975, 3.569931, 3.822683, 4.045778, 4.239305, 4.403849, 4.540447
1440 DATA 4.650522, 4.735827, 4.798367, 4.840344, 4.864085, 4.871982, 4.866445, 4.84984
1450 DATA 4.824465, 4.792504, 4.756006, 4.716868, 4.676813, 4.637392, 4.599973, 4.565743
1460 DATA 4.535709, 4.510699, 4.491374, 4.478221, 4.471564, 4.471564, 4.478221, 4.491374
1470 DATA 4.5107, 4.535708, 4.565743, 4.599974, 4.637392, 4.676813, 4.716869, 4.756007
1480 DATA 4.792504, 4.824465, 4.84984, 4.866445, 4.871983, 4.864084, 4.840344, 4.798368
1490 DATA 4.735827, 4.650522, 4.540447, 4.403848, 4.239304, 4.045777, 3.822682, 3.569931
1500 DATA 3.287974, 2.977833, 2.641093, 2.279923, 1.897015, 1.495574, 1.079232, .6519791
1510 DATA .2180707
1520 END
```

Program 3
BASIC Listing of BEAMPAT8 Program

[The ARRL-supplied disk filename for this program is FRELICH3.BAS—*Ed.*]

```
10 REM [BEAMPAT8] PROGRAM  PAUL D. FRELICH  W1ECO  DATE 12-08-1987  REV. 6-17-88
20 REM NULL ANGLES, PATTERN & DI; ELEMENT CURRENTS, LENGTHS AND IMPEDANCES.
30 PI=3.141592654#
40 DIM ANG(32),PHI(32),RESP(181),AX(32),AY(32),BX(32),BY(32),MAG(32),PHASE(32),MX(41),MY(41),
      ZMX(32,33),ZMY(32,33),ZSELFA(32),ZSELFB(32),ZDRIVEA(32),ZDRIVEB(32),VDA(32),VDB(32),
      XTUN(32),LENGTH(32),DLENGTH(32),HF(181),ACC(181),ZDSELFA(32),ZDSELFB(32)
50 DIM DELTAL(32),DELTADL(32)
60 PRINT "SCHEME, PATTERN NULL METHOD FOR DESIGNING YAGI ANTENNAS"
70 INPUT "ENTER NUMBER OF ELEMENTS          ELNUM = ",ELNUM
80 INPUT "ENTER UNIFORM ELEMENT SPACING    SPACING = ",SPACING
90 INPUT "ENTER FRACTIONAL NULL LOCATION     FRC = ",FRC
100 BOOMLENGTH=(ELNUM-1)*SPACING/360
110 PRINT
120 PRINT "ELNUM" ELNUM "SPACING" SPACING "BOOMLENGTH"BOOMLENGTH "FRC" FRC
130 PRINT
140 PRINT "REM--PATTERN DESIGN FACTORS-- PDF Null Angles & Phase Angles."
150 PRINT
160 PRINT "PDF","NULL ANGLE","PHASE, PHI"
170 S=SPACING*PI/360
```

```
180 FOR I = 1 TO (ELNUM-1)
190 ANG(I)=(PI/(ELNUM))*(I+FRC)
200 PHI(I)=PI/2+S*COS(ANG(I))
210 PRINT I,ANG(I)*(180/PI),PHI(I)*(180/PI)
220 NEXT I
230 ALPHA=0
240 GOSUB 650
250 R0=R
260 PRINT
270 PRINT "R0= "R0"Main lobe Resp.(numeric)"
280 PRINT
290 PRINT "Use PDF's to compute Pattern Responses, Directivity & half-Power half-Beamwidth."
300 PRINT
310 PRINT "N","BEARING","RESPONSE-DB","RESP"
320 DALPHA=PI/72
330 ACCUM=0
340 FOR N=0 TO 72
350 READ HF(N)
360 ALPHA=N*DALPHA
370 GOSUB 650
380 GOSUB 710
390 GOSUB 730
400 ALPHA= (N-.5)*DALPHA
410 GOSUB 650
420 ACCUM=ACCUM+HF(N)*R^2
430 ACC(N)=ACCUM
440 NEXT N
450 ACCUM=ACCUM*DALPHA/20
460 DIRECTIVITY=R0^2/ACCUM
470 DI=4.343*LOG(DIRECTIVITY)
480 DI=INT(DI*10^6)/10^6
490 DBD=DI-2.150341
500 DB3=-3.010299956#
510 FOR N=0 TO 72
520 ON SGN(DB3-RESP(N))+2 GOTO 600, 580, 530
530 D1=RESP(N-1)-RESP(N)
540 D2=RESP(N+1)-RESP(N)
550 D3=DB3-RESP(N)
560 BW=DALPHA*(D3*(D3*(D1+D2)-D1^2-D2^2)/(D1*D2^2-D2*D1^2)+N)*180/PI
570 GOTO 610
580 BW=DALPHA*N*180/PI
590 GOTO 610
600 NEXT N
610 PRINT
620 PRINT "DIRECTIVITY"DIRECTIVITY"Numeric"SPC(2)"DBI"DI"DBD"DBD"BW"BW"Deg."
630 PRINT
640 GOTO 750
650 R=2^(ELNUM-1)
660 A=S*COS(ALPHA)
670 FOR I = 1 TO (ELNUM-1)
680 R=R*COS(PHI(I)-A)
690 NEXT I
700 RETURN
710 RESP(N)=8.686*LOG(ABS(R/R0)+10^-30)
720 RETURN
730 PRINT N,ALPHA*180/PI,RESP(N),R
740 RETURN
750 PRINT
760 PRINT
770 AX(1)=1
780 FOR M=1 TO (ELNUM-1)
790 C=COS(2*PHI(M))
800 S=SIN(2*PHI(M))
810 FOR N=1 TO M
820 BX(N)=AX(N)*C-AY(N)*S
830 BY(N)=AX(N)*S+AY(N)*C
840 NEXT N
850 FOR N=1 TO M
860 AX(N+1)=AX(N+1)+BX(N)
870 AY(N+1)=AY(N+1)+BY(N)
880 NEXT N
890 NEXT M
900 GOSUB 2250
910 FOR I=0 TO 40
920 READ MX(I),MY(I)
930 NEXT I
940 PRINT
```

```
950 PRINT "REM--MUTUAL IMPEDANCE TABLE-- PRINT? YES OR NO, ENTER 1, OR  Ø"
960 INPUT "TABLE PRINT OPTION ",OPT1
970 PRINT
980 IF OPT1=Ø THEN 1050
990 PRINT " SPACING, Ø    MUTUAL IMPEDANCE          SPACING 1 TO 2 LAMBDA"
1000 PRINT " TO 1 LAMBDA    REAL            IMAG.        REAL            IMAG."
1010 FOR I=Ø TO 20
1020 PRINT I/20,MX[I],MY[I],MX[I+20],MY[I+20]
1030 NEXT I
1040 PRINT
1050 NSPACE=SPACING/360
1060 I=1
1070 FOR J=1 TO ELNUM
1080 L=(J-I)*NSPACE
1090 IF L>2 THEN 1170
1100 FOR K=1 TO 39 STEP 2
1110 U=20*L-K
1120 IF ABS(U)<=1 THEN 1140
1130 NEXT K
1140 ZMX[I,J]=([(MX[K+1]+MX[K-1])/2-MX[K])*U+(MX[K+1]-MX[K-1])/2)*U+MX[K]
1150 ZMY[I,J]=([(MY[K+1]+MY[K-1])/2-MY[K])*U+(MY[K+1]-MY[K-1])/2)*U+MY[K]
1160 GOTO 1200
1170 A=2*PI*L
1180 ZMX[I,J]=120*(COS(A)/A^2+SIN(A)*(1/A-1/A^3))
1190 ZMY[I,J]=120*(-SIN(A)/A^2+COS(A)*(1/A-1/A^3))
1200 NEXT J
1210 FOR I=2 TO ELNUM
1220 FOR J=I TO ELNUM
1230 ZMX[I,J]=ZMX[1,J-I+1]
1240 ZMY[I,J]=ZMY[1,J-I+1]
1250 NEXT J
1260 FOR J=1 TO I-1
1270 ZMX[I,J]=ZMX[1,I-J+1]
1280 ZMY[I,J]=ZMY[1,I-J+1]
1290 NEXT J
1300 NEXT I
1310 PRINT "REM--MUTUAL IMPEDANCE MATRIX-- PRINT? YES OR NO, ENTER  1, OR  Ø"
1320 INPUT "MUT. IMP. MATRIX PRINT OPTION ",OPT2
1330 PRINT
1340 IF OPT2=Ø THEN 1420
1350 PRINT "MUTUAL IMPEDANCE MATRIX--ZMX[I,J], ZMY[I,J], WHERE I=ROW AND J=COLUMN"
1360 FOR I=1 TO ELNUM
1370 PRINT I
1380 FOR J=1 TO ELNUM
1390 PRINT ,J,ZMX[I,J],ZMY[I,J]
1400 NEXT J
1410 NEXT I
1420 IF ELNUM=1 THEN 1430 ELSE 1490
1430 ZDRIVEA[1]=MX[Ø]
1440 ZDSELFA[1]=MX[Ø]
1450 ZDRIVEB[1]=MY[Ø]
1460 ZDSELFB[1]=MY[Ø]
1470 DLENGTH[1]=PI/2
1480 GOTO 1930
1490 PRINT
1500 PRINT "SELF IMPEDANCES for PARASITIC ELEMENTS, from Linear Equations and Currents."
1510 PRINT
1520 PRINT "J","ZSELFA","ZSELFB"
1530 REM AX, AY Real and Imaginary parts of Element Currents.
1540 REM VA, VB Real and Imaginary parts of BACK VOLTAGES.
1550 FOR J=1 TO ELNUM
1560 VA=Ø
1570 VB=Ø
1580 FOR K=1 TO ELNUM
1590 IF K=J THEN 1620
1600 VA=VA+AX[K]*ZMX[J,K]-AY[K]*ZMY[J,K]
1610 VB=VB+AX[K]*ZMY[J,K]+AY[K]*ZMX[J,K]
1620 NEXT K
1630 REM ZSELF Required Self Impedance that makes Drive Voltage = ZERO.
1640 ZSELFA[J]=-(VA*AX[J]+VB*AY[J])/MAG[J]^2
1650 ZSELFB[J]=-(VB*AX[J]-VA*AY[J])/MAG[J]^2
1660 PRINT J,ZSELFA[J],ZSELFB[J]
1670 NEXT J
1680 PRINT
1690 REM R(KL)=104.093(KL)^2-190.970(KL)+116.256
1700 REM X(KL)=29.263(KL)^2-26.862(KL)+12.452
1710 A=104.093
```

```
1720 B=-190.97
1730 C=116.256
1740 ELRAD=PI/400
1750 PRINT "PARASITIC ELEMENT LENGTH, & PARASITIC ELEMENT TUNING, XTUN"
1760 PRINT
1770 PRINT "J","LENGTH","XTUN","% DELTA-L"
1780 FOR J=1 TO ELNUM
1790 C1=C-ZSELFA(J)
1800 ON SGN(B^2-4*A*C1)+2 GOTO 1810, 1830, 1830
1810 PRINT J SPC(2) "SQUARE ROOT OF NEGATIVE NUMBER"
1820 GOTO 1920
1830 KL=(-B+(B^2-4*A*C1)^.5)/(2*A)
1840 XKL=(29.263*KL-26.862)*KL+12.452
1850 LENGTH(J)=KL
1860 DELTAL(J)=100*(2*LENGTH(J)/PI-1)
1870 IF KL=PI/2 THEN 1880 ELSE 1900
1880 XTUN(J)=ZSELFB(J)-XKL
1890 GOTO 1910
1900 XTUN(J)=ZSELFB(J)-XKL+120*(LOG(2*KL/ELRAD)-1)/TAN(KL)
1910 PRINT J,LENGTH(J),XTUN(J),DELTAL(J)
1920 NEXT J
1930 PRINT "DRIVEN ELEMENT DRIVE IMPEDANCE, SELF IMPEDANCE and LENGTH—required to make Gain equal
        to the Directivity."
1940 PRINT
1950 PRINT "ELNUM","ZDRIVE-REAL","ZSELF-REAL","ZDSELF-REAL","DLENGTH"
1960 FOR J=1 TO ELNUM
1970 IF ELNUM=1 THEN 2120 ELSE 1980
1980 ZDRIVEA(J)=(120/DIRECTIVITY)*R0^2/MAG(J)^2
1990 ZDSELFA(J)=ZDRIVEA(J)+ZSELFA(J)
2000 C1=C-ZDSELFA(J)
2010 ON SGN(B^2-4*A*C1)+2 GOTO 2020, 2040, 2040
2020 PRINT J SPC(2) "SQUARE ROOT OF NEGATIVE NUMBER"
2030 GOTO 2130
2040 KL=(-B+(B^2-4*A*C1)^.5)/(2*A)
2050 XKL=(29.263*KL-26.862)*KL+12.452
2060 DLENGTH(J)=KL
2070 DELTADL(J)=100*(2*DLENGTH(J)/PI-1)
2080 IF KL=PI/2 THEN 2090 ELSE 2110
2090 ZDSELFB(J)=XKL
2100 GOTO 2120
2110 ZDSELFB(J)=XKL-120*(LOG(2*KL/ELRAD)-1)/TAN(KL)
2120 PRINT J,ZDRIVEA(J),ZSELFA(J),ZDSELFA(J),DLENGTH(J)
2130 NEXT J
2140 PRINT
2150 PRINT "ELNUM","ZDRIVE-IMAG","ZSELF-IMAG","ZDSELF-IMAG","% DELTA-DL"
2160 FOR J=1 TO ELNUM
2170 ZDRIVEB(J)=ZDSELFB(J)-ZSELFB(J)
2180 PRINT J,ZDRIVEB(J),ZSELFB(J),ZDSELFB(J),DELTADL(J)
2190 NEXT J
2200 PRINT
2210 INPUT "PLOT OPTION; 1=YES,0=NO. ENTER 1 OR 0.  ",OPT3
2220 PRINT
2230 IF OPT3=0 THEN 2550
2240 GOTO 2790
2250 PRINT "Compute ELEMENT CURRENTS from PDF's and Spacing."
2260 PRINT
2270 PRINT "ELNUM","REAL","IMAGINARY","MAGNITUDE","PHASE, DEG."
2280 FOR N=1 TO ELNUM
2290 ON SGN(AX(N))+2 GOTO 2300, 2310, 2320
2300 ON SGN(AY(N))+2 GOTO 2330, 2350, 2370
2310 ON SGN(AY(N))+2 GOTO 2390, 2410, 2430
2320 ON SGN(AY(N))+2 GOTO 2450, 2470, 2490
2330 PHASE=PI+ATN(AY(N)/AX(N))
2340 GOTO 2500
2350 PHASE=PI
2360 GOTO 2500
2370 PHASE=PI+ATN(AY(N)/AX(N))
2380 GOTO 2500
2390 PHASE=3*PI/2
2400 GOTO 2500
2410 PHASE=0
2420 GOTO 2500
2430 PHASE=PI/2
2440 GOTO 2500
2450 PHASE=2*PI+ATN(AY(N)/AX(N))
2460 GOTO 2500
2470 PHASE=0
```

```
2480 GOTO 2500
2490 PHASE=ATN(AY(N)/AX(N))
2500 PHASE(N)=PHASE
2510 MAG(N)=(AX(N)^2+AY(N)^2)^.5
2520 PRINT N,AX(N),AY(N),MAG(N),PHASE(N)*180/PI
2530 NEXT N
2540 RETURN
2550 PRINT "FRC"FRC "ELNUM"ELNUM "SPACING"SPACING "BOOMLENGTH"(ELNUM-1)*SPACING/360
2560 PRINT "R0"R0 "DIR"DIRECTIVITY "DBI"DI "DBD"DBD "BW/2"BW SPC(2)"L XT LD RD XD DL DDL"
2570 FOR I=1 TO ELNUM
2580 PRINT LENGTH(I) SPC(0)
2590 NEXT I
2600 PRINT
2610 FOR I=1 TO ELNUM
2620 PRINT XTUN(I) SPC(0)
2630 NEXT I
2640 PRINT
2650 FOR I=1 TO ELNUM
2660 PRINT DLENGTH(I) SPC(0)
2670 NEXT I
2680 PRINT
2690 FOR I=1 TO ELNUM
2700 PRINT ZDRIVEA(I) SPC(0)
2710 NEXT I
2720 PRINT
2730 FOR I=1 TO ELNUM
2740 PRINT ZDRIVEB(I) SPC(0)
2750 NEXT I
2760 PRINT
2761 FOR I=1 TO ELNUM
2762 PRINT DELTAL(I) SPC(0)
2763 NEXT I
2764 PRINT
2765 FOR I=1 TO ELNUM
2766 PRINT DELTADL(I) SPC(0)
2767 NEXT I
2768 PRINT
2770 PRINT "END OF ROUTINE"
2780 END
2790 GOSUB 2960
2800 FOR N=0 TO 72
2810 GOSUB 2860
2820 NEXT N
2830 GOSUB 2960
2840 GOSUB 2930
2850 GOTO 2550
2860 DB=RESP(N)
2870 ON SGN(70+DB)+2 GOTO 2880, 2880, 2890
2880 DB=-70
2890 C=INT(DB+.5)
2900 DBTR=INT(10*DB+.5)/10
2910 PRINT TAB(70+C) "*" 2.5*N DBTR
2920 RETURN
2930 PRINT TAB(1)"-69"TAB(9)"-60"TAB(19)"-50"TAB(29)"-40"TAB(39)"-30"TAB(49)"-20"TAB(59)"-10"
     TAB(70)"0"
2940 PRINT TAB(26)"RELATIVE DB"
2950 RETURN
2960 FOR I=0 TO 70 STEP 10
2970 PRINT TAB(I) "L";
2980 NEXT I
2990 RETURN
3000 DATA 0,.2180718,.6519813,1.079234,1.495576,1.897017,2.279923,2.641095
3010 DATA 2.977833,3.287975,3.56993,3.822683,4.045778,4.239305,4.403849,4.540447
3020 DATA 4.650522,4.735827,4.798367,4.840344,4.864085,4.871982,4.866445,4.84984
3030 DATA 4.824465,4.792504,4.756006,4.716868,4.676813,4.63739,4.599974,4.565743
3040 DATA 4.535709,4.5107,4.491374,4.478221,4.471564,4.471564,4.478221,4.491374
3050 DATA 4.5107,4.535708,4.565743,4.599974,4.637392,4.676813,4.716869,4.756007
3060 DATA 4.792504,4.824465,4.84984,4.866445,4.871983,4.864084,4.840344,4.798368
3070 DATA 4.735827,4.650522,4.540447,4.403848,4.239304,4.045777,3.822682,3.56993
3080 DATA 3.287974,2.977833,2.641093,2.279923,1.897015,1.495574,1.079232,.651980
3090 DATA .2180717
3100 SP=SPACING*PI/180
3110 DALPHA=PI/60
3120 DEL=DELTA*PI/180
3130 FOR N=0 TO 60
3140 PH=-SP*COS(N*DALPHA)
3150 X1=0
```

```
3160 Y1=0
3170 X2=0
3180 Y2=0
3190 ON SCHEME GOTO 3200,3250,3480
3200 R=2^[ELNUM-1]
3210 FOR M=1 TO ELNUM-1
3220 R=R*COS[PHI[M]+PH/2]
3230 NEXT M
3240 GOTO 3300
3250 R=2^FAMNO
3260 COV=1-COS[N*DALPHA]
3270 FOR M=1 TO FAMNO
3280 R=R*COS[M*SP*COV/2+DEL]
3290 NEXT M
3300 FOR M= 1 TO ELNUM
3310 C=COS[[M-1]*PH]
3320 S=SIN[[M-1]*PH]
3330 X1=X1+AX[M]*C-AY[M]*S
3340 Y1=Y1+AY[M]*C+AX[M]*S
3350 PHASE=[M-1]*PH+PHASE[M]
3360 X2=X2+MAG[M]*COS[PHASE]
3370 Y2=Y2+MAG[M]*SIN[PHASE]
3380 NEXT M
3390 RSP1=[X1^2+Y1^2]^.5
3400 RSP2=[X2^2+Y2^2]^.5
3410 IF N=0 THEN 3420 ELSE 3460
3420 R0=R
3430 R01=RSP1
3440 R02=RSP2
3450 PRINT "R0"R "R01"RSP1 "R02"RSP2
3460 PRINT 3*N,8.686*LOG[ABS[R]/R0+10^-30],8.686*LOG[RSP1/R01+10^-30],8.686*LOG[RSP2/R02+10^-30]
3470 NEXT N
3480 END
3490 DATA 73.12,42.46,71.7,24.3,67.3,7.5,60.4,-7.1,51.4,-19.2,40.8,-28.4
3500 DATA 29.3,-34.4,17.5,-37.4,6.2,-37.4,-4.0,-34.8,-12.5,-29.9
3510 DATA -19.1,-23.4,-23.3,-15.9,-25.2,-7.9,-24.9,-0.3,-22.5,6.6
3520 DATA -18.5,12.3,-13.3,16.3,-7.5,18.6,-1.6,19.0,4.0,17.7
3530 DATA 8.8,15.0,12.3,11.2,14.5,6.7,15.3,1.9,14.6,-2.7
3540 DATA 12.6,-6.7,9.6,-9.8,6.0,-11.9,2.0,-12.7,-1.9,-12.3
3550 DATA -5.4,-10.8,-8.2,-8.4,-10.0,-5.3,-10.9,-2.0,-10.6,1.4
3560 DATA -9.4,4.5,-7.4,7.0,-4.8,8.7,-1.9,9.5,1.1,9.4
```

Program 4
BASIC Listing of BEAMPAT9 Program

[The ARRL-supplied disk filename for this program is FRELICH4.BAS—Ed.]

```
10 REM [BEAMPAT9] PROGRAM  PAUL D. FRELICH  W1ECO  DATE 12-08-1987   REV. 6-21-88
20 REM NULL ANGLES, PATTERN & DI; ELEMENT CURRENTS, LENGTHS AND IMPEDANCES.
30 PI=3.141592654#
40 DIM ANG[32],PHI[32],RESP[181],AX[32],AY[32],BX[32],BY[32],MAG[32],PHASE[32],MX[41],MY[41],
       ZMX[32,33],ZMY[32,33],ZSELFA[32],ZSELFB[32],ZDRIVEA[32],ZDRIVEB[32],VDA[32],VDB[32],
       XTUN[32],LENGTH[32],DLENGTH[32],HF[181],ACC[181],ZDSELFA[32],ZDSELFB[32]
50 DIM DELTAL[32],DELTADL[32]
60 PRINT "SCHEME, PATTERN NULL METHOD FOR DESIGNING YAGI ANTENNAS"
65 LPRINT "SCHEME, PATTERN NULL METHOD FOR DESIGNING YAGI ANTENNAS"
70 INPUT "ENTER NUMBER OF ELEMENTS          ELNUM  =  ",ELNUM
80 INPUT "ENTER UNIFORM ELEMENT SPACING    SPACING =  ",SPACING
90 INPUT "ENTER FRACTIONAL NULL LOCATION     FRC  =  ",FRC
100 BOOMLENGTH=[ELNUM-1]*SPACING/360
110 PRINT
115 LPRINT
120 PRINT "ELNUM" ELNUM "SPACING" SPACING "BOOMLENGTH"BOOMLENGTH "FRC" FRC
130 LPRINT "ELNUM" ELNUM "SPACING" SPACING "BOOMLENGTH"BOOMLENGTH "FRC" FRC
140 PRINT
145 LPRINT
150 PRINT "REM--PATTERN DESIGN FACTORS-- PDF Null Angles & Phase Angles."
155 LPRINT "REM--PATTERN DESIGN FACTORS-- PDF Null Angles & Phase Angles."
160 PRINT
165 LPRINT
170 PRINT "PDF","NULL ANGLE","PHASE, PHI"
```

```
175 LPRINT "PDF","NULL ANGLE","PHASE, PHI"
180 S=SPACING*PI/360
190 FOR I = 1 TO (ELNUM-1)
200 ANG(I)=(PI/(ELNUM))*(I+FRC)
210 PHI(I)=PI/2+S*COS(ANG(I))
220 PRINT I,ANG(I)*(180/PI),PHI(I)*(180/PI)
225 LPRINT I,ANG(I)*(180/PI),PHI(I)*(180/PI)
230 NEXT I
240 ALPHA=0
250 GOSUB 660
260 R0=R
270 PRINT
275 LPRINT
280 PRINT "R0= "R0"Main lobe Resp.(numeric)"
285 LPRINT "R0= "R0"Main lobe Resp.(numeric)"
290 PRINT
295 LPRINT
300 PRINT "Use PDF's to compute Pattern Responses, Directivity & half-Power half-Beamwidth."
305 LPRINT "Use PDF's to compute Pattern Responses, Directivity & half-Power half-Beamwidth."
310 PRINT
315 LPRINT
320 PRINT "N","BEARING","RESPONSE-DB","RESP"
325 LPRINT "N","BEARING","RESPONSE-DB","RESP"
326 LPRINT "SEE PATTERN PLOT"
330 DALPHA=PI/72
340 ACCUM=0
350 FOR N=0 TO 72
360 READ HF(N)
370 ALPHA=N*DALPHA
380 GOSUB 660
390 GOSUB 720
400 GOSUB 740
410 ALPHA= (N-.5)*DALPHA
420 GOSUB 660
430 ACCUM=ACCUM+HF(N)*R^2
440 ACC(N)=ACCUM
450 NEXT N
460 ACCUM=ACCUM*DALPHA/20
470 DIRECTIVITY=R0^2/ACCUM
480 DI=4.343*LOG(DIRECTIVITY)
490 DI=INT(DI*10^6)/10^6
500 DBD=DI-2.150341
510 DB3=-3.010299956#
520 FOR N=0 TO 72
530 ON SGN(DB3-RESP(N))+2 GOTO 610, 590, 540
540 D1=RESP(N-1)-RESP(N)
550 D2=RESP(N+1)-RESP(N)
560 D3=DB3-RESP(N)
570 BW=DALPHA*(D3*(D3*(D1+D2)-D1^2-D2^2)/(D1*D2^2-D2*D1^2)+N)*180/PI
580 GOTO 620
590 BW=DALPHA*N*180/PI
600 GOTO 620
610 NEXT N
620 PRINT
625 LPRINT
630 PRINT "DIRECTIVITY"DIRECTIVITY"Numeric"SPC(2)"DBI"DI"DBD"DBD"BW"BW"Deg."
635 LPRINT "DIRECTIVITY"DIRECTIVITY"Numeric"SPC(2)"DBI"DI"DBD"DBD"BW"BW"Deg."
640 PRINT
645 LPRINT
650 GOTO 760
660 R=2^(ELNUM-1)
670 A=S*COS(ALPHA)
680 FOR I = 1 TO (ELNUM-1)
690 R=R*COS(PHI(I)-A)
700 NEXT I
710 RETURN
720 RESP(N)=8.686*LOG(ABS(R/R0)+10^-30)
730 RETURN
740 PRINT N,ALPHA*180/PI,RESP(N),R
750 RETURN
760 PRINT
770 PRINT
780 AX(1)=1
790 FOR M=1 TO (ELNUM-1)
800 C=COS(2*PHI(M))
810 S=SIN(2*PHI(M))
820 FOR N=1 TO M
```

```
830 BX(N)=AX(N)*C-AY(N)*S
840 BY(N)=AX(N)*S+AY(N)*C
850 NEXT N
860 FOR N=1 TO M
870 AX(N+1)=AX(N+1)+BX(N)
880 AY(N+1)=AY(N+1)+BY(N)
890 NEXT N
900 NEXT M
910 GOSUB 2260
920 FOR I=0 TO 40
930 READ MX(I),MY(I)
940 NEXT I
950 PRINT
955 LPRINT
960 PRINT "REM---MUTUAL IMPEDANCE TABLE-- PRINT? YES OR NO, ENTER 1, OR  0"
965 LPRINT "REM---MUTUAL IMPEDANCE TABLE-- PRINT? YES OR NO, ENTER 1, OR  0"
970 INPUT "TABLE PRINT OPTION ",OPT1
980 PRINT
985 LPRINT
990 IF OPT1=0 THEN 1060
1000 PRINT " SPACING, 0    MUTUAL IMPEDANCE          SPACING 1 TO 2 LAMBDA"
1005 LPRINT " SPACING, 0    MUTUAL IMPEDANCE           SPACING 1 TO 2 LAMBDA"
1010 PRINT " TO 1 LAMBDA    REAL         IMAG.        REAL         IMAG."
1015 LPRINT " TO 1 LAMBDA    REAL         IMAG.        REAL         IMAG."
1020 FOR I=0 TO 20
1030 PRINT I/20,MX(I),MY(I),MX(I+20),MY(I+20)
1035 LPRINT I/20,MX(I),MY(I),MX(I+20),MY(I+20)
1040 NEXT I
1050 PRINT
1055 LPRINT
1060 NSPACE=SPACING/360
1070 I=1
1080 FOR J=1 TO ELNUM
1090 L=(J-I)*NSPACE
1100 IF L>2 THEN 1180
1110 FOR K=1 TO 39 STEP 2
1120 U=20*L-K
1130 IF ABS(U)<=1 THEN 1150
1140 NEXT K
1150 ZMX(I,J)=(((MX(K+1)+MX(K-1))/2-MX(K))*U+(MX(K+1)-MX(K-1))/2)*U+MX(K)
1160 ZMY(I,J)=(((MY(K+1)+MY(K-1))/2-MY(K))*U+(MY(K+1)-MY(K-1))/2)*U+MY(K)
1170 GOTO 1210
1180 A=2*PI*L
1190 ZMX(I,J)=120*(COS(A)/A^2+SIN(A)*(1/A-1/A^3))
1200 ZMY(I,J)=120*(-SIN(A)/A^2+COS(A)*(1/A-1/A^3))
1210 NEXT J
1220 FOR I=2 TO ELNUM
1230 FOR J=I TO ELNUM
1240 ZMX(I,J)=ZMX(1,J-I+1)
1250 ZMY(I,J)=ZMY(1,J-I+1)
1260 NEXT J
1270 FOR J=1 TO I-1
1280 ZMX(I,J)=ZMX(1,I-J+1)
1290 ZMY(I,J)=ZMY(1,I-J+1)
1300 NEXT J
1310 NEXT I
1320 PRINT "REM---MUTUAL IMPEDANCE MATRIX-- PRINT? YES OR NO, ENTER  1, OR  0"
1325 LPRINT "REM---MUTUAL IMPEDANCE MATRIX-- PRINT? YES OR NO, ENTER  1, OR  0"
1330 INPUT "MUT. IMP. MATRIX PRINT OPTION ",OPT2
1340 PRINT
1350 IF OPT2=0 THEN 1430
1360 PRINT "MUTUAL IMPEDANCE MATRIX--ZMX(I,J), ZMY(I,J), WHERE I=ROW AND J=COLUMN"
1370 FOR I=1 TO ELNUM
1380 PRINT I
1385 LPRINT I
1390 FOR J=1 TO ELNUM
1400 PRINT ,J,ZMX(I,J),ZMY(I,J)
1405 LPRINT ,J,ZMX(I,J),ZMY(I,J)
1410 NEXT J
1420 NEXT I
1430 IF ELNUM=1 THEN 1440 ELSE 1500
1440 ZDRIVEA(1)=MX(0)
1450 ZDSELFA(1)=MX(0)
1460 ZDRIVEB(1)=MY(0)
1470 ZDSELFB(1)=MY(0)
1480 DLENGTH(1)=PI/2
1490 GOTO 1940
```

```
1500 PRINT
1505 LPRINT
1510 PRINT "SELF IMPEDANCES for PARASITIC ELEMENTS, from Linear Equations and Currents."
1515 LPRINT "SELF IMPEDANCES for PARASITIC ELEMENTS, from Linear Equations and Currents."
1520 PRINT
1525 LPRINT
1530 PRINT "J","ZSELFA","ZSELFB"
1535 LPRINT "J","ZSELFA","ZSELFB"
1540 REM AX, AY Real and Imaginary parts of Element Currents.
1550 REM VA, VB Real and Imaginary parts of BACK VOLTAGES.
1560 FOR J=1 TO ELNUM
1570 VA=0
1580 VB=0
1590 FOR K=1 TO ELNUM
1600 IF K=J THEN 1630
1610 VA=VA+AX[K]*ZMX[J,K]-AY[K]*ZMY[J,K]
1620 VB=VB+AX[K]*ZMY[J,K]+AY[K]*ZMX[J,K]
1630 NEXT K
1640 REM ZSELF Required Self Impedance that makes Drive Voltage = ZERO.
1650 ZSELFA[J]=-(VA*AX[J]+VB*AY[J])/MAG[J]^2
1660 ZSELFB[J]=-(VB*AX[J]-VA*AY[J])/MAG[J]^2
1670 PRINT J,ZSELFA[J],ZSELFB[J]
1675 LPRINT J,ZSELFA[J],ZSELFB[J]
1680 NEXT J
1690 PRINT
1695 LPRINT
1700 REM R(KL)=104.093(KL)^2-190.970(KL)+116.256
1710 REM X(KL)=29.263(KL)^2-26.862(KL)+12.452
1720 A=104.093
1730 B=-190.97
1740 C=116.256
1750 ELRAD=PI/400
1755 LPRINT "ELRAD=PI/400"
1760 PRINT "PARASITIC ELEMENT LENGTH, & PARASITIC ELEMENT TUNING, XTUN"
1765 LPRINT "PARASITIC ELEMENT LENGTH, & PARASITIC ELEMENT TUNING, XTUN"
1770 PRINT
1775 LPRINT
1780 PRINT "J","LENGTH","XTUN","% DELTA-L"
1785 LPRINT "J","LENGTH","XTUN","% DELTA-L"
1790 FOR J=1 TO ELNUM
1800 C1=C-ZSELFA[J]
1810 ON SGN(B^2-4*A*C1)+2 GOTO 1820, 1840, 1840
1820 PRINT J SPC(2) "SQUARE ROOT OF NEGATIVE NUMBER"
1825 LPRINT J SPC(2) "SQUARE ROOT OF NEGATIVE NUMBER"
1830 GOTO 1930
1840 KL=(-B+(B^2-4*A*C1)^.5)/(2*A)
1850 XKL=(29.263*KL-26.862)*KL+12.452
1860 LENGTH[J]=KL
1870 DELTAL[J]=100*(2*LENGTH[J]/PI-1)
1880 IF KL=PI/2 THEN 1890 ELSE 1910
1890 XTUN[J]=ZSELFB[J]-XKL
1900 GOTO 1920
1910 XTUN[J]=ZSELFB[J]-XKL+120*(LOG(2*KL/ELRAD)-1)/TAN(KL)
1920 PRINT J,LENGTH[J],XTUN[J],DELTAL[J]
1925 LPRINT J,LENGTH[J],XTUN[J],DELTAL[J]
1930 NEXT J
1940 PRINT "DRIVEN ELEMENT DRIVE IMPEDANCE, SELF IMPEDANCE and LENGTH---required to make Gain equal
        to the Directivity."
1945 LPRINT "DRIVEN ELEMENT DRIVE IMPEDANCE, SELF IMPEDANCE and LENGTH---required to make Gain equal
        to the Directivity."
1950 PRINT
1955 LPRINT
1960 PRINT "ELNUM","ZDRIVE-REAL","ZSELF-REAL","ZDSELF-REAL","DLENGTH"
1965 LPRINT "ELNUM","ZDRIVE-REAL","ZSELF-REAL","ZDSELF-REAL","DLENGTH"
1970 FOR J=1 TO ELNUM
1980 IF ELNUM=1 THEN 2130 ELSE 1990
1990 ZDRIVEA[J]=(120/DIRECTIVITY)*R0^2/MAG[J]^2
2000 ZDSELFA[J]=ZDRIVEA[J]+ZSELFA[J]
2010 C1=C-ZDSELFA[J]
2020 ON SGN(B^2-4*A*C1)+2 GOTO 2030, 2050, 2050
2030 PRINT J SPC(2) "SQUARE ROOT OF NEGATIVE NUMBER"
2035 LPRINT J SPC(2) "SQUARE ROOT OF NEGATIVE NUMBER"
2040 GOTO 2140
2050 KL=(-B+(B^2-4*A*C1)^.5)/(2*A)
2060 XKL=(29.263*KL-26.862)*KL+12.452
2070 DLENGTH[J]=KL
```

```
2080 DELTADL(J)=100*(2*DLENGTH(J)/PI-1)
2090 IF KL=PI/2 THEN 2100 ELSE 2120
2100 ZDSELFB(J)=XKL
2110 GOTO 2130
2120 ZDSELFB(J)=XKL-120*(LOG(2*KL/ELRAD)-1)/TAN(KL)
2130 PRINT J,ZDRIVEA(J),ZSELFA(J),ZDSELFA(J),DLENGTH(J)
2135 LPRINT J,ZDRIVEA(J),ZSELFA(J),ZDSELFA(J),DLENGTH(J)
2140 NEXT J
2150 PRINT
2155 LPRINT
2160 PRINT "ELNUM","ZDRIVE-IMAG","ZSELF-IMAG","ZDSELF-IMAG","% DELTA-DL"
2165 LPRINT "ELNUM","ZDRIVE-IMAG","ZSELF-IMAG","ZDSELF-IMAG","% DELTA-DL"
2170 FOR J=1 TO ELNUM
2180 ZDRIVEB(J)=ZDSELFB(J)-ZSELFB(J)
2190 PRINT J,ZDRIVEB(J),ZSELFB(J),ZDSELFB(J),DELTADL(J)
2195 LPRINT J,ZDRIVEB(J),ZSELFB(J),ZDSELFB(J),DELTADL(J)
2200 NEXT J
2210 PRINT
2215 LPRINT
2220 INPUT "PLOT OPTION; 1=YES,0=NO. ENTER 1 OR 0.   ",OPT3
2230 PRINT
2235 LPRINT
2240 IF OPT3=0 THEN 2560
2250 GOTO 2880
2260 PRINT "Compute ELEMENT CURRENTS from PDF's and Spacing."
2265 LPRINT "Compute ELEMENT CURRENTS from PDF's and Spacing."
2270 PRINT
2275 LPRINT
2280 PRINT "ELNUM","REAL","IMAGINARY","MAGNITUDE","PHASE, DEG."
2285 LPRINT "ELNUM","REAL","IMAGINARY","MAGNITUDE","PHASE, DEG."
2290 FOR N=1 TO ELNUM
2300 ON SGN(AX(N))+2 GOTO 2310, 2320, 2330
2310 ON SGN(AY(N))+2 GOTO 2340, 2360, 2380
2320 ON SGN(AY(N))+2 GOTO 2400, 2420, 2440
2330 ON SGN(AY(N))+2 GOTO 2460, 2480, 2500
2340 PHASE=PI+ATN(AY(N)/AX(N))
2350 GOTO 2510
2360 PHASE=PI
2370 GOTO 2510
2380 PHASE=PI+ATN(AY(N)/AX(N))
2390 GOTO 2510
2400 PHASE=3*PI/2
2410 GOTO 2510
2420 PHASE=0
2430 GOTO 2510
2440 PHASE=PI/2
2450 GOTO 2510
2460 PHASE=2*PI+ATN(AY(N)/AX(N))
2470 GOTO 2510
2480 PHASE=0
2490 GOTO 2510
2500 PHASE=ATN(AY(N)/AX(N))
2510 PHASE(N)=PHASE
2520 MAG(N)=(AX(N)^2+AY(N)^2)^.5
2530 PRINT N,AX(N),AY(N),MAG(N),PHASE(N)*180/PI
2535 LPRINT N,AX(N),AY(N),MAG(N),PHASE(N)*180/PI
2540 NEXT N
2550 RETURN
2560 PRINT "FRC"FRC "ELNUM"ELNUM "SPACING"SPACING "BOOMLENGTH"(ELNUM-1)*SPACING/360
2565 LPRINT "FRC"FRC "ELNUM"ELNUM "SPACING"SPACING "BOOMLENGTH"(ELNUM-1)*SPACING/360
2570 PRINT "R0"R0 "DIR"DIRECTIVITY "DBI"DI "DBD"DBD "BW/2"BW SPC(2)"L XT LD RD XD DL DDL"
2575 LPRINT "R0"R0 "DIR"DIRECTIVITY "DBI"DI "DBD"DBD "BW/2"BW SPC(2)"L XT LD RD XD DL DDL"
2580 FOR I=1 TO ELNUM
2590 PRINT LENGTH(I) SPC(0)
2595 LPRINT LENGTH(I) SPC(0)
2600 NEXT I
2610 PRINT
2615 LPRINT
2620 FOR I=1 TO ELNUM
2630 PRINT XTUN(I) SPC(0)
2635 LPRINT XTUN(I) SPC(0)
2640 NEXT I
2650 PRINT
2655 LPRINT
2660 FOR I=1 TO ELNUM
2670 PRINT DLENGTH(I) SPC(0)
```

```
2675 LPRINT DLENGTH(I) SPC(Ø)
2680 NEXT I
2690 PRINT
2695 LPRINT
2700 FOR I=1 TO ELNUM
2710 PRINT ZDRIVEA(I) SPC(Ø)
2715 LPRINT ZDRIVEA(I) SPC(Ø)
2720 NEXT I
2730 PRINT
2735 LPRINT
2740 FOR I=1 TO ELNUM
2750 PRINT ZDRIVEB(I) SPC(Ø)
2755 LPRINT ZDRIVEB(I) SPC(Ø)
2760 NEXT I
2770 PRINT
2775 LPRINT
2780 FOR I=1 TO ELNUM
2790 PRINT DELTAL(I) SPC(Ø)
2795 LPRINT DELTAL(I) SPC(Ø)
2800 NEXT I
2810 PRINT
2815 LPRINT
2820 FOR I=1 TO ELNUM
2830 PRINT DELTADL(I) SPC(Ø)
2835 LPRINT DELTADL(I) SPC(Ø)
2840 NEXT I
2850 PRINT
2855 LPRINT
2860 PRINT "END OF ROUTINE"
2865 LPRINT "END OF ROUTINE"
2870 END
2880 GOSUB 3Ø5Ø
2890 FOR N=Ø TO 72
2900 GOSUB 295Ø
2910 NEXT N
2920 GOSUB 3Ø5Ø
2930 GOSUB 3Ø2Ø
2940 GOTO 256Ø
2950 DB=RESP(N)
2960 ON SGN(7Ø+DB)+2 GOTO 297Ø, 297Ø, 298Ø
2970 DB=-7Ø
2980 C=INT(DB+.5)
2990 DBTR=INT(1Ø*DB+.5)/1Ø
3000 PRINT TAB(7Ø+C) "*" 2.5*N DBTR
3005 LPRINT TAB(7Ø+C) "*" 2.5*N DBTR
3010 RETURN
3020 PRINT TAB(1)"-69"TAB(9)"-6Ø"TAB(19)"-5Ø"TAB(29)"-4Ø"TAB(39)"-3Ø"TAB(49)"-2Ø"TAB(59)"-1Ø"
       TAB(7Ø)"Ø"
3025 LPRINT TAB(1)"-69"TAB(9)"-6Ø"TAB(19)"-5Ø"TAB(29)"-4Ø"TAB(39)"-3Ø"TAB(49)"-2Ø"TAB(59)"-1Ø"
       TAB(7Ø)"Ø"
3030 PRINT TAB(26)"RELATIVE DB"
3035 LPRINT TAB(26)"RELATIVE DB"
3040 RETURN
3050 FOR I=Ø TO 7Ø STEP 1Ø
3060 PRINT TAB(I) "L";
3065 LPRINT TAB(I) "L";
3070 NEXT I
3080 RETURN
3090 DATA Ø,.218Ø718,.6519813,1.Ø79234,1.495576,1.897Ø17,2.279923,2.641Ø95
3100 DATA 2.977833,3.287975,3.56993,3.822683,4.Ø45778,4.2393Ø5,4.4Ø3849,4.54Ø447
3110 DATA 4.65Ø522,4.735827,4.798367,4.84Ø344,4.864Ø85,4.871982,4.866445,4.84984
3120 DATA 4.824465,4.792504,4.75600 6,4.716868,4.67 6813,4.63739,4.599974,4.565743
3130 DATA 4.535709,4.51Ø7,4.491374,4.478221,4.471564,4.471564,4.478221,4.491374
3140 DATA 4.51Ø7,4.535708,4.565743,4.599974,4.637392,4.67 6813,4.716869,4.756ØØ7
3150 DATA 4.792504,4.824465,4.84984,4.866445,4.871983,4.864084,4.84Ø344,4.798368
3160 DATA 4.735827,4.65Ø522,4.54Ø447,4.4Ø3848,4.2393Ø4,4.Ø45777,3.822682,3.56993
3170 DATA 3.287974,2.977833,2.641Ø93,2.279923,1.897Ø15,1.495574,1.Ø79232,.651980
3180 DATA .218Ø717
3190 SP=SPACING*PI/18Ø
3200 DALPHA=PI/6Ø
3210 DEL=DELTA*PI/18Ø
3220 FOR N=Ø TO 6Ø
3230 PH=-SP*COS(N*DALPHA)
3240 X1=Ø
3250 Y1=Ø
3260 X2=Ø
3270 Y2=Ø
```

```
3280 ON SCHEME GOTO 3290,3340,3570
3290 R=2^(ELNUM-1)
3300 FOR M=1 TO ELNUM-1
3310 R=R*COS(PHI(M)+PH/2)
3320 NEXT M
3330 GOTO 3390
3340 R=2^FAMNO
3350 COV=1-COS(N*DALPHA)
3360 FOR M=1 TO FAMNO
3370 R=R*COS(M*SP*COV/2+DEL)
3380 NEXT M
3390 FOR M= 1 TO ELNUM
3400 C=COS((M-1)*PH)
3410 S=SIN((M-1)*PH)
3420 X1=X1+AX(M)*C-AY(M)*S
3430 Y1=Y1+AY(M)*C+AX(M)*S
3440 PHASE=(M-1)*PH+PHASE(M)
3450 X2=X2+MAG(M)*COS(PHASE)
3460 Y2=Y2+MAG(M)*SIN(PHASE)
3470 NEXT M
3480 RSP1=(X1^2+Y1^2)^.5
3490 RSP2=(X2^2+Y2^2)^.5
3500 IF N=0 THEN 3510 ELSE 3550
3510 R0=R
3520 R01=RSP1
3530 R02=RSP2
3540 PRINT "R0"R "R01"RSP1 "R02"RSP2
3550 PRINT 3*N,8.686*LOG(ABS(R)/R0+10^-30),8.686*LOG(RSP1/R01+10^-30),8.686*LOG(RSP2/R02+10^-30)
3560 NEXT N
3570 END
3580 DATA 73.12,42.46,71.7,24.3,67.3,7.5,60.4,-7.1,51.4,-19.2,40.8,-28.4
3590 DATA 29.3,-34.4,17.5,-37.4,6.2,-37.4,-4.0,-34.8,-12.5,-29.9
3600 DATA -19.1,-23.4,-23.3,-15.9,-25.2,-7.9,-24.9,-0.3,-22.5,6.6
3610 DATA -18.5,12.3,-13.3,16.3,-7.5,18.6,-1.6,19.0,4.0,17.7
3620 DATA 8.8,15.0,12.3,11.2,14.5,6.7,15.3,1.9,14.6,-2.7
3630 DATA 12.6,-6.7,9.6,-9.8,6.0,-11.9,2.0,-12.7,-1.9,-12.3
3640 DATA -5.4,-10.8,-8.2,-8.4,-10.0,-5.3,-10.9,-2.0,-10.6,1.4
3650 DATA -9.4,4.5,-7.4,7.0,-4.8,8.7,-1.9,9.5,1.1,9.4
```

Quad and Loop Antennas

Half-Loop Antennas

By Bob Alexander, W5AH

2720 Posey Drive
Irving, TX 75062

The half-loop antenna is a vertically polarized, more or less omnidirectional radiator that offers a reasonable match to 50-ohm transmission line and has maximum radiation at low angles. As its name implies, the half loop is one half of a full-wave loop. The ground plane below the antenna provides the return path for the loop (see Fig 1). The antenna is fed at the base of either leg with 50-ohm line.

The feed-point impedance, radiation pattern and an equation for element length were derived through the use of a model constructed on an aluminum ground plane and tested at UHF. The antenna element consisted of a no. 18 wire 12 inches in length from feed point to ground. Both an inverted-U and an inverted-V shape were tested.

Twelve inches was chosen for the element length because the resonant frequency of the antenna, once determined, can be used to calculate the length required to resonate the antenna at any given frequency. For example, if the antenna resonated at 502.5 MHz, 502.5/f, where f is in MHz, would yield the length required for any other frequency.

A network analyzer initially calibrated for 450 to 510 MHz was used to measure the impedance and resonant frequency of the model. The results were a surprise. The plots obtained were nowhere near the real axis of the Smith Chart. After recalibrating the analyzer for 450-550 MHz the plots shown in Figs 2 and 3 were obtained. Resonance—in other words, the point where the plot crosses the real axis of the chart—occurs approximately 10% higher in frequency than anticipated. Impedance at resonance ranged from 62 ohms for the inverted-V shape to 66 ohms for the inverted-U shape. It should be noted that the antenna element was completely self-supporting and was over a near-perfect ground plane.

The radiation pattern of the half loop is the same as that of a pair of quarter-wavelength verticals spaced ¼ wavelength and fed in phase. It has a slight gain in the broadside direction and about 3-dB front-to-side ratio (see Fig 4).

During testing of the model it was found that nearby objects have an effect on both the resonant frequency and impedance of the half loop. When a central support was added for the inverted-V shaped element, the resonant frequency dropped to within the range originally expected and the measured impedance was 49 ohms.

A 30-Meter Version

A half loop cut for the center of the 30-meter band, using 530/f for the element length in feet, made the importance of a good ground plane very evident. The loop was built in an inverted-U

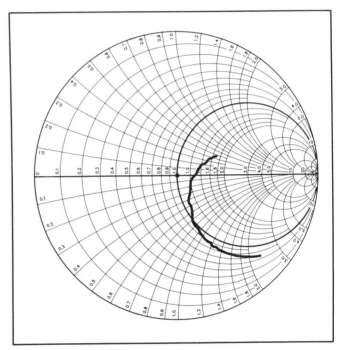

Fig 2—450- to 550-MHz impedance plot of the inverted-U half loop model over a near-perfect ground plane. Resonance is at 532.5 MHz, where the impedance is 65.8 ohms. This chart and that of Fig 3 are normalized to 50 ohms.

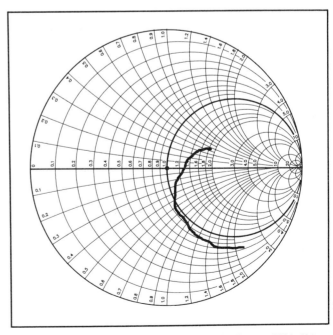

Fig 3—450- to 550-MHz impedance plot of the inverted-V half loop model over a near-perfect ground plane shows 62.6 ohms at a resonant frequency of 528 MHz.

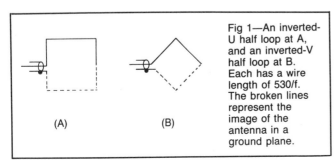

(A)   (B)

Fig 1—An inverted-U half loop at A, and an inverted-V half loop at B. Each has a wire length of 530/f. The broken lines represent the image of the antenna in a ground plane.

shape using fiberglass poles to support the vertical legs. The ground connections were made to 4-foot ground rods placed at the base of each leg.

The best SWR recorded was over 6:1, indicating very high ground losses. After adding four radials to the base of each leg the SWR was measured as 2.8:1. Eight radials brought the SWR down to 2:1.

At this point I decided to determine if the frequency with the best SWR and the resonant frequency were the same. A good SWR does not mean that an antenna is resonant. Since the network analyzer previously used was unavailable, another method was used to calculate the exact impedance of the antenna.[1] The resulting plot on the Smith Chart was very near the real axis, indicating that resonance and best SWR were occurring together.

Going back to the antenna, a ground bus was added between the two radial sets. The bus consisted of a double length of braid that had been removed from a piece of RG-8 coax. An SWR of 1.5:1 was then recorded. That was still a bit higher than the model but more than acceptable, considering the initial values. The element length, which had not been altered during the measurements, was then adjusted but

no improvement in SWR was noted.

On-the-air tests were made after stringing up an inverted-V dipole as a comparison antenna. On receive, the half loop outperformed the dipole. The noise from thunderstorms in the area as well as powerline noise was two to three S units below the inverted-V dipole. The improved signal-to-noise ratio of the loop made it possible to copy several stations that could not be heard on the dipole. On transmit, neither antenna proved better than the other.

Further experimentation with the UHF model indicated that a unidirectional array of two half loops in a phased array should be possible. However, the ground requirements make construction difficult.

Conclusion

The low noise characteristics of the half loop antenna as well as its physical size make it ideal for the low bands. Being about ¼ wavelength long, it can be erected where space limitations prevent use of a ½-wavelength antenna. Its low height can be used to advantage in areas having severe height limitations. One added bonus is that the half loop can be used on its harmonic frequencies.[2]

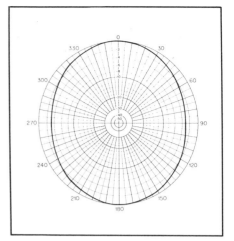

Fig 4—The azimuthal plot of the half-loop antenna radiation pattern indicates slight directivity broadside to the wire.

Notes

[1]R. Rhea, "Measuring Complex Impedance with an SWR Bridge," *Ham Radio*, May 1975, p 46.
[2]J. S. Belrose and D. DeMaw, "The Half-Delta Loop: A Critical Analysis and Practical Deployment," *QST*, Sep 1982, pp 28-32.

Coil Shortened Quads— A Half-Size Example on 40 Meters

By Kris Merschrod, KA2OIG/TI2

Apartado 104-2070
San Jose, Costa Rica

T he general antenna rule is the more wire you have in the air the greater the efficiency. Thus, it is better to have four quarter waves than a dipole or a vertical. The advantages of a vertically hung loop over a dipole are numerous: angle of radiation is lower for the same height, greater gain and greater bandwidth.

The disadvantage is mechanical. It has a third dimension, which means that construction and support present serious design challenges. The ice load can be a problem in temperate zones.

However, here in Costa Rica there are no ice problems and the bamboo grows tall and straight. My first attempt was a four-element quad for 10 meters made of bamboo and macrame cord, which was strung up in a eucalyptus tree at a height of about 45 feet. It has been a great weapon for pileups with only 80 watts going out to it. The macrame cord rotted after six months and the dozens of clove hitches had to be retied with plastic cord. After this success I turned to 40 meters. First I used the vertical, then the dipole and now the quad. There are limits to the size and stress which bamboo and plastic cord can handle. A technique had to be found to reduce the elements to a manageable size.

Literature Review

In the literature on reduced size quads there are examples of top-hat or capacitive loading,[1] linear loading,[2] stub loading,[3] trap and coil loading,[4] a folded-mini[5] and finally coil loading.[6] The folded, capacitive, linear and stub-loaded quads all present mechanical problems, that is, they tend to be more complex than a simple quad, except for size.

Orr and Cowan state, "Attempts have been made to 'shrink' the quad elements by the use of loading coils placed in the loops; the results have been inconclusive, and little specific comparative data has been found on the performance of the loaded quad antenna. A coil-loaded mini-

loop element was tried at W6SAI for a period of time to determine if it was competitive with a half-wave dipole mounted at the approximate mid-height of the element. After a period of tests, it was concluded that the loaded loop compared poorly with the dipole as far as signal strength reports were concerned and furthermore, the bandwidth of the loaded loop was quite restrictive compared to that of the reference dipole."[7]

They reported that the stub-loaded variety was superior in gain and bandwidth, but I believe that the 0.14-wave dangling stub is a problem, especially when one plans a rotating beam on 40 meters. I've used coil-loaded dipoles before with good results and felt that this route to a reduced-size quad should be explored.

Coil Loading

In the *ARRL Antenna Anthology*[8] there is a section on off-center-loaded dipole antennas. The formula presented therein can be used in a computer program, or the chart provided can be used to estimate the approximate inductance required to shorten a dipole for a given frequency and coil position. The *1989 ARRL Handbook*[9] provides a formula for coil designs. Sander[10] provides a computer program to

combine these formulas. The idea is to place a loading coil near the center of each leg of the dipole. If you load toward the end you approach an infinite amount of inductance needed, and if you are close to the feed point the bandwidth is narrow (as those of you who have tuned base-loaded verticals know). Also, the high current near the feed point requires a heavier coil. Program 1 is a BASIC computer program listing which can be used to calculate the dimensions of a coil-loaded loop. [The program is available on diskette for the IBM PC and compatibles; see information on an early page of this book.—Ed.]

Hypothesis

If a loop can be assumed to be made of two dipoles, then the position and the calculation of the coils needed to reduce its size would be similar to that of a coil-loaded dipole. That is, place four coils of equal inductance and Q (same size form and number of turns) 1/8 wavelength on each side of the feed point and 1/8 wavelength on each side of the mid point of the other half of the loop as shown in Fig 1. If this would work, then the reduced size would be easier to build. The antenna would be more stable with the coils placed on (around) the struts because

[1]Notes appear on page 92.

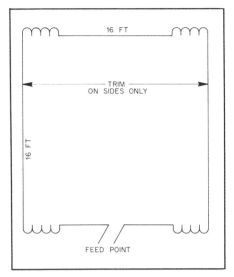

Fig 1—Coil-loaded 40-meter loop. Physical size is one half as large as a conventional 40-meter loop.

of the decreased wind area.

40-Meter Half-Size Loop Example

In Costa Rica one can operate single sideband below 7.1 MHz. There is plenty of DX action in this section of the band for both SSB and CW. I picked 7.15 for the center frequency because I hoped that it would be broad enough to work the whole band. An existing full-sized loop in the form of a trapezoid (55 feet at the top and 10 feet off the ground) was the reference antenna. It was decided that a loop of 16 feet on a side (half size) would be ideal because it would be similar to a full-size 20-meter loop, which is manageable in a multielement-quad formation.

The formulas in the references cited were used.[11] The results called for 22 μH of inductance in each coil. Plastic water pipe 2-3/8 inches OD and 4 inches long was used for the forms. The winding of 29 turns was of the same wire used for the elements, no. 14 multi-stranded copper with insulation. Whatever was readily available was used.

Instead of the usual crossed-bamboo struts, two 17-foot bamboo poles were used horizontally. The coils were slipped over the poles and tied at each end. The bottom pole hangs from the side wires of the loop connecting the coils, and the top pole was tied to a rope and pulled up a straight eucalyptus tree inside of the existing full quad. Both had the same center of gravity or effective height above ground.

Results

From the shape of SWR curve A in Fig 2 it appeared that the shortened quad was too long. Accordingly, the sides were shortened to bring the low point up to the band segment I wished to use without a tuner.[12] This is a 100-kHz segment between 7.05 and 7.175 MHz with an SWR of less than 2:1. As shown in Fig 2, 1¾ feet was trimmed off each side of the driven element to bring the low SWR point up to 7.125 MHz. This was a substantial improvement over the full quad (lowest SWR was 4:1). The rig used was a Drake TR4-CWx, which tolerates this SWR easily.

So much for the electrical results; how did the half-size loop compare with the full loop? The two were compared across the band using the antenna tuner where necessary.

Reception showed the full loop to be 1 to 2 dB stronger, but when there was QRN, intelligence was better on the half-size loop. The half size was quieter, so much quieter

that stations lost in the noise with the full loop were copied. It seemed to have a better signal-to-noise ratio.

Transmitting was done with 100 watts output through almost equal lengths of Belden 9913 (practically no loss at 7 MHz) fed directly to the loops with no balun or coils. TI2ZM and HJ4OIG gave a "no difference" report. The same was true from KA9WIE on CW. DL8PC gave a "slight gain" report for the full loop, and UR1RWW gave the full loop a 1.5-dB advantage. This was encouraging information because it meant that a small sacrifice was all that was needed to make the radical leap to a two-element rotating quad on 40 meters. Little time was wasted pulling down the full loop to make way for a 24-foot boom of bamboo and another half-size loop for a reflector. Reflectors are generally 5% longer than the driven elements. Thus, without trimming, another half-size loop based on the original specifications should be a good reflector.

Two-Element Results

The first question was element spacing; according to various sources, anything from 0.12 to 0.15 wavelength will do. Fig 3 shows the SWR curves for approximately 0.12, 0.13 and 0.14 wavelengths in reference to the single element at 7.125 MHz. The shapes of the curves are similar, but the 0.13 spacing gave the lowest curve. Reception and signal reports on the air did not seem to show a difference between spacing, but the idea was to have a low SWR and bring in the QSOs. I checked into the Century Net and the OMISS nets from time to time to see how stateside reports were at that end of the band (7.272 and 7.234 MHz). Also, other stateside contacts were consistently

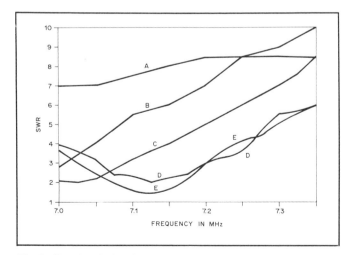

Fig 2—Results of trimming the half-size coil-loaded loop for 40 meters. The sides of the loop were shortened by equal amounts.
Curve A—16-foot sides.
Curve B—Sides shortened 1 foot.
Curve C—Sides shortened 1½ feet.
Curve D—Sides shortened 2 feet.
Curve E—Sides shortened 1¾ feet.

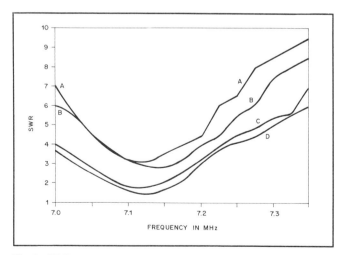

Fig 3—SWR curves for a single half-size 40-m coil-loaded loop and various element spacings for a similar two-element quad.
Curve A—19.5 feet.
Curve B—16.5 feet.
Curve C—17.5 feet.
Curve D—A single coil-loaded loop.

five or better for stations using loaded verticals, mobiles and dipoles. The "big guns" using Yagis and used to copying linears give me five nine plus reports, but "stateside" is close by.

From Europe, I1KFB came in at 20 dB over S9 and gave me a five eight. Similar results came from DL4RS, LA7DFA, DJ6BN, GØCNG and YU4AU. From ZL2APW came a five five. These reports are from different nights during May and June 1988.

Carlos and Sofia (TI2KD and TI2IY of TI9 DX fame) live about 200 meters away and have a fixed two-element full-size quad on 40 meters. Their effective height is about the same as mine. Carlos and I spent an evening with 100 watts output, roaming the band doing "invidious" comparisons. The results are shown in Table 1.

It is clear the full loops are better than the reduced size. The front-to-back ratio and side ratio have not been easy to measure. At this point all I have is "slightly better" for front to back and "that's better" for the front-to-side report. Copying off the side is weaker and the audio seems garbled. When TI2KD is on the air, the RF gain and the audio gain must be turned all the way down or the noise is painful for either front or back. But to the side only the audio gain has to be backed off.

Table 1
Signal Report Comparisons of Half-size and Full-size Quads

Station	Comments	Report Half-size	Full-size
IK6BOB	4-el Yagi, European QRM	3 × 5	5 × 8
W5ILR	Was copied 5 × 8	4 × 4	5 × 6
K5HWA	Was 20 over 9	5 × 8	25 over 9
YS1MAE	Was 25 over 9	5 over 9	20 over 9
F6ARC	Was 10 over 9	5 × 7	5 × 9

Conclusions

Coil loading is a practical way to reduce the size of quads. Even though the bandwidth is not too wide (100 kHz for less than 2:1 SWR), the SWR is lower than a full-size quad (apparently because the impedance is lower than a full loop because of the coil loading). Because of the smaller size, a multiple-element rotating quad is possible without expensive structural elements. I'm searching for a 40-foot bamboo boom to try a three-element model. In the meanwhile the two-element model, up about 50 feet in a eucalyptus, is doing a fine job.

Notes and References

[1]L. A. Moxon, *HF Antennas For All Locations* (Potters Bar, England: Radio Society of Great Britain. 1986), p 178.
[2]M. S. Anderson, Ed., *The ARRL Antenna Anthology* (Newington: ARRL, 1978), p 73.
[3]W. I. Orr and S. D. Cowan, *The Radio Amateur Antenna Handbook* (Wilton, CT: Radio Publications Inc, 1978), p 103.
[4]See ref 3, p 125.
[5]See ref 2, p 69.
[6]See ref 3, p 127.
[7]See ref 3, p 103.
[8]See ref 2, pp 107-112.
[9]B. S. Hale, Ed., *The 1989 ARRL Handbook for the Radio Amateur* (Newington: ARRL, 1988), p 2-17.
[10]D. Sander, "A Computer Designed Loaded Dipole Antenna," *CQ*, Dec 1981, p 44.
[11]Program 1 is a printout of the program used. It is basically (no pun intended) Sander's program with a few adjustments.
[12]The sides of the "loaded" quad are easiest to adjust. If one uses tuning stubs, they should be placed on each side and not between the coils and feed point. Avoid placing the coils at the center. Recall the hypothesis that these are two dipoles with their ends connected. Each side (the "end") should be adjusted by equal amounts.

Program 1
BASIC Program to Calculate Coil Dimensions for Reduced-Size Quad Antennas
ARRL-supplied disk filename is MERSCHR.BAS.

```
100 CL=15:GOSUB 10000
160 PRINT
170 PRINT"                                    ? FT."
175 PRINT
178 PRINT" I[[[[———————————————— 'W' ————————————————]]]]I"
179 PRINT" I                                                   I"
180 PRINT" I                                                   I"
181 PRINT" I                                                   I"
182 PRINT" I                                                   I"
183 PRINT" I                                                   I"
184 PRINT" I              PROGRAM FOR REDUCED SIZE             I"
185 PRINT" I                 QUAD ANTENNAS                     I"
186 PRINT"'H' ? FT.                                            I"
187 PRINT" I                   KA2OIG/TI2                      I"
189 PRINT" I                      1988                         I"
190 PRINT" I                                                   I"
191 PRINT" I                                                   I"
192 PRINT" I                                                   I"
193 PRINT" I                                                   I"
194 PRINT" I                                                   I"
195 PRINT" I[[[[—————————————————/ /—————————————————]]]]I"
200 PRINT
210 INPUT" WIDTH OF QUAD IN FEET ";A
290 PRINT
300 INPUT" HEIGHT OF QUAD IN FEET ";H
301 TEST = A/(H+A)
302 IF TEST <.5 OR TEST =.5 GOTO 310
303 PRINT" QUAD IS TOO WIDE FOR HEIGHT "
304 INPUT"  PRESS RETURN TO CONTINUE ";YN
305 GOTO 160
```

```
310 GOSUB 10000
311 PRINT" ENTER DIAMETER OF ELEMENT IN INCHES OR":PRINT
312 INPUT" GA. (8 TO 18) WIRE ";D
313 IF D <8 GOTO 330
314 IF D = 8 THEN D=.1285
315 IF D = 10 THEN D=.1019
316 IF D = 12 THEN D=.0808
317 IF D = 14 THEN D=.0641
318 IF D = 16 THEN D=.0508
319 IF D = 18 THEN D=.0403
320 IF D > 18 THEN GOTO 311
330 GOSUB 10000
340 INPUT"FREQUENCY IN MHz. ";F
341 TEST2 = 1005/F
342 IF TEST2 > (A*2+H*2)   GOTO 350
343 PRINT " THIS" ;(A*2+H*2);" Ft. LOOP DOES NOT NEED LOADING"
344 PRINT " FOR ";F;" MHz. OPERATION"
345 INPUT " PRESS RETURN TO BEGIN AGAIN ";YN
346 GOTO 100
350 GOSUB 10000
351 A1=A
352 B=A/2
354 A=(A+H)
360 F1=10^6/(68*(3.14159^2)*F^2)
370 F2=LOG(24*((251/F)-B)/D)-1
380 F3=((1-(F*B/251))^2)-1
390 F4=(251/F)-B
400 F5=LOG((((24*A/2)-B)/D)-1
410 F6=((((F*A/2)-F*B)/251)^2-1
420 F7=A/2-B
430 LMH=F1*((F2*F3/F4)-(F5*F6/F7))
450 PRINT"    THE INDUCTANCE NEEDED IS:"
460 PRINT
470 PRINT"               ";LMH;" uH"
480 PRINT
490 PRINT
500 PRINT"                -)))))-"
510 CL=2:GOSUB 10000
520 GOSUB 5000
750 INPUT" DO YOU WISH A PRINTOUT ";PO$
755 IF PO$ <>"Y" AND PO$ <> "y" AND PO$ <> "n" AND PO$ <>"N" GOTO 750
780 IF PO$="N" OR PO$="n" THEN 900
790 CL=15: GOSUB 10000
800 PRINT"          NOW PRINTING"
805 LPRINT:LPRINT:LPRINT:LPRINT:LPRINT
810 CL=7:GOSUB 10000
811 LPRINT"                              ";A1;" Ft."
812 LPRINT"    I[[[[---------------------- 'W' ----------------------]]]]I"
813 LPRINT"    I                                                       I"
814 LPRINT"    I                                                       I"
815 LPRINT"    I                                                       I"
816 LPRINT"    I                                                       I"
817 LPRINT"    I                                                       I"
818 LPRINT"    I                PROGRAM FOR REDUCED SIZE               I"
819 LPRINT"    I                    QUAD ANTENNAS                      I"
820 LPRINT"    'H' = ";H;"Ft.                                        I"
821 LPRINT"    I                    KA2OIG/TI2                        I"
823 LPRINT"    I                       1988                           I"
824 LPRINT"    I                                                       I"
825 LPRINT"    I                                                       I"
826 LPRINT"    I            ELEMENT DIAMETER =";D;" INCHES        I"
827 LPRINT"    I                                                       I"
828 LPRINT"    I                                                       I"
829 LPRINT"    I[[[[----------------------/ /----------------------]]]]I"
830 LPRINT
831 LPRINT"                         ";F;"MHz":LPRINT
832 LPRINT"                        ********":LPRINT
836 LPRINT"             USE ONE LOADING COIL"
837 LPRINT"               AT EACH CORNER"
838 LPRINT"       DO NOT CHANGE OVERALL WIDTH OR COIL SIZE"
839 LPRINT"       TRIM BY CHANGING HEIGHT ON EACH SIDE"
866 LPRINT"             COIL SPECIFICATIONS"
868 LPRINT"             ***************"
872 LPRINT
874 LPRINT"             INDUCTANCE IS ";LMH;" uH"
876 LPRINT
878 LPRINT"             DIAMETER IS   ";DIA;" INCHES"
880 LPRINT
882 LPRINT"             LENGTH IS     ";L;" INCHES"
```

```
884 LPRINT
886 LPRINT"                          ";N;"TURNS NEEDED"
890 IF WGA < 10 GOTO 893
891 LPRINT:LPRINT"                     USE ";WGA;" Ga. INSULATED WIRE FOR COIL"
892 GOTO 900
893 LPRINT:PRINT" USE ";WGA;"INCH DIAMETER WIRE --";(1/WGA)/2;"TURNS PER INCH"
900 CL=2:GOSUB 10000
920 INPUT" DO YOU WISH TO DESIGN ANOTHER REDUCED QUAD ";YN$
925 IF YN$ <>"Y" AND YN$ <> "y" AND YN$ <> "N" AND YN$ <> "n" GOTO 920
930 IF YN$ = "Y" OR YN$ = "y" GOTO 100
950 PRINT:PRINT"   73 DE KA2OIG/TI2"
960 PRINT:PRINT
990 END
5000 '      COIL DESIGN SUBROUTINE
5002 '
5004 PRINT"        COIL DESIGN SPECIFICATIONS"
5005 REIT = 0
5010 CL=2:GOSUB 10000
5020 PRINT"        ************************"
5515 GOSUB 10000
5600 INPUT" DIAMETER OF THE COIL IN INCHES ";DIA :R=DIA/2
5700 GOSUB 10000
5800 INPUT" LENGTH OF THE COIL IN INCHES ";L
5900 GOSUB 10000
5910 TEST3 = L/DIA
5920 IF TEST3 > .4 GOTO 6600
5930 PRINT"        **********************"
5940 PRINT:PRINT" THE PROPORTION OF LENGTH TO DIAMETER"
5950 PRINT" SHOULD BE AT LEAST .4"
5960 PRINT" THE PROPOSED PROPORTION IS ";L/DIA
5980 PRINT:PRINT"        **********************
5990 PRINT:INPUT" PRESS RETURN TO CONTINUE";YN
6500 GOSUB 10000
6600 ' CALCULATE THE NUMBER OF TURNS
6700 N=(LMH*(9*R+10*L)/R^2)^.5
6810 CL=3:GOSUB 10000
6815 IF REIT >0 THEN 6829
6819 PRINT" ENTER DIAMETER OF COIL-WIRE IN INCHES OR":PRINT
6820 INPUT" GA. (10 TO 16) INSULATED HOUSE WIRE = ";WGA:GOSUB 10000
6821 IF WGA<10 THEN WW=1/(WGA*2)
6824 IF WGA=10 THEN WW=5.0
6825 IF WGA=12 THEN WW=6.5
6826 IF WGA=14 THEN WW=7.5
6827 IF WGA=16 THEN WW=9.5
6828 IF WGA>16 THEN 6819
6829 LN=N/WW
6831 IF LN < L GOTO 6900
6834 L = LN + .25:REIT = 1: GOTO 5910
6835 PRINT:PRINT:INPUT" PRESS RETURN TO CONTINUE ";YN
6838 L = LN+1:GOTO 5910
6900 CL=1:GOSUB 10000
6905 REM NOTICE OF AUTOCALCULATION OF COIL LENGTH
6910 IF REIT <=0 GOTO 7500
6915 PRINT" TOO MANY TURNS FOR GIVEN FORM LENGTH":PRINT
6920 PRINT" LENGTH AUTOMATICALLY RECALCULATED !":PRINT:PRINT
7500 REM SCREEN OUTPUT FOR COIL SPECS
8350 PRINT"        *****  COIL SPECS  *****"
8360 GOSUB 10000
8400 PRINT"          INDUCTANCE IS ";LMH;"uH"
8450 GOSUB 10000
8500 PRINT"          DIAMETER IS   ";DIA;" INCHES"
8550 GOSUB 10000
8600 PRINT"          LENGTH IS     ";L;"INCHES"
8650 GOSUB 10000
8800 PRINT"          USE           ";N;" TURNS"
8900 GOSUB 10000
8905 IF WGA < 10 GOTO 8930
8910 PRINT"          USE ";WGA;" Ga. INSULATED WIRE FOR COIL"
8915 GOTO 8935
8930 PRINT:PRINT" USE ";WGA;"INCH DIAMETER WIRE --";(1/WGA)/2;"TURNS PER INCH"
8935 GOSUB 10000
9910 INPUT " DO YOU WISH TO DESIGN ANOTHER COIL ";YN$
9915 IF YN$ <> "Y" AND YN$ <> "y" GOTO 9998 ELSE 5000
9998 RETURN
9999 END
10000 FOR ICL = 1 TO CL
10001 PRINT
10002 NEXT ICL
10003 RETURN
10004 END
```

Multiband and Broadband Antenna Systems

A 14-30 MHz LPDA for Limited Space

By Fred Scholz, K6BXI

3035 Dickens Ct
Fremont, CA 94536

Here is a 14- to 30-MHz log-periodic dipole array that has a 13.5-foot turning radius, weighs less than 25 pounds, and fits on the roof of my city-lot house. This array is made primarily of wire with a minimal but rugged supporting structure and is therefore low cost. It is also relatively easy to transport, folding into itself after loosening just a few bolts. These features will allow anyone to install the antenna single-handedly with ease. Low wind resistance is also a feature of this array, but the real key to the success of this design is its size. To achieve this goal, a small compromise in front-to-back ratio is made on the 20-meter band. This is done by resonating element 2 at 14.1 MHz, leaving only element 1 to act as a "reflector." The net effect on array gain can be determined from Fig 1.

According to Isbell, the radiation efficiency of the LPDA with one element behind the resonant radiator is about 0.83 (assuming a τ of 0.9).[1] If a second element is added to the rear, the radiation efficiency is about 0.86. In dB this is 10 log (0.86/0.83) or only 0.15 dB (indicating practically no loss of gain if this element is eliminated), and a tremendous reduction in size compared to an antenna with element 3 resonated for 20 meters. It can be seen that the single "reflector" element 1 adds 10 log (0.83/0.5) or 2.2 dB additional gain to the array—therefore this element is included in the design. Additionally, all elements are swept forward by varying amounts, reducing the size of the array but having the positive effect of an increase in gain and bandwidth for comparable size.

The boom and spreaders are supported from a single quad-element "spider mount" hub, available at flea markets or from a supplier of quad antennas. The hub is used as the mast mount in the horizontal, rather than the vertical plane. The boom is constructed of telescoping sections of 1-inch-diameter and 7/8-inch-diameter aluminum tubing. The tubing is slotted at the ends and clamped at the proper length with stainless-steel hose clamps. If you wish, you can join the boom sections with screws instead of hose clamps. Two 13-foot fiberglass spreaders are used as the center support arms for the array (see Fig 2). My spreaders were too thin and weak at the

[1]Notes appear on page 99.

ends, so I cut off the outer 18 inches and extended their length to 13 feet using ¾- inch-diameter Plexiglas® rod, available from most plastic shops. This extension was made at the hub end of the spreader.

Building the LPDA

Initial assembly can start with individual parts in your workshop. First, cut the boom-mounted insulators (Fig 3) to size, and drill the holes for the brass screws, wire feed-through holes, and mounting holes. Then mount the brass screws and tighten them down. Build the boom and measure the locations for the insulators. Drill the holes in the boom for insulator mounting and mount the insulators. Cut the element wires to length plus about six inches, label and roll up. Cut sections of 450-ohm ladder line to length plus about three inches and label. Then mount and solder the ladder line and the element wires directly to the brass screws. Be sure to put a 180-degree twist in the ladder line so that each succeeding element is 180 degrees out of phase with the next.

The last piece to prepare before final assembly is the forward spreader (Fig 4). This piece is unique in that it provides the main forward structural strength of the array and also forms element 10, which is

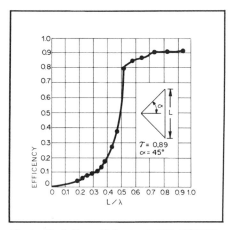

Fig 1—Radiating efficiency versus antenna size of a log-periodic dipole array. (After Isbell; see note 1)

resonant at 33 MHz. This element is cut about 3% shorter than it would be for a wire element because of its diameter. First, cut an 8-foot section of 3/4-inch and 5/8-inch diameter aluminum tubing. Cut the 6-foot 1/2-inch-diameter fiberglass rod in half. Slot one end of each 4-foot section

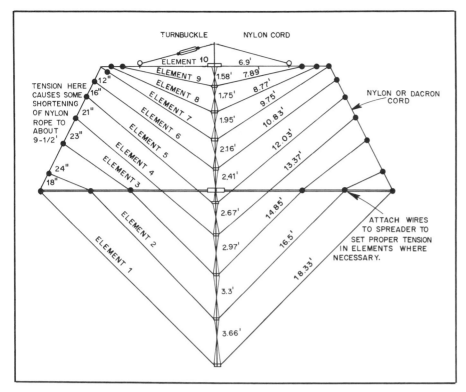

Fig 2—A top view of the LPDA shows that elements 1, 2 and 3 are attached (with insulators) to the spreader arms.

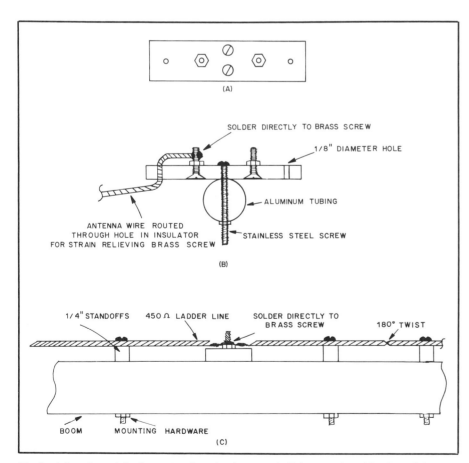

Fig 3—A top view of the boom insulator is shown at A. B is a section side view of the insulator mounted on the aluminum tubing. In C, the stand-off insulators and boom insulator support the feed line.

of tubing. Insert the 5/8-inch tubing and clamp it to the proper length. Extend the elements with the fiberglass rods, clamping so that 2-1/2 feet extend beyond the ends of the 5/8-inch tubing. (I first tried Plexiglas rod in this application and it broke under tension.) Cut the fiberglass-bar mounting insulators to size and mount them to the aluminum-tube elements. Angle each half a few degrees forward, to help tension the wire support. Last, mount hose clamps on each fiberglass-rod extender. The clamps are used to support the ends of elements 7, 8, and 9.

The final assembly can be done at ground level and then the array hoisted and put in place. If your yard is too small to complete the assembly, it is not difficult to do on the roof prior to final installation. I used an 8-foot section of 2-inch PVC pipe as the mast support for the array. Mount the U bolts or guy-support rings as shown in Fig 5 and the photos. My 2-meter beam is mounted at the top of this section, just above the strut supports for the boom. Finally, locate and drill holes for the boom-mounted eyebolts that are used to support the boom. These bolts can be positioned at the holes in the 450-ohm ladder-line insulation.

Mount the forward-element/spreader combination with the ¼-inch bolt. Then string insulators on the dacron or nylon lines and run these lines from the ends of the forward spreaders to the ends of the central spreaders. This is the proper time to mount the forward-tension truss (Fig 4). Do not apply final tension until element one is cut and secured in place. Unroll the wire elements one at a time, starting with the number-1 element. Cut it to length after attaching to the ends of the center spreaders. Now use the turnbuckle to apply final tension to the forward spreader. Proceed with the number-2 element, attaching it to the first insulator on the nylon or plastic line. Elements 3, 4, 5 and 6 are done in like manner. Elements 7, 8 and 9 can now be attached to the forward spreader using the three hose clamps on each end.

The weight of the forward spreader assembly unbalances the boom because most of the weight of the antenna is in this member. To counterbalance this, tie lead weights along the rear section of the boom until the antenna is balanced. I used several 6- and 8-ounce fishing weights. To transport the LPDA (to the roof, or new QTH) undo two bolts on each forward spreader and disconnect the center spreaders. The entire assembly will now fold up along the boom. Wrap the wire elements carefully along the boom, and secure with electricians tape. For a list of the parts used in building this antenna, see Table 1.

My roof peaks at a point 23 feet above the ground. To be effective as a DX antenna on 20 meters, I wanted the antenna at a height of 35 feet, yet wanted access to work on it while standing on the roof without using a ladder. I used a 5-foot tri-

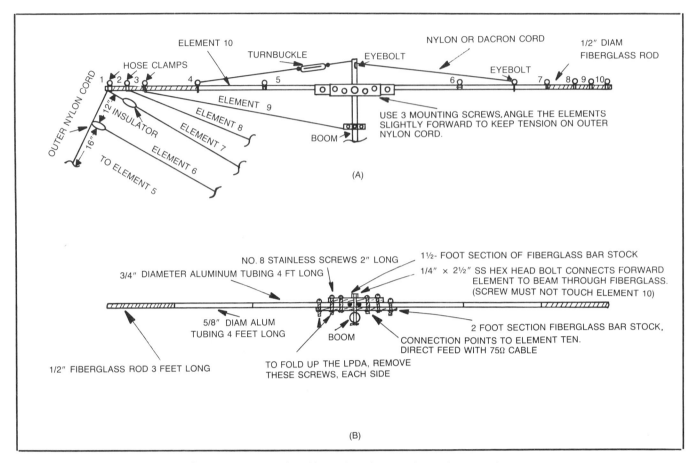

(A)

(B)

Fig 4—A gives a detailed top view of the forward spreader with spacings between hose clamps as follows: element 1 and 2 = 9 in., element 2 and 3 = 4 in., element 3 and 4 = 12 in., element 4 and 5 = 3 ft. Adjust element 10 to a length of 6 ft, 10½ in. These are approximate dimensions. At B the details of mounting and tubing dimensions are shown (minus hose clamps) for the forward spreader.

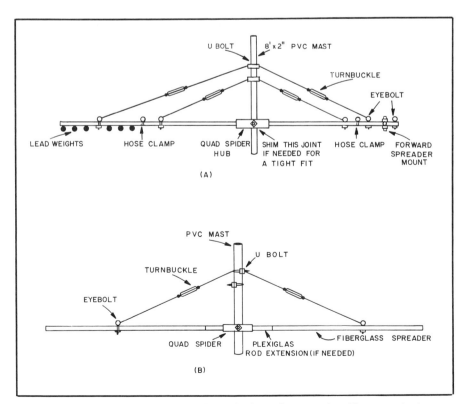

(A)

(B)

Fig 5—A shows how to add support to keep the boom from sagging. The spreader is supported in the same manner at B.

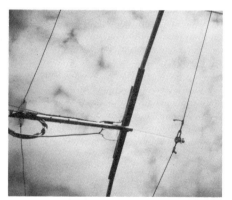

Fig 6—A view of the feed arrangement. The author found fiberglass to be more serviceable than Plexiglas® (used here to brace element 10) for extensions.

pod mount with an 11-foot telescoping mast made by cutting a 20-foot telescoping mast into two 5½-foot sections. This allowed the antenna to be raised to 35 feet for operation, and lowered to 6 feet above the roof for adjustments.

Using the LPDA

Nothing I found in my references deals

Table 1

14- to 30-MHz LPDA Parts List

10 Plexiglas insulators ¾ in. × 3 in. × ¼ in. cut from sheet
12 antenna end insulators (Budwig HYE-QUE 2 or similar)
8 sets flat-head brass screws, nuts, washers: no. 8 × ¾ in.
9 no. 6 × 1½ in. stainless screws, with washers and nuts to mount boom insulators
1 each ¼ in. × 2½ in. stainless-steel hex-head bolt, washer, and locking nut
6 no. 8 × 2 in. stainless-steel screws, washers, and nuts
9 no. 10 eyebolts
7 turnbuckles
Aluminum tubing:
    8-ft sections 1-in. OD, 2 req'd
    6-ft sections 7/8-in. OD, 2 req'd
    8-ft section 3/4-in. OD
    8-ft section 5/8-in. OD
2 fiberglass spreader rods, 13 ft long
1 fiberglass rod (solid) 6 ft × ½-in. diameter (Tap Plastics)
1 fiberglass bar 6 ft × 1-3/16 × 1/2 in. thick (Tap Plastics, comes in 6 ft lengths, 3-1/2 ft needed)
1 spider hub piece from quad element
30 feet of 3/16-in. nylon rope (or 1/8-in. dacron cord)
25 ft of 450-ohm ladder line
260 ft of heavy duty copper antenna wire (Radio Shack 278-1329)
50 ft of guy wire for strut supports (nylon rope could also be used)
30 spacers ¼-in. long with no. 6 × 1½-in. stainless hardware to secure the ladder line to the boom
6 small and 6 medium hose clamps
2 each 2-in. U bolts or one 2-in. diameter guy ring to tie guy wires from support strut
Several small fishing weights to balance boom

Table 2

SWR Data—Direct 75-Ohm Coax Feed

Figures represent frequency and SWR. For example, the first 20-meter listing gives an SWR of 1.4:1 at 14.05 MHz.

20 meters		10 meters	
14.05	1.4	28.05	2.8
14.15	1.8	28.15	2.8
14.25	2.4	28.25	2.8
14.35	2.8	28.35	2.7
		28.45	2.7
17 meters		28.55	2.7
18.07	1.3	28.65	2.6
18.12	1.4	28.75	2.6
18.16	1.4	28.85	2.5
		28.95	2.4
15 meters		29.05	2.3
21.05	1.1	29.15	2.0
21.15	1.2	29.25	1.7
21.25	1.4	29.35	1.7
21.35	1.8	29.45	1.6
21.45	2.1	29.55	1.4
		29.65	1.3
12 meters			
24.90	3.0		
24.95	3.2		
24.99	3.2		

with what happens to the feed-point impedance of an LPDA when the elements are swept forward. There are even some differences on how to calculate the impedance of a conventional array among the references. I experimented with various angles and element spacings. Every change I made had an effect on feed-point impedance. Direct feed with 75-ohm coax (no transformer) provided the best results with the dimensions given here. SWR readings range from 1:1 to 2.8:1 (for 20, 15, and 10 meters; see Table 2). A Transmatch located at the input end of the transmission line can be used to improve the match if desired. The coax is wound into an 8-inch-diameter three-turn choke at the point of connection (element 10) to the antenna. This is to keep current off the outside of the transmission line braid. If desired, a 1:1 balun can be used instead of the choke.

In the seven months of on-the-air experience so far, this antenna has captured 100 countries and produced more than 450 QSOs. Working the pileups is now more of a pleasure since my success rate is much higher. Signal reports have compared favorably with others in the pileups, with many QSOs "barefoot." When the linear is needed, the LPDA takes a full 1500 watts. The LPDA has also survived a winter of storms with winds gusting to 55 mph.

Many thanks to Emil, W1EJI (my father) for his help and suggestions in the contruction of the LPDA.

Notes and References

[1]D. E. Isbell, "Log Periodic Dipole Arrays," *IRE Transactions on Antennas and Propagation*, Vol AP-8, no. 3, 1960; Fig 7.
Reference Data for Radio Engineers, 6th ed. (Indianapolis: Howard W. Sams and Co, subsidiary of ITT, 1977), pp 18-20, 27.
G. L. Hall, Ed., *The ARRL Antenna Book*, 15th ed. (Newington: ARRL, 1988), Chap 10.
W. I. Orr, *Radio Handbook*, 22nd ed. (Indianapolis: Howard W. Sams and Co, 1981), sections 28.5 and 30.3.
L. A. Moxon, *HF Antennas for All Locations* (Potters Bar, Herts: RSGB, 1982).
H. Jasik, *Antenna Engineering Handbook* (New York: McGraw Hill, 1961), Chap 18.
A. Eckols, "The Telerana—A Broadband 13-30 MHz Directional Antenna," *QST*, Jul 1981, pp 24-27.
D. A. Mack, "A Second-Generation Spiderweb Antenna," *The ARRL Antenna Compendium, Vol 1* (Newington: ARRL, 1985), pp 55-59.

Antenna Trap Design Using a Home Computer

By Larry East, W1HUE
119-7 Buckland St
Plantsville, CT 06479

This article describes a BASIC program that will allow you to quickly and accurately design traps for multiband antennas on your home computer. The program is an adaptation of one originally written by Andy Griffith, W4ULD, for a Timex/Sinclair computer.[1] It allows you to design antenna traps from common coaxial cable (or any two-conductor cable for which the capacitance per foot is known) for any desired frequency. Such traps have the general configuration shown in Fig 1 and have been described in several Amateur Radio publications.[2-5]

Several changes were made to Griffith's original program to speed up computation and to improve accuracy. The present program, listed as Program 1, is written in "standard BASIC" and can be used with BASICA or GWBASIC on IBM PCs or clones. It can be easily adapted to other implementations of BASIC with few, if any, changes. (A version is also available from the author in APPLESOFT for the Apple II series of computers.)

When the program is executed, you will be asked to supply the following information: coax cable outer diameter in inches, coax capacitance per foot in pF, form diameter in inches and frequency in MHz at which the trap is to be resonant.

The program will then calculate and display

1) The inductive reactance of the resulting trap at the specified design frequency,

2) The capacitive reactance of the resulting trap (should be equal to the inductive reactance, but may be slightly different due to the iterative method of calculation used),

3) The inductance of the trap,

4) The capacitance of the trap,

5) The number of turns required,

6) The winding length in inches,

7) The total length of cable required for the trap, and

8) The effective length of the trap (explained later).

Different parameter values may be entered without restarting the program—just follow the instructions on the screen!

Design Method

The method of calculation is based on

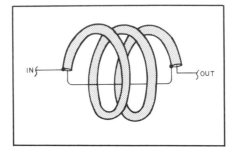

Fig 1—Preferred method of constructing a trap from coaxial cable.

Griffith's original iteration method to determine the number of turns required for a resonant trap at a given frequency. This is done by progressively increasing the number of turns until the capacitive reactance approximately equals the inductive reactance (the requirement for resonance in a parallel L-C circuit). The inductance of the coil is calculated using the usual approximate equation for a single-layer air-core coil:

$$L = \frac{A^2 \times N^2}{18A + 40B} \qquad \text{(Eq 1)}$$

where

L = inductance, μH
A = mean coil diameter, inches
B = coil length, inches
N = number of turns

If the coil is tightly wound from wire (in our case, coax) with diameter D, then the coil length, B (for an integer number of turns, at least), is given by $B = D \times N$.

The parallel capacitance is assumed to consist entirely of the capacitance of the length of cable. This capacitance is given by

$$C = \frac{A \times \pi \times N \times e}{12} \qquad \text{(Eq 2)}$$

where

C = total capacitance, pF
π = 3.14159...
e = capacitance per foot of the cable, pF
A = mean diameter of the coil, inches
N = number of turns

This assumes that the cable ends exactly at the coil ends. Coils will normally have

about $\frac{1}{2}$ inch of stripped cable extending through holes in the coil form at each end. The capacitance of this extra cable should also be included, as pointed out in an excellent article by Robert Sommer.[6] Assuming that there is 1 inch of extra cable, Eq 2 becomes

$$C = \frac{(A \times \pi \times N + 1)e}{12} \qquad \text{(Eq 3)}$$

The basic idea behind Griffith's iterative procedure is to pick a value of N and calculate the inductive reactance, X_L, and the capacitive reactance, X_C, for the design frequency of the trap. If these values are not equal, then make an adjustment in N (up or down, as indicated by the sign of the difference) and try again. The actual method used by Griffith was to start with $N = 1$ and increase it by one full turn until X_L was greater than or equal to X_C. At this point, the value of N was decreased by one and then increased in steps of 1/10 turn until again X_L was greater than or equal to X_C. The value of N at this point was assumed to be the desired value, correct to the nearest 1/10 turn. Doing the calculation in one-turn steps followed by 1/10-turn steps drastically reduces the number of calculations over what would be required if 1/10-turn steps had been used from the beginning.

This is certainly a reasonable approach; however, there is one obvious and one less-obvious improvement that can be made. The obvious one is to include the capacitance of all of the stripped cable, as noted above. The less-obvious one has to do with the way Griffith calculated the length of the coil for use in Eq 1. He used $B = D \times (N + 1)$, where N is the value at the end of the first iteration loop, in other words, to the next full turn. He chose to use $N + 1$ rather than N because this gave better results. However, it turns out that the value of N after the first loop can be as much as 9/10 of a turn greater than the final value. For a coil of only a few turns, this can cause a large error!

The current program therefore uses Eq 3 for the capacitance calculation, and uses $B = D \times N1$ in Eq 1 for the inductance calculation in the second loop. The program also "fine tunes" N1 in steps of 0.05 instead of 0.1.

Program 1
Program Listing for Coax Trap Calculations

ARRL-supplied disk filename is EAST.BAS.

```
1    REM   COAX TRAP PROGRAM - IBM-PC Version
2    REM   WRITTEN NOV. 1984 - LAST MOD 11/7/87
3    REM   ADAPTED FROM PROGRAM BY ANDY GRIFFITH, W4ULD
4    CLS : PRINT
5    PRINT TAB(25);"*************************"
6    PRINT TAB(25);"*   Coax Trap Program   *"
7    PRINT TAB(25);"*    Larry East, W1HUE   *"
8    PRINT TAB(25);"*************************"
9    PRINT
10   PRINT TAB(6);"This program will calculate the turns required for an antenna"
11   PRINT TAB(6);"trap made from coaxial cable. It also calculates and displays"
13   PRINT TAB(6);"XL, XC,  L and C values.  Inputs are the coax cable diameter,"
15   PRINT TAB(6);"capacitance per foot, coil form diameter, and trap frequency."
27   PRINT
30   PRINT TAB(8); : INPUT "Enter 0 for SCREEN, 1 for PRINTER ";PT
33   IF PT < 0 GOTO 30 : IF PT > 1 GOTO 30
35   PRINT
40   PRINT TAB(8); : INPUT "Coax Diameter (in) ";CD
50   PRINT TAB(8); : INPUT "Coax Cap./ft. (pF) ";CC
55      CC = CC / 12 : REM - capacitance/inch
60   PRINT TAB(8); : INPUT "Form Diameter (in) ";D
65      A = D + CD : REM - mean winding diameter
70      ASQ = A * A : A18 = 18 * A : C40 = 40 * CD
80   PRINT TAB(8); : INPUT "Frequency in MHz ";F
90      PI = 3.14159265 : RAD = 2 * PI * F
100  REM - INITIAL ESTIMATE OF N
110     N0 = INT ((570 / F) * SQR (CD / (A * ASQ * CC)) + 0.5)
120  REM - FIRST CALCULATION LOOP
130  FOR N = N0 TO N0+20 STEP 0.5
140     L = ASQ * N * N / (A18 + C40 * N)
150     XL = RAD * L
160     C = (A * PI * N + 1) * CC
170     XC = 1E6 / (RAD * C)
180        IF XL >= XC THEN GOTO 210
190  NEXT N
200  REM - SECOND CALCULATION LOOP
210  FOR N1=(N-0.5) TO (N+0.5) STEP 0.05
220     L =  ASQ * N1 * N1 / (A18 + C40 * N1)
230     XL = RAD * L
240     C = (A * PI * N1 + 1) * CC
250     XC = 1E6 / (RAD * C)
260        IF XL >= XC THEN GOTO 310
270  NEXT N1
300  REM - NOW DISPLAY RESULTS
310  IF PT = 0 THEN OPEN "SCRN:" FOR OUTPUT AS #1 : GOTO 340
315     OPEN "LPT1:" FOR OUTPUT AS #1 : REM - STANDARD PRINTER OUTPUT
320     PRINT #1, : PRINT #1,
325     PRINT #1,TAB(8);"Coax:  ";CD;" in. OD, ";CC*12;" pF/ft"
330     PRINT #1,TAB(8);"Coil Form Diameter = ";D;" in."
340  PRINT #1,
345  PRINT #1,TAB(8);"For Frequency of ";F;" MHz:"
350  PRINT #1,TAB(8);"XL = ";INT(XL * 10 + .5)/10, "XC = ";INT(XC * 10 + .5)/10
360  X1 = 100 : IF L < 1 THEN X1 = 1000
365  X2 = 100 : IF C >= 10 THEN X2 = 10
370  PRINT #1,TAB(8);"L = ";INT(L*X1+.5)/X1;" uH", "C = ";INT(C*X2+.5)/X2;" pF"
375  PRINT #1,TAB(8);"No. of turns = ";INT (N1 * 100 + .5) / 100
380  W = INT (INT (N1 + .5) * CD * 100 + .5) / 100
385  CL = INT ((A * PI * N1 + 1.5 * W) * 10 + .5) / 10
390  PRINT #1,TAB(8);"Winding Length = ";W;" in."
395  PRINT #1,TAB(8);"Coax Length = ";CL;" in."
397  PRINT #1,TAB(8);"Effective Length = ";INT (CL * 8.8 + .5) / 10;" in."
400  PRINT #1, : CLOSE #1
410     IF PT <> 0 THEN PRINT
500  PRINT TAB(4);"1 - Repeat from Beginning."
510  PRINT TAB(4);"2 - Repeat for New Frequency Only."
520  PRINT TAB(4);"3 - New Coil Diameter & Frequency."
530  PRINT TAB(4);"4 - Exit Program."
535  PRINT
540  PRINT TAB(8); : INPUT "Your Choice (1 - 4) ";X1
545  PRINT
550     IF X1 = 1 THEN GOTO 40
560     IF X1 = 2 THEN GOTO 80
570     IF X1 = 3 THEN GOTO 60
580     IF X1 = 4 THEN GOTO 600
590  GOTO 540
600  END
```

With these simple changes, the calculated number of turns agrees much better with as-built traps using high-quality coax. The results for four sample trap designs are summarized in Table 1. The calculated number of turns is "right on the money" for the four designs tested, and considerably closer than calculated from Griffith's original program.

The actual (measured) characteristics of the cables used in these examples, capacitance and diameter in particular, were very close to the specified values. This contributed to much of the good agreement between the number of turns calculated and those actually required. This is not always the case with some of the cheaper "generic" cables that are available.

The number of turns is more sensitive than one might expect to very small changes in the diameter of the coil form (note the squared value in Eq 1). For example, rounding off the form diameter in Table 1 from 1.68 to 1.70 inches reduces the calculated number of turns by 0.05 for the 10- and 15-meter traps, and by 0.1 for the 20- and 30-meter traps.

As mentioned in the introduction, this design method can be used for traps made from any parallel-conductor cable, not just coax. For example, speaker cable has been suggested for use in lightweight traps.[7] All that is required is to determine the capacitance per foot of the cable to be used and plug it into the program. The method of construction would be essentially the same as with coax.

Program Description

A complete listing of the program is shown in Program 1. [The program is available on diskette for the IBM PC and compatibles; see information on an earlier page of this book.—*Ed*.] The comments within the program listing should give you a pretty clear idea of what is going on. However, a few points might need further clarification.

The "initial estimate," N0, of the number of turns calculated in line 110 is based on a simplification of the equation that is obtained by substituting Eqs 1 and 2 into the formula for the resonant frequency of a parallel-resonant tuned circuit. The original equation cannot be solved directly for N; if it could, the design problem would be much simpler! (See the references of notes 4, 5 or 6 for details.) However, by ignoring a couple of terms, it can be solved for an approximate value of N (the N0 value calculated in line 110). The nature of the simplification guarantees that N0 will always be less than the required number of turns, so it is a good starting point for the first iteration loop beginning at line 130. One could use N0 = 1 as the starting point, as Griffith did in his program, but an initial value closer to the required value reduces the number of times through the first iteration loop. This can result in a significant saving in calculation time for traps having many turns.

Results of the calculation are displayed on the CRT screen or printed, as selected in line 30, starting at line 300. Sample output from the program is shown in Table 2. The values are for the 10.1-MHz trap in Table 1.

The total length of the piece of coax needed for the trap is calculated in line 385 and printed in line 395. An extra amount sufficient to make the required connection from the inner conductor of the input end to the shield of the output end, as shown in Fig 1, is included.

The "effective length" printed in line 397 is an estimate of the amount that the next-lower-frequency leg of the trap antenna must be shortened to obtain resonance. For example, if you are making a 10- and 15-meter trap dipole, place 10-meter traps at the ends of a normal-length 10-meter dipole. The amount of wire to be connected to the trap outputs to make the 15-meter dipole will be reduced by the "effective length" of each trap from that calculated using the usual half-wave-dipole formula,

$$\ell = \frac{468}{f} \qquad \text{(Eq 4)}$$

where

ℓ = length, feet
f = frequency, MHz

The effective length is taken as 88% of the length of the coax in the trap. This is a conservative estimate; the actual effective length could be as low as 80% but probably no more than 90%. You should therefore expect to do a little pruning to bring the antenna to resonance at the desired frequency. More on this later.

A definite attempt was made to avoid the use of nonstandard BASIC statements. Also, no PEEKs or POKEs are used in the program. Only a few lines should require any changes for other versions of BASIC. You may have to change (or just delete) line 4, which clears the display screen (the "clear screen" command might be HOME). If your version of BASIC happens to internally define PI as 3.14159, then remove the definition from line 90.

Multiple statements on a line are separated by colons (:). Some BASICs may require a different separator, or may require a separate line for each command. Also, some versions (early implementations in particular) allow variable names to consist only of a letter followed by an optional number; if yours is one of these, then some variable names will require changing.

Table 1

Comparison of Design and Measured Values

Type	Coax Cap, pF/ft	Coax Diam, in.	Form Diam, in.	Design Freq, MHz	W4ULD Program	This Program	Actual
RG-58C	30.8	0.195	1.68	21.1	4.5	4.10	4.20*
RG-58C	30.8	0.195	1.68	28.2	3.6	3.25	3.25
RG-59A	21.0	0.242	1.68	14.1	7.4	6.90	6.85
RG-174	30.8	0.100	1.68	10.1	7.2	6.85	6.85

*Loosely spaced turns.

Table 2

Sample Program Output

Results shown are for the 10.1-MHz trap in Table 1. The format is not precisely duplicated here. Coax: 0.1 in. OD, 30.8 pF/ft.
Coil form diameter = 1.68 in.

X_L = 158.7 Ω
X_C = 156.2 Ω
L = 2.5 μH
C = 100.9 pF
No. of turns = 6.85
Winding length = 0.7 in.
Coax length = 39.4 in.
Effective length = 34.7 in.

Trap Construction

Designing the traps is only part of the fun—the rest is putting them together and making the antenna. This is when most of the work (and fun?) really begins!

Methods for coax trap construction are described in the references of notes 2, 3 and 6. Additional useful information is given in a 1983 *QST* article by Doug DeMaw.[8] With so much information already available, it is probably not necessary to give construction details here. However, there are a few things that are important enough to be repeated.

First of all, it is absolutely necessary to have some means of checking the resonant frequency of antenna traps. These are relatively narrow-band devices, and should be resonant at least some place in the band for which they are designed! Probably the most readily available method of checking the resonant frequency is with a dip oscillator. Remember, however, to check the indicated resonant frequency with a receiver or frequency counter, since the calibration of most grid or gate dippers is only approximate at best. The "Beyond the Dipper" device described in *QST* a while back would appear to be ideal for measuring trap

resonance, if you are big on construction projects.[9] A simple method using an ordinary transceiver is described in the reference of note 8.

It is also necessary to fix the trap in some way to keep the turns from shifting. For a trap of a few turns, a slight shift in the spacing can result in a significant change in resonant frequency. Weatherproof silicone sealant can be used, but you should be careful to use a noncorrosive type (if it smells like vinegar, don't use it!). I have had very good success with the type sold for automotive gasket use. Remember to seal the ends of the cable against moisture as well.

When constructing a multiband trap antenna, always start by making sure that the highest frequency antenna is pruned to the desired resonant frequency before adding traps. The traps will pull the resonant frequency of the system toward the resonant frequency of the trap, making it almost impossible to properly trim an antenna with traps present for the same band. After the highest frequency antenna has been properly tuned, add the traps and the additional wire (mast, or whatever) for the next highest frequency antenna to the trap outputs. Again, prune to resonance before adding the next set of traps.

Notes
[1] A. S. Griffith, "A Coax-Antenna Trap Program for the Timex/Sinclair 1000," *QEX*, August 1984, p 8.
[2] R. H. Johns, "Coaxial Cable Antenna Traps," *QST*, May 1981, p 15.
[3] G. E. O'Neil, "Trapping the Mysteries of Trapped Antennas," *Ham Radio*, October 1981, p 10.
[4] C. Nouel, "Exploring the Vagaries of Traps," *CQ*, August 1984, p 32.
[5] F. Noble "Coaxial Antenna Trap Design," Technical Correspondence, *QST*, March 1984, p 47.
[6] R. Sommer, "Optimizing Coaxial-Cable Traps," *QST*, December 1984, p 37.
[7] See note 4.
[8] D. DeMaw, "Lightweight Trap Antennas—Some Thoughts," *QST*, June 1983, p 15.
[9] W. Hayward, "Beyond the Dipper," *QST*, May 1986, p 14.

The Suburban Multibander

By Charles A. Lofgren, W6JJZ
1934 Rosemount Ave
Claremont, CA 91711

No room for an antenna farm? The Suburban Multibander, a simple 80-10 meter antenna, may be the solution. It allows Transmatch-free operation on 40 and 20 meters while avoiding the complications and losses of traps or resonators. On the remaining HF bands, it has the versatility of a center-fed Zepp (to use the common misnomer), but greater convenience. On the higher frequencies it even shows some directivity and gain. Not least, the multibander fits on a small lot. Others with limited space may find it just as satisfactory as I have.

The antenna, Fig 1, resembles a standard center-fed Zepp, especially the G5RV version, but its lineage and operation are different. The design draws on Taft Nicholson's compact multiband antenna for tube-type transmitters.[1] But special attention is given to the need of modern solid-state transmitters to see low standing wave ratios. The place to begin is with a bit of theory.

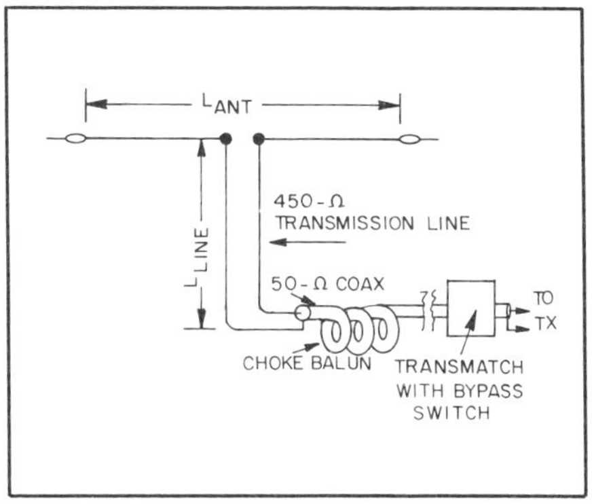

Theory

According to Nicholson, the design can be understood by considering the antenna and transmission line as open and closed line segments that are joined together to form a resonant circuit. (For the open segment, the length equals one side or leg of the dipole.) Resonance occurs at all frequencies where the reactances of the segments have the same absolute value and are opposite in polarity. As a result, when the open and closed segments are each $60°$ long, that is, one-sixth wavelength, the circuit is resonant on most harmonics, for segments with lengths equal to discrete-angle multiples of $60°$, that is, $120°$, $240°$, $300°$ and so on, display a uniform absolute value of reactance, with alternating polarity at each successive interval. It can be shown mathematically that these harmonic frequencies are the 2nd, 4th, 5th, 7th, 8th, etc, and exclude the 3rd harmonic and its multiples (the 6th, 9th, etc).[2]

As we shall see, this analysis is useful. But, in fact, the antenna itself does not behave like an open line segment. The

[1]Notes appear on page 104.

Fig 1—Construction details of the Suburban Multibander, a simple 80-10 meter antenna. See text for details and length formulas. For "fundamental" operation on 7.1 MHz, approximate lengths are: $L_{ANT} = 88.8$ feet, $L_{LINE} = 46.2 \times$ VF.

reason is that while a line segment is a pure reactance, ignoring losses, this is not true for an antenna, which consumes real power through radiation. A resulting problem is that the feed-point impedance of the antenna is *not* neatly uniform in absolute value at the fundamental frequency and harmonics described above.[3] This difficulty is compounded by end effects, variations in antenna height, and other environmental factors.

In practice, however, the analysis based on line segments and hence on pure reactances works in one instance and comes close in a second. The first is when a "thin" two-thirds-wave ($240°$) dipole is fed with a one-third-wave ($120°$) segment of 450-Ω transmission line, and the near fit is when a thin four-thirds-wave ($480°$) dipole is fed with a two-thirds-wave ($240°$) segment of 450-Ω line. Moreover, the two instances are particularly interesting because they describe the same antenna, the second case merely representing harmonic operation.

The explanation is straightforward. To begin with, even though the actual feed-point impedance of a thin dipole is not iterative at all the $60°$-per-leg intervals listed above, the impedances for dipole lengths of $240°$ and $480°$ are fairly close to one

another. The dipole in Fig 1, when constructed as described below, displays a 40-meter feed-point impedance of about $190 + j740 \ \Omega$.[4] The 20-meter figure is near that, with the polarity of the reactive component reversed.

Given the similarity of these two feed-point impedances in absolute value, the 450-Ω feed line enters the picture. Because of its characteristic or surge impedance (Z_0), and because of the iterative property of discrete-angle $60°$ intervals along a transmission line, the input impedance of the 450-Ω feed line on 40 meters is approximately $50 + j0 \ \Omega$. (In technical terms, a conjugate match occurs at the antenna feed point.[5]) On 20 meters, the figure at the input of 450-Ω line is close enough to 50 Ω to give an SWR on the coaxial line of less than 2:1.

It is also easy to see why on 80 and 10 meters the design fails to produce a match that is suitable for modern solid-state transmitters. On each of these bands, the feed line remains an appropriate length ($60°$ or a discrete-angle multiple of $60°$), but in each case the antenna feed-point impedance shifts further away from the 40-meter figure. On 80 meters, the dipole length is enough *under* a half wave to significantly lower the feed-point resistance, resulting in an estimated feed-point impedance of approximately $25 - j \ 700 \ \Omega$.[6]

On 10 meters, one factor is the *increased* overall antenna length in terms of wavelength (approximately two and two-thirds wavelengths, versus two-thirds wavelength on 40 meters). The major culprit, however, is probably the decreased impact of end-effect loading, which makes an antenna cut for 40 meters electrically somewhat shorter on 10 meters than the desired length of $960°$.[7] (These two factors also shift the impedance on 20 meters, but not as much.) On both 80 and 10 meters, though, sufficient matching occurs to allow reasonable *reception* without use of a Transmatch, which, as explained later, is itself a useful feature.

Construction

In building and installing the antenna shown in Fig 1, there are several factors to keep in mind. Two are the gauge of the wire in the dipole and its height, both of which affect the antenna surge impedance and thus its feed-point impedance. The surge impedance shifts slowly, however, because the change is logarithmic with respect to the ratio of wire diameter and height. Basically, then, wire gauge and height are not overly critical. My version uses no. 18 solid copper wire, with the antenna supported at the center, about 30 feet high, and the ends drooping to about 20 feet. No. 16 wire should be suitable, particularly at greater heights. The resulting surge impedance should be in the vicinity of 1200 to 1300 Ω. If in doubt, you can calculate it.[8]

As for the antenna length, the best approach is to begin with an approximation slightly on the long side. Use the formula

$$L_{ANT} = \frac{630}{f} \text{ feet}$$

where f is in MHz

After installation, check the antenna center frequency with an SWR meter in the coax line and trim the ends for the desired frequency. (Ideally, the SWR reading should be taken at the junction of the balun and the coax to the rig, but measurement at the rig will do.) Use temporary end clips while making the adjustments. You may take several inches off each leg, depending on height, apex angle, nearby objects, etc.

By contrast, cut the 450-Ω transmission line exactly to an electrical third of a wavelength at the desired center frequency and then *leave it alone* when you make the final adjustments. The formula here is

$$L_{LINE} = \frac{328 \times VF}{f} \text{ Feet}$$

where VF = velocity factor of the line (discussed below)

The approximate dimensions given in Fig 1 are intended for Transmatch-free operation primarily on 40 meters. The foregoing formulas also allow cutting a shorter antenna for similar operation on a higher band, along with Transmatch operation elsewhere.

When cut for 40 meters, the antenna dimensions also provide operation without a Transmatch on 20 meters. Note, however, that the two resonant frequencies may not have a precise harmonic relationship, which is not surprising in view of the matching system and the impact of end effects. In addition, while the SWR on 40 meters will be nearly 1:1, the 20-meter figure will be higher, between 1.5:1 and 1.8:1 at resonance. Although not ideal, these latter figures are adequate for modern equipment. At the same time, the antenna is only a little longer on 20 meters than an extended double Zepp, being 0.67 wavelength per leg versus the prescribed

0.64 wavelength, and in the real world it should show about the same directivity and gain.

In any event, room remains for further investigation and experimentation on 20 meters. You may wish to try lowering the SWR figure on the band by slightly varying the length of both the antenna *and* the 450-Ω transmission line. The same approach could be used to nudge the exact harmonic relationship one way or the other.

For the transmission line, use 450-Ω line. *This is important.* Do not use 300-Ω twin-lead, which seriously alters the impedance match, besides causing higher losses.[9] In determining the physical length of the 450-Ω line, be sure to take the velocity factor into account. The safest approach is to measure it instead of relying on the manufacturer's nominal figure.[10] For the line that I'm using (which resembles wide twin-lead with windows cut in it), the velocity factor measures 90.5%.

The choke balun in Fig 1 is inserted at the junction of the 450-Ω balanced line and the 50-Ω coax. It consists of 14 turns of RG-58 coax on an Amidon FT 240-43 core (μ = 850). While the balun might be eliminated altogether, it ensures against feeding RF onto the outside of the coax at the junction of the two transmission lines. With the FT-240-43 core, the balun is heftier in terms of power-handling capability than is needed in most installations, but the very high reactance obtained with this core seems advisable for operation of the antenna on 30 and 15 meters, where the junction with the coax is a point of high impedance.[11]

Because the SWR on the coax line is especially high on 30 and 15 meters, use low-loss coax (RG-8X or better, except for the RG-58 in the choke balun) and keep the run to the shack as short as possible. By comparison, the SWR on the coax is lower on other bands (and flat or nearly so on a couple of them), but purists may still object to using coax at all in a feed system where the SWR is sometimes high. The combination of balanced line and coax is admittedly not the best way to center-feed a wire in such circumstances. It bears mention, though, that the same method is widely employed with the popular G5RV antenna.[12]

Operation

Once trimmed to frequency, the antenna is as easy to use on 40 meters as a half-wave dipole. Just switch your Transmatch to the Transmatch-bypass position. Depending on your installation and equipment requirements, "straight-through" operation may also work on 20 meters. In any case, observe the usual precautions about harmonic attenuation. On 40 meters, where

the SWR is close to 1:1 at the selected center frequency, nearly the whole band will fall within the 2:1 SWR points, and about 200 kHz within the 1.5:1 points. Although the center-frequency SWR is higher on 20 meters, about 150 kHz falls within the 2:1 points. (If you trim the antenna primarily for 40 meters, the actual usable coverage on 20 meters will depend on the harmonic relationship displayed by the antenna.)

For the other bands, simply switch in the Transmatch and operate the antenna as you would a center-fed Zepp or G5RV antenna. In fact, because the Transmatch is unnecessary on 40 and perhaps 20 meters, you can leave it adjusted for another band. A further bonus comes from the "near-miss" match on 80 and 10 meters. This feature allows monitoring to check band conditions, or to read the mail, without switching in the Transmatch. On 30 and 15 meters, the Transmatch is necessary for reception as well as regular operation. (My rig does not cover the 12-meter band, but the same should hold true for it.)

In short (so to speak!), if you are restricted to about 90 feet, here is an antenna that's efficient, simple to build, versatile in operation and convenient to use. You can still dream of that antenna farm, but meanwhile, why wait?

Notes
[1] T. Nicholson, "Compact Multiband Antenna Without Traps," *QST*, Nov 1981, pp 26-27.
[2] This brief analysis only summarizes the long appendix (with diagrams) in Nicholson, p 27 in the reference of note 1.
[3] As the electrical length increases at successive harmonics, coupling between adjoining half-wave segments and radiation from the segments cause progressive shifts in current magnitude and in phase angle at the antenna feed point. See *The ARRL Antenna Book*, 15th ed. (1988), pp 2-8 to 2-10 (and especially Fig 10 on p 2-10).
[4] This figure, based on moving "backwards" from the input of the 450-Ω line, can be worked out on a Smith Chart, but can be calculated more easily with the computer program FEEDLINE TRANSFORMER, which is part of the software package designed to accompany J. Devoldere, *Low Band DXing*, 2nd ed. (1987). (The package may be purchased through the ARRL.) Anyone interested in antennas should note that many of the programs in the package, and especially those which computerize the Smith Chart, are useful in antenna design generally and not just on the low bands. For example, either FEEDLINE TRANSFORMER or IMPEDANCE ITERATION can be used to check the iterative property of transmission lines at 60° intervals.
[5] See the useful discussion in *The ARRL Antenna Book*, 15th ed., p 25-2. (This discussion is in the context of transmitter-to-line coupling, but the principle applies to line-to-antenna coupling.) To elaborate, consider the situation at 40-meter resonance. Viewed from the junction of the feed line and the antenna, the 450-Ω transmission line is terminated 120° away in 50 + $j0$ Ω, and it accordingly displays an impedance of approximately 190 $-j740$ Ω at the antenna feed point. At the same feed point, the antenna impedance is approximately 190 +

$j740\ \Omega$. The reactances thus cancel each other at the feed point, the resultant feed-point impedance is purely resistive (approximately $1540 + j0\ \Omega$), and resonance is established and matching occurs throughout the system. (The resultant feed-point impedance, although not really necessary to know, can easily be calculated with the program FEEDLINE T-JUNCTION in the Devoldere software package, note 4 above.) On 20 meters, the polarities of the reactive components of feed line and antenna impedances at the feed point are reversed, which produces some cancellation of reactances. The absolute values, however, are not equal, hich accounts for the higher SWR on 20 meters.

[6]See pp II-9 and II-13 in. Devoldere, note 4 above. Owing to ground losses from the low antenna height, the actual resistive component is doubtless higher than in the estimate. Not surprisingly, Nicholson (note 1) had to lengthen the feed line with a loading arrangement in order to obtain a relatively low SWR below 3.750 MHz.

[7]As the antenna becomes longer in terms of a wavelength, the impact of end effects becomes proportionately lower because only the outermost quarter-wave segments are affected. The basic phenomenon is familiar to anyone who has used a 40-meter half-wave dipole on 15 meters, where the resonant frequency is higher than the third harmonic of 40-meter resonance. But unlike a 40-meter dipole on 15 meters, the antenna described here depends on a feed-line transformer, which does not show a comparable skewing of electrical length on harmonics. In other words, at the frequency where the antenna electrical length on 10 meters is 960° (toward the high end of the band), the feed-line electrical length is not the necessary 480°. Hence, the impact of end effects is magnified.

[8]Nicholson, note 1, contains an approximate formula for dipole surge impedance, and a more precise formula is in Devoldere, *Low Band DXing*, note 4, p II-13. Either one is adequate.

[9]However, if you are inclined toward calculations and experiments, you can try *low-loss* 300-Ω line. If you do, tests indicate (as expected) that the length of the antenna and/or of the line will require substantial adjustment, which in turn will significantly change the harmonic relationship. As starting points on 40 meters, keep the *electrical* length of the 300-Ω line at 120°, and for the antenna use the formula $L_{ANT} = 582/f_{MHz}$. Be prepared to look around for the harmonic resonance.

[10]See G. Downs, "Measuring Transmission Line Velocity Factor," *QST*, June 1979, pp 27-28.

[11]Because of the low current flowing on the outside of the coax, core saturation is typically not a problem with a choke balun. See *The ARRL Antenna Book*, 15th ed., p 25-11.

[12]For the originator's own assessment, see L. Varney, "The G5RV Multiband Antenna . . . Up-to-Date," in *The ARRL Antenna Compendium, Volume 1* (1985), pp 89-90. Varney recommends against use of a balun at the junction of the balanced line and coax, but his precaution applies to the typical transmission-line balun, and he in fact suggests a choke.

Fat Dipoles

By Robert C. Wilson

c/o ARRL
225 Main St.
Newington, CT 06111

A ntennas without problems make radio communications enjoyable. I design overseas radio stations for a living, so I'd rather not have to fight my own ham station when just relaxing and rag chewing. Fat dipoles do the things I want. They match the coax line well over a wide band, and they launch the signal remarkably well.

Theory

Making a dipole conductor thicker than normal with respect to wavelength will increase the bandwidth and modify the working impedance of the antenna. The trick is to make a dipole "fat" in such a way that it may be easily constructed from cheap materials, be highly efficient and at the same time arrange things so that it will match the transmission line from the lower band edge to the upper band edge.

I started with the assumption that my band of interest would be the 80/75-meter band. From end to end, this requires a 13% bandwidth to the 2:1 SWR points for my broadband-solid-state final. I also assumed

Table 1

Dimensions and Bandwidths on Various Bands

Frequency MHz	Length, ft	Spacing (A), ft	Bandwidth, MHz
1.9	233	5.9	0.252
3.75	118	3	0.5
7.15	62	1.6	0.951
14.175	31.2	0.8	1.885
21.225	20.85	0.53	2.823
28.850	15.34	0.39	3.837

that this antenna was going to be at a nominal height of 30 feet or 0.11 wavelength above ground. The calculations indicated that a dipole built of four quarter-wavelength no. 14 wires (0.064 inch) with a spacing of 0.0114 wavelength would produce the necessary results. The correct length would have to be 0.45 wavelength to match a 50-ohm line.

Length = 442.5/f feet
Width = 11.25/f feet
Height = 112.5/f feet

where f = center frequency in MHz

Construction

Very few problems will be encountered in building this simple fat dipole if you follow the drawing (Fig 1). First you will need five good insulators. I prefer egg type insulators but there is no critical problem

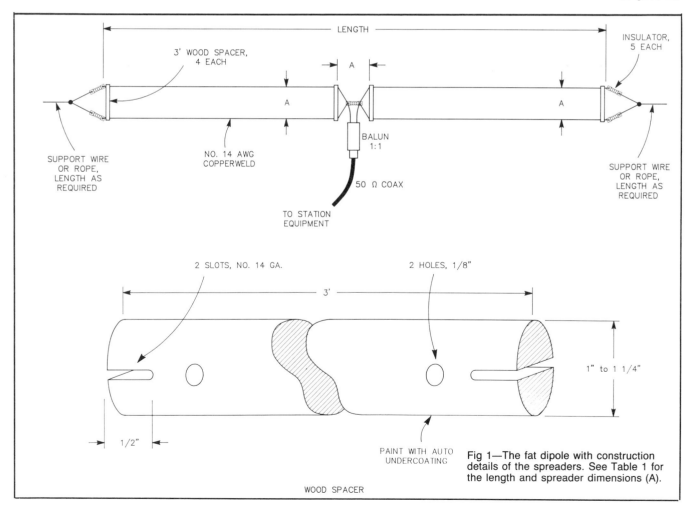

Fig 1—The fat dipole with construction details of the spreaders. See Table 1 for the length and spreader dimensions (A).

here. One insulator is for the center of the dipole and the others are for the four ends. You will need four 3-foot-long broom sticks or 1-inch wood or plastic rods with good weatherproofing. I painted mine with auto undercoating but outdoor paint or varnish should also work. Copperweld wire is very desirable because it won't stretch and change the tuning of your antenna. This type of wire is available through advertisements in *QST*. The same source may also be able to supply the essential wide-band balun transformer and coax. Either the RG-58 or RG-8 types of coax cable are satisfactory but the latter requires more support because of its greater weight.

Measure your wire carefully and leave enough extra so that the insulators can be attached. The final length will need to be the calculated value, from insulator wire end to insulator wire end. After building the four-wire section, attach dowel rods as shown to act as spreaders. Fasten each rod in place with pieces of wire threaded through the holes and then wrapped around the antenna wire. Wrap these spreader wires tight enough so that the rods will not slip out of place. Snip off projecting wire ends wherever they occur to prevent RF corona power loss. Then, using either wire or rope, make a bridle to hold the ends of the antenna.

Last, solder the balanced end of the balun transformer to the dipole. Each wire from the balun should go to the pair of wires on the same side of the dipole. The solder job should be of the best quality and permanent because it is hard to repair later. The coax needs to be connected to the unbalanced side of the balun. If you use large-diameter coax (3/8 inch) then think about ways to support the weight. Perhaps a piece of nylon rope from the dipole center insulator to the coax will help take the load, but I'll leave the details of the problem up to you. After this final construction step, haul the antenna up in the trees, using care that no twists are allowed.

Operation

For once I had a 75-meter antenna that worked better than predicted. The SWR was 1.6:1 or better from 3.5 to 4.0 MHz. Better yet, reports received were excellent with my old 100-watt solid-state transceiver. Moving up and down the band gave no loading problems from the broadband final. The fat dipole is just what I needed for a good, relaxing rag chew after a hard day with the 500-kW rig at the office.

Swallow Tail Antenna Tuner

By Dave Guimont, WB6LLO

5030 July St
San Diego, CA 92110

Build the antenna tuner for less than a dollar. The complete antenna will cost you from ten to fifteen dollars, depending on your junk box. It will fit on a 40 foot × 40 foot lot. The SWR is less than 2:1 from 3.5 MHz to 4.0 MHz and from 7.0 MHz to 7.3 MHz.

The antenna itself is a classic design that has been around for years. It is popular because it is inexpensive. It allows an 80-meter antenna on a small lot, and it is easily constructed by a novice builder. About the only disadvantage of the classic design is the narrow bandwidth abhored by today's transciever.

This simple device eliminates that disadvantage. It accomplishes that by actually tuning the antenna. Most so-called antenna tuners tune the antenna system, but do nothing to the antenna itself.

General Design

For those unfamiliar with the design, I will first describe the antenna. It is a combination 40/80-meter antenna that resonates at 80 meters because of a lumped inductance at the end of the 40-meter section. The 80-meter section is tuned by opening and closing a 50-inch (approximately) tail, on one end only, that is outboard of the lumped inductance. Fig 1 illustrates the optimized antenna. By formula the 40-meter inverted-V dipole would be 32 feet 6 inches on a side. It was discovered many years ago that this section has to be lengthened to 33 feet 2 inches to provide resonance near the center of the 40-meter band. It was first assumed that the center section would have to be shortened, but this is not the case. I am not an engineer and can only assume this lengthening is required because of the capacitance between turns on the lumped inductance. I would appreciate information on the reason this section has to be lengthened.

The original antenna covered 40 meters adequately, and the swallow tail has little or no effect on 40 meters. Cut the center section initially to approximately 33 feet 8 inches and prune to resonance in the desired portion of 7 MHz. Proximity to other conductors, height above ground and the apex angle will all have an effect on the desired frequency. The peak of my antenna

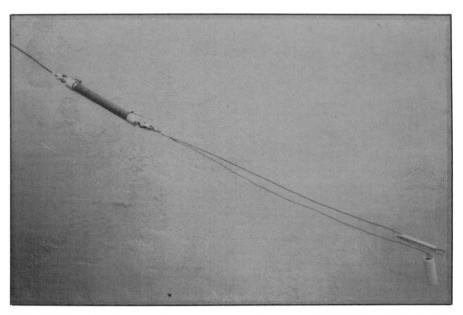

Fig 1—With the tail pulled up as shown, the antenna is resonant at 3.995 with an SWR of 1.4:1.

is 36 feet high and the ends are about 12 feet off the ground. I am using a 1:1 balun from an old Hy-Gain beam for the present antenna. It has worked as well with no balun, and a 10-turn coil of coax served nicely at a recent Field Day exercise.

The lumped inductance (about 120 μH) is made by winding 50 feet (about 150 turns) of no. 18 wire on a piece of ¾-inch schedule 40 PVC that is 10 inches long. Two ¾-inch PVC end caps and suitable hardware complete the coil portion. The coil occupies about 8 inches on the form. Make two. For those who are wary about using PVC as the coil form, the antenna has been used at over a kilowatt. Pieces of old garden hose have been used successfully in the past. The end insulators are 4-inch ceramics, and I use 60-pound nylon monofilament line to support the ends. See Fig 2.

Construction Details

The antenna wire itself is no. 16 stranded copper. Most any size near that should work as well. Reasonable substitution can

be made in the parts list shown in Table 1.

Start out with the 80-meter outboard portion approximately 56 inches long. Be certain to have this portion installed before attempting to tune the 40-meter section. After tuning at 40 meters, prune the outboard sections (equally) for resonance at 3.99 MHz. As mentioned before, antenna height, apex angle and other conductors will all have an effect on individual installations. Some additional pruning may be necessary after adding the swallow-tail tuner. Element length is critical, as is proximity to other objects. Bandwidth is extremely narrow at this point, which is the reason we add the swallow tail.

As the swallow tail is opened, the resonant frequency goes down. Remember that altering the apex angle also has an effect on the resonant frequency.

The tuning portion is simplicity in itself. A length of the same size wire as the outboard section is soldered to a doubled piece of RG-8 outer braid (one inside the other) 2½ inches long. Use a heat sink on the center portion of the braid to prevent solder wicking into the braid. This piece of braid is the hinge upon which the tuning wire will flex. Solder the other end of the braid to the outboard end of one of the coils. See Fig 2.

I use a large ceramic insulator as the counterweight to pull the end of the tuning wire down. Any nonmetallic weight providing minimum wind resistance will work as well. Sixty pound nylon fish line is reeved through the end insulator and is routed to my operating position. Various electromechanical devices could obviously be used to position the wire; I keep it simple, and merely pull on the fish line to tune the antenna. The fish line is routed

Table 1
Parts List

78 ft no. 16 stranded copper wire
100 ft no. 18 enameled copper wire
Center insulator and/or balun
4-in. end insulators (2 req'd)
20-in. schedule 40 ¾-in. PVC tubing
¾-in. PVC cap (4 req'd)
50-ohm coaxial line as needed
60-lb-test nylon fish line as needed
2½ in. piece of RG-8 outer braid (2 req'd)
Bob weight (see text)
Miscellaneous hardware, weather proof coating for coils, suitable coax connectors

through the same entrance as the coaxial cable and a counterweight in the shack maintains its position. No indicating device of any kind is required other than the in-line SWR meter. Tune-up requires only seconds to go from the bottom to the top of the band. Your QTH may require other devices.

My antenna tunes from 3.525 to 3.995 MHz at an SWR of less than 2:1 across that entire range. The "footprint" is approximately 57 feet by 1 inch, and will fit on a 40 foot × 40 foot lot. Though I have not tried, I assume an 80/160-meter combination, appropriately scaled, would work as well.

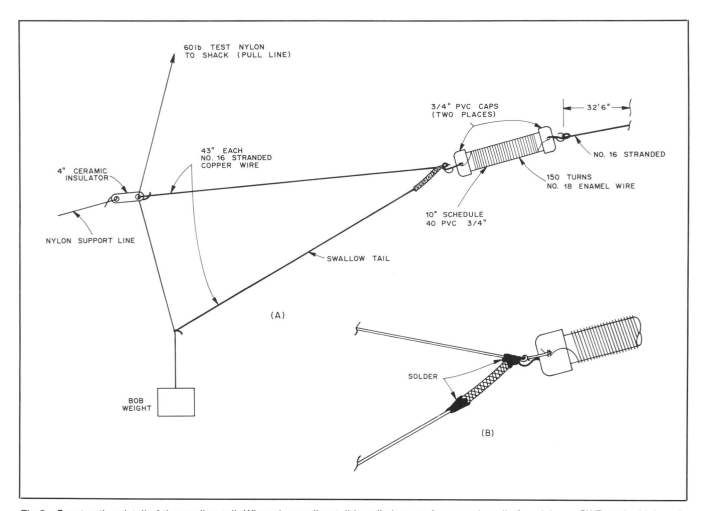

Fig 2—Construction detail of the swallow tail. When the swallow tail is pulled up, at A, prune the tails for minimum SWR at the high end of the 75-meter phone band. The nylon pull line will tune the antenna for minimum SWR. When the tail is vertical, the SWR will be less than 2:1 at 3.5 MHz. At B, 2½ in. of RG-8 coaxial braid (doubled, one inside the other). When soldering, use a heat sink at the braid center to prevent wicking. This is the swallow-tail hinge.

The Coaxial Resonator Match

By Frank Witt, AI1H
20 Chatham Rd
Andover, MA 01810

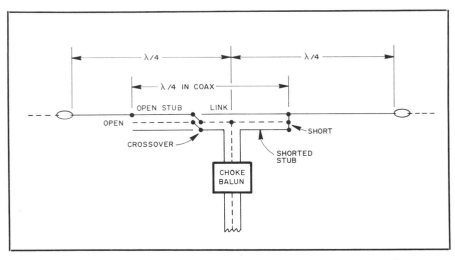

Fig 1—The coaxial-resonator-matched dipole. The ¼-λ resonator provides the elements needed to broaden the match of the dipole to the transmission line.

hrough the use of the coaxial-resonator match, one can realize an antenna which has the radiating properties of a ½-λ dipole and achieves a broadband match to a transmission line. Typically, the 2:1 SWR bandwidth of the antenna is almost three times that of a conventional ½-λ dipole. Physically, the antenna takes up the same space as a ½-λ dipole.

The coaxial-resonator match technique was introduced in April 1989 *QST*.[1] In that article, the concept was described, and details for two specific 80-meter antennas were presented. This paper presents the design equations which are required to extend the coaxial-resonator match technique to other applications. Also, results of a computer simulation of the resulting antenna are given. Further, the measure of tradeoff which exists between bandwidth and matching-network loss is highlighted.

Broadband Dipole Structure

In Fig 1 is shown a broadband antenna system made up of a coaxial transmission line, a choke balun, and a ½-λ dipole antenna that contains a coaxial-resonator match. The balun is recommended in order to prevent radiation from the feed line. The coaxial-resonator match is made up of three segments of cable that make a total length equal to ¼ λ. There is a short at one end, an open at the other, and a crossover; it is fed at a T junction. Wires are connected as shown in Fig 1 to build out the total antenna length to be ½ λ. The lengths may be chosen so that the feed line is connected to the physical center of the antenna, and variations of the design allow the feed line to be connected off center.

The coaxial-resonator match performs the same functions as its predecessors —the delta match, the T match and the gamma match, ie, that of matching a transmission line to a resonant dipole. The coaxial-resonator match has some similarity to the gamma match in that it allows connection of the shield of the coaxial feed line to the center of the dipole, and it feeds the dipole off-center. (The effective feed point of the dipole occurs at the position of the crossover.) The coaxial-resonator match has a further advantage, in that it

[1]Notes appear on page 116.

110

may be used to broadband the antenna system while it is providing an impedance match. Fig 2 clearly shows the value of the coaxial-resonator match as a means for enhancing the bandwidth over which the SWR is satisfactory.

What is the performance of this broadband antenna relative to that of a conventional ½-λ dipole? Aside from the loss in the matching network, which is quantified later, the broadband version will behave essentially the same as a dipole cut for the frequency of interest. That is, the radiation patterns for the two cases will be virtually the same. In reality, the dipole itself is not made "broadband" by the coaxial-resonator match; it is inherently already a broadband radiator, as long as energy is delivered to it. It is more accurate to say that the coaxial-resonator match provides a broadband match between the transmission line and the dipole antenna. Consequently, it would be better to think of the term "bandwidth" used in this paper as "match bandwidth," rather than "dipole bandwidth."

Later it is shown how the coaxial segment lengths should be chosen to optimize the bandwidth and efficiency. In the next two sections, the principles of the resonant transformer and off-center feed are described. Then the coaxial-resonator-match broadband dipole, which combines

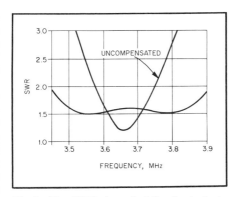

Fig 2—The SWR characteristic of a typical coaxial-resonator-match dipole. Shown for comparison is the SWR characteristic of an uncompensated dipole made from the same materials and fed with 50-Ω coax.

these two principles, is presented.

Resonant Transformer

The coaxial-resonator match is a resonant transformer made from a ¼-λ piece of coaxial cable. It is based on a technique used at VHF and UHF to realize a low-loss impedance transformation. Fig 3A shows how a ¼-λ transmission line with a short at one end and an open at the other may be used to provide transformer action over a limited band. Note that the equivalent circuit of Fig 3B consists of a transformer with a parallel-tuned circuit connected across its secondary. Such a topology has been used in the past to achieve broadbanding.[2-6] The equivalent resonator has a Q which is related to the

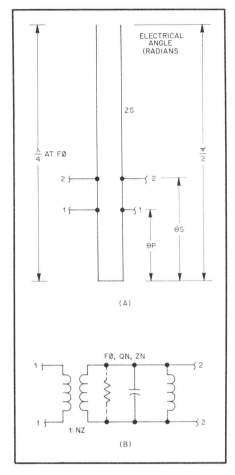

Fig 3—The ¼-λ resonator used as a transformer. Notice from the equivalent circuit of B that a simple piece of transmission line can provide the function of a tuned transformer.

loss of the coax at the frequency of interest. It is the value of the resonator Q which will determine the matching-network loss. The higher the Q, the lower the loss.

The approximate impedance transformation ratio is given by

$$NZ = \frac{\sin^2\theta S}{\sin^2\theta P}$$

where θS and θP are the electrical angles of the secondary and primary taps, respectively, measured from the shorted end of the resonator.

For the practical application of matching the coaxial cable to a dipole, the desired impedance transformation ratio may be readily obtained. The resonator-transformer impedance transformation ratio is analogous to a conventional transformer, where

$$NZ = \frac{NS^2}{NP^2}$$

where NS and NP are the number of secondary and primary turns, respectively.

Off-Center Feed

The reason for the use of coaxial cable for the resonator will be seen later. But first, the concept of off-center dipole feed is covered. The effective feed point of the dipole using the coaxial-resonator match is at the location of the crossover, which may not necessarily be at the physical center of the dipole. Hence, the effect of off-center feed must be dealt with.

In most cases, ½-λ dipoles are split and fed at the center. However, off-center feed is possible and has been used before. Two examples are the so-called Windom antenna and the dipole using the gamma match. Fig 4 shows a dipole with off-center feed. If one assumes (and this is usually a very good assumption) that the current distribution over the dipole is sinusoidal in shape, with zero current at the ends and maximum current at the center, then the radiation resistance at resonance is

modified as follows.

$$RAF = \frac{RA}{\cos^2\theta D}$$

where
  RAF = the radiation resistance at the feed point
  RA = the radiation resistance at the center of the dipole
  θD = the electrical angular distance off of center

The change in antenna feed-point impedance arising from off-center feed must be taken into account for best results.

The Coaxial-Resonator-Match Broadband Dipole

All of the necessary elements of the broadband dipole have been described. It remains to assemble them into an antenna system. The transformer and resonant circuit realized by the resonant transformer are the necessary elements for matching and broadbanding. The off-center feed concept provides the finishing touch.

Fig 5 shows the evolution of the broadband dipole for the case when the transmission line is connected to the physical center of the dipole. Now it becomes clear why coaxial cable is used for the ¼-λ resonator-transformer; interaction between the dipole and the matching network is minimized. The effective dipole feed point is located at the crossover. In effect, the match is physically located "inside" the dipole. Currents flowing on the inside of the shield of the coax are associated with the resonator; currents flowing on the outside of the shield of the coax are the usual dipole currents. Skin effect provides a degree of isolation and allows the coax to perform its dual function. The wire extensions at each end make up the remainder of the dipole, making the overall length equal to ½ λ. The design equations which follow apply to the structure of Fig 5A, which is not the coaxial-resonator match of Fig 5D, but experience has demonstrated that the simple model is sufficiently accurate for practical purposes.

The coaxial-resonator match, like the gamma match, allows one to connect the shield of the coaxial feed line to the center of the dipole. If the antenna were completely symmetrical, then the RF voltage would be zero (relative to ground) at the center and no balun would be required. In the real situation, some voltage (again referred to ground) does exist at the dipole midpoint (as it does with the gamma match), and a balun should be used. It should be a choke balun, such as a longitudinal choke made by threading several turns of coax through a ferrite toroid, or a commercial variety, such as the W2DU balun.[7]

A useful feature of an antenna using the coaxial-resonator match is that the entire antenna is at the same dc potential as the

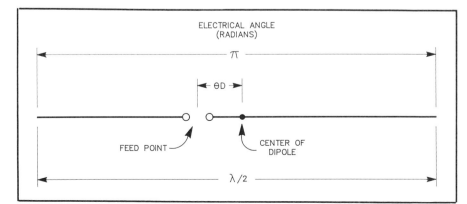

Fig 4—The dipole with off-center feed.

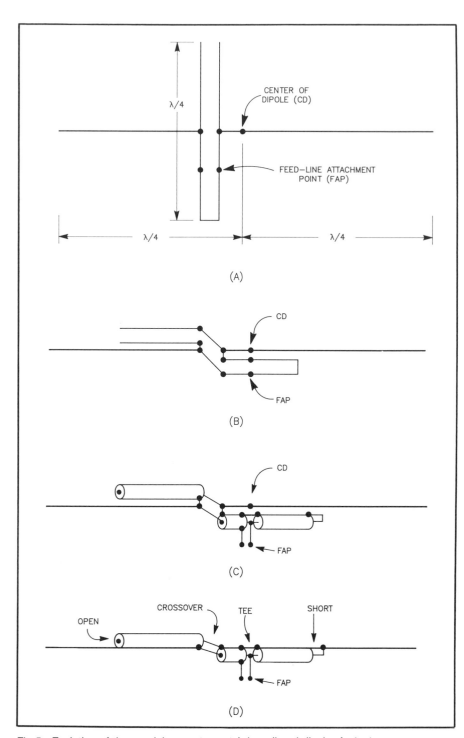

Fig 5—Evolution of the coaxial-resonator-match broadband dipole. At A, the resonant transformer is used to match the feed line to the off-center-fed dipole. At B, the match and the dipole are made collinear. At C, the balanced transmission-line resonator-transformer of A and B is replaced by a coaxial version. Since the shield of the coax can serve the role of a part of the dipole radiator, the wire adjacent to the coax match may be eliminated, as shown at D.

effect on the shape of the SWR characteristic. On the other hand, the availability of the resonator may be used to broadband the match to the antenna by providing partial cancellation of the reactance of the antenna away from resonance.

Determination of Optimum Matching-Network Parameters

This section shows how the optimum segment lengths may be determined, optimum in the sense that maximum match bandwidth is achieved. It is important to realize, however, that many other segment length combinations will yield an improved, though not necessarily an optimum, match.

In order to analyze and design the broadband antenna described in this paper, some algebra, calculus and trigonometry are required. The following sections summarize the results of this analysis. The antenna system model that makes the analysis tractable is shown in Fig 6. The matching network topology is a transformer and a parallel resonant circuit. By comparing Fig 6 with the resonant transformer of Fig 3B, one can see how the coaxial-resonator match provides the network elements necessary to achieve broadbanding. Since the resonator has a finite Q, the matching network will exhibit loss which must be taken into account in the design.

Refer to Table 1 for the definition of all the terms used below. The definitions of SM (and SMMIN), as well as bandwidth, BW (and BWMAX), are given in Fig 7. Note the W-shaped SWR-versus-frequency characteristic.

Illustrative Example

In order to add to the understanding of the analytical results that follow, a specific example is considered. This example is the 80-Meter DX Special described in April 1989 *QST*.[8] This design provides an antenna with a good match from the low end of the 80-meter band up to 3.85 MHz. The following input parameters apply.

FL = 3.5 MHz
FH = 3.85 MHz
QA = 13
RA = 60 ohms
SM = 1.6:1
Coaxial cable type: RG-213
ZT = ZS = 50 ohms
VF = 0.66
A = 0.345 dB per 100 feet

The dipole Q and radiation resistance are generally not known precisely. One may make an assumption based on past experience or published information, however, simple SWR measurements may be used to derive values for QA and RA. See Appendix 1 for methods for finding QA and RA.

Center Frequency

The center frequency is the geometric mean of the band edge frequencies.

feed-line potential, thereby avoiding charge buildup on the antenna. Hence, noise and the potential of damage from nearby lightning strikes are reduced.

The coaxial-resonator match enables one to install at the antenna a transformer that has enough bandwidth to enhance the match of the transmission line to the antenna. Hence, through the use of the coaxial-resonator match, one can obtain with wire antennas the same impedance-transformation function one can get with a T or gamma match on antennas made from metal tubing. The transformer realized with the coaxial-resonator match has associated with it a parallel resonant circuit; its impedance level may be selected to be high enough so that it has a negligible

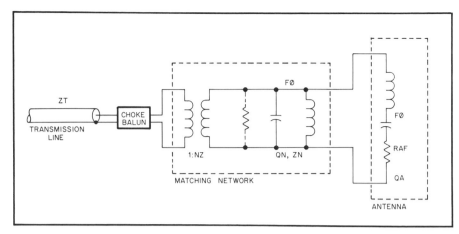

Fig 6—The broadband antenna system. In the matching network, the resonant circuit provides broadbanding by compensating for the reactance of the dipole, while the transformer adjusts the impedance level of the antenna feed to an optimum value.

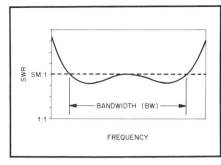

Fig 7—The definition of SM, the maximum bandwidth over the band, and BW, the bandwidth.

Table 1

Definition of Terms

Input Data:

FL	= lower band-edge frequency, MHz
FH	= upper band-edge frequency, MHz
QA	= dipole Q
RA	= dipole radiation resistance (at center), ohms
SM	= maximum SWR over band, achieved at band center and at band edges
ZT	= transmission-line characteristic impedance, ohms
ZS	= resonator characteristic impedance, ohms
VF	= velocity factor
A	= resonator attenuation per 100 feet at frequency F0, dB

Unknown Parameters:

F0	= dipole resonant frequency, MHz
BW	= SM:1 SWR bandwidth of compensated dipole, MHz
BWD	= SM:1 SWR bandwidth of uncompensated dipole, MHz
QN	= resonator Q
BWMAX	= maximum SM:1 SWR bandwidth of compensated dipole, MHz
SMMIN	= minimum allowable value SM for a given bandwidth
ZN	= matching-network impedance level, ohms
NZ	= transformer impedance ratio
LOC	= matching-network loss at band center, dB
LOE	= matching-network loss at band edges, dB
θS	= electrical angle of secondary stub measured from shorted end, radians
θP	= electrical angle of primary stub measured from shorted end, radians
θD	= electrical angle of feed measured from center of antenna, radians
LO	= length of open stub, feet
LS	= length of shorted stub, feet
LL	= length of link, feet
RAF	= dipole radiation resistance at off-center feed point, ohms

$$F0 = \sqrt{FL \times FH} = \sqrt{3.5 \times 3.85}$$
$$= 3.671 \text{ MHz}$$

Desired Bandwidth

The bandwidth of the 80-Meter DX Special coaxial-resonator match is

$$BW = FH - FL = 3.85 - 3.5$$
$$= 0.350 \text{ MHz}$$

where
   BW = bandwidth for SWR less than 1.6:1.

Bandwidth of Uncompensated Dipole

For reference, the uncompensated SM:1 SWR bandwidth of a dipole which is perfectly matched at band center (RA = ZT) is given by

$$BWD = \frac{F0 (SM - 1)}{QA \sqrt{SM}}$$

$$= \frac{3.671 \times (1.6 - 1)}{13 \sqrt{1.6}} = 0.134 \text{ MHz}$$

The uncompensated dipole in the example has a bandwidth of only 134 kHz (SWR less than 1.6:1). Thus the desired bandwidth is 350/134 = 2.6 times the uncompensated bandwidth.

Resonator Q

Before designing the matching network parameters, the Q of the coaxial ¼-λ resonator must be determined

$$QN = \frac{2.774 \text{ F0}}{A \times VF} = \frac{2.774 \times 3.671}{0.345 \times 0.66}$$
$$= 44.7$$

Although 44.7 is fairly high, it is close enough to the Q of the uncompensated dipole that it must be taken into account in determining the matching-network parameters. Note that if the coaxial cable had twice the loss, QN would drop by half.

Maximum Bandwidth

For a given dipole, allowable SWR and resonator Q, there is a maximum achievable bandwidth. Fortunately, though somewhat tedious to derive, the result is fairly simple.

$$BWMAX = \frac{F0}{QA}\sqrt{(SM + \triangle)^2 - 1}$$

where $\triangle = \frac{QA}{2QN}\left(SM - \frac{1}{SM}\right)$

In the example,

$$\triangle = \frac{13}{2 \times 44.7}\left(1.6 - \frac{1}{1.6}\right) = 0.142$$

and

$$BWMAX = \frac{3.671}{13}\sqrt{(1.6 + 0.142)^2 - 1}$$

$$= 0.403 \text{ MHz}$$

or 403 kHz. Fortunately, this is more than the 350 kHz desired bandwidth. Note that the potential bandwidth improvement factor is 403/134 = 3.0.

Minimum SWR

Similarly, for a given dipole Q, bandwidth and resonator Q, there is a minimum limit to SM. See Fig 7. This minimum limit, SMMIN, is useful when one is interested in achieving the best match over a given band. This formula is approximate, but close enough for practical purposes.

$$SMMIN = \sqrt{1 + \frac{\left(\frac{BW \times QA}{F0}\right)^2}{\left(1 + \frac{QA}{2\,QN}\right)^2}}$$

Plugging the values in for the example,

$$SMMIN = \sqrt{1 + \frac{\left(\frac{0.350 \times 13}{3.671}\right)^2}{\left(1 + \frac{13}{2 \times 44.7}\right)^2}}$$

$$= 1.47{:}1$$

By choosing a higher SM than SMMIN, the matching-network loss is reduced.

Matching Network Impedance Level

Before the coaxial-resonator-match segment lengths can be calculated, the desired impedance level, ZN, must be determined. Impedance level is simply the reactance of the equivalent tank circuit inductance (or capacitance) at resonance. The equation for ZN that follows is very general. It applies to any type of resonator used for broadbanding the dipole. In the case of the coaxial-resonator match, the value of ZN establishes the location of the crossover.

$$ZN = \frac{RAF \times SM}{QA} \times$$

$$\left[SM + \triangle \pm \sqrt{(SM + \triangle)^2 - 1 - \left(\frac{BW \times QA}{F0}\right)^2} \right]$$

where ZN is in ohms. Because of the possibility of off-center feed, this equation uses RAF, the off-center radiation resistance, rather than RA, the radiation resistance at the dipole center. It is used in the following equations as well. However, RAF cannot be calculated until the coaxial stub lengths are determined. Consequently, an iterative process is used. First, all calculations are made using RA in place of RAF. The value of RAF so derived is used to recalculate all parameters. This process is repeated until the value of RAF converges to a nonchanging value. The use of a spreadsheet for this analysis is described later. When a spreadsheet is used, this iterative process is carried out very quickly.

Incidentally, the use of RA for RAF as above yields a viable antenna design. It means that the crossover is at the physical center of the antenna and the T is located off-center, as shown in Fig 8. In the rest of the calculations associated with the 80-Meter DX Special, RA is used in place of RAF.

Note that the equation for ZN has a plus or minus sign, each of which will yield the same bandwidth. The plus sign solution should be chosen since this gives the higher of the two allowable values for ZN, and hence the lower matching-network loss. The effect of this choice is shown later when matching network loss is discussed. For the 80-Meter DX Special, when the plus sign is used and RA for RAF,

$$ZN = \frac{60 \times 1.6}{13} \times$$

$$\left[1.6 + 0.142 \pm \sqrt{(1.6 + 0.142)^2 - 1 - \left(\frac{0.350 \times 13}{3.671}\right)^2} \right]$$

$$= 18.10\ \Omega$$

If the minus sign had been used, ZN = 7.65 Ω.

Here, the formula for impedance level applies for the case when the SWR bandwidth, BW, is specified. Of course, the specified bandwidth must be less than the maximum possible bandwidth, BWMAX, given in the previous section. It is important to present this general case, since one can take advantage of the situation to obtain higher antenna system efficiency when the absolute maximum bandwidth is not required.

For the special case when the maximum bandwidth is desired, the equation for ZN simplifies to

$$ZN = \frac{RAF \times SM}{QA}\,(SM + \triangle)$$

Transformer Impedance Ratio

The expression for the transformer impedance ratio, NZ, is very general and applies to any matching network with the transformer/resonant circuit topology. In the case of the coaxial-resonator match, the value of NZ establishes the location of the T.

$$NZ = \frac{RAF \times SM \times ZN \times QN}{ZT\,(RAF + ZN \times QN)}$$

In the case of the example,

$$NZ = \frac{60 \times 1.6 \times 18.1 \times 44.7}{50\,(60 + 18.1 \times 44.7)}$$

$$= 1.79{:}1$$

Matching Network Loss

The matching-network loss will vary over the band and is worst at the band edges. At band center,

$$LOC = 10 \log\left(1 + \frac{RAF}{QN \times ZN}\right)$$

and at the edges,

$$LOE = 10 \log$$
$$\left\{ 1 + \frac{RAF}{QN \times ZN}\left[1 + \left(\frac{BW \times QA}{F0}\right)^2 \right] \right\}$$

For the 80-Meter DX Special,

$$LOC =$$
$$10 \log\left(1 + \frac{60}{44.7 \times 18.1}\right) = 0.31\ dB$$

and

$$LOE = 10 \log\left\{ 1 + \frac{60}{44.7 \times 18.1} \times \right.$$

$$\left. \left[1 + \left(\frac{0.350 \times 13}{3.671}\right)^2 \right] \right\} = 0.75\ dB$$

These results apply when the plus sign is used, but LOC = 0.70 dB and LOE = 1.60 dB when the negative sign is used. This illustrates the importance of using the plus sign in the equation for ZN, since the loss is more than doubled when the negative sign is used.

Electrical Angles of the Resonator-Transformer

In Fig 3, the approximate equivalent circuit of a ¼-λ resonator-transformer was given. Equations for the electrical angles shown in the figure are given below. The angle associated with the transformer "secondary" is given by

$$\theta S = \arcsin\sqrt{\frac{\pi \times ZN}{4\,ZS}} = \sqrt{\frac{\pi \times 18.1}{4 \times 50}}$$

$$= 0.562\ radian$$

For the transformer "primary,"

$$\theta P = \arcsin\sqrt{\frac{\pi \times ZN}{4\,ZS \times NZ}}$$

$$= \sqrt{\frac{\pi \times 18.1}{4 \times 50 \times 1.79}} = 0.410\ radian$$

Coaxial Stub Lengths

The above equations are preparation for computing the sought-after result, the lengths of the coax segments (in feet) which make up the coaxial-resonator match. See Fig 1 for physical definitions of the shorted (LS), link (LL), and open (LO) segments.

$$LS = \frac{492 \ VF \times \theta P}{\pi \times F0}$$

$$= \frac{492 \times 0.66 \times 0.410}{\pi \times 3.671} = 11.5 \text{ feet}$$

$$LL = \frac{492 \ VF \times \theta S}{\pi \times F0} - LS$$

$$= \frac{492 \times 0.66 \times 0.562}{\pi \times 3.671} - 11.5 = 4.3 \text{ ft}$$

$$LO = \frac{246 \ VF}{F0} - LS - LL$$

$$= \frac{246 \times 0.66}{3.671} - 11.5 - 4.3 = 28.4 \text{ ft}$$

Off-Center Feed Impedance

All of the above calculations assume that the crossover is located at the physical center of the dipole, as shown in Fig 8. This implies that the T is off-center. One can modify the design slightly and allow the T to be located at the dipole center. In order to do this, RAF must be calculated and the iterative process mentioned earlier must be employed. The value of RAF depends on the electrical angle of the offset.

$$\theta D = \frac{\pi \times F0 \times LL}{492} = \frac{\pi \times 3.671 \times 4.3}{492}$$

$$= 0.101 \text{ radian}$$

$$RAF = \frac{RA}{\cos^2 \theta D}$$

$$= \frac{60}{\cos^2 0.101 \text{ radians}} = 60.6 \text{ ohms}$$

The value of RAF is used in an iterative process as described earlier in the section entitled ''Matching Network Impedance Level.'' In the next section, the final values for the example with the T at the center are given.

Design Spreadsheet

I used the above equations to develop a spreadsheet for the design of broadband dipoles with the coaxial-resonator match. The values shown in the spreadsheet after the iteration is performed do not differ much from those calculated above. The values obtained were LS = 11.54 feet, LL

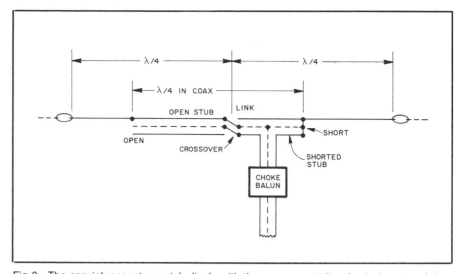

Fig 8—The coaxial-resonator-match dipole with the crossover at the physical center of the dipole. For this case, RAF = RA.

= 4.37 feet, and LO = 28.32 feet. One should use these values for constructing the antenna when the coax feed line is connected to the physical center of the dipole, as shown in Fig 1. The calculated SWR characteristic of the antenna is shown in Fig 2. Some tips for tuning the antenna are given in Appendix 2.

In the spreadsheet, I enter the loss per 100 feet at 1 MHz. The program computes the loss at F0 with the following formula.

$$A = A_{1MHz} \sqrt{F0}$$

Simulation Results

I have developed a simulation program that uses more accurate models for the dipole and coaxial resonator. The spreadsheet results were used in an initial simulation run. Then the values were manually changed to optimize the SWR-versus-frequency result. Relatively small differences between the calculated lengths and the optimized lengths were obtained with the aid of the simulation program. It is not necessary to use a simulation program to design a coaxial-resonator match. The equations given earlier provide segment lengths which are satisfactory for most practical applications.

Of interest is the electrical stress on the coaxial cable in the coaxial-resonator match application. The simulation program has a feature which determines the maximum equivalent power in the coax as well as the peak voltage at the open when the total power into the antenna is one kilowatt. The maximum equivalent power is 10.2 kilowatts. This occurs at 3.5 MHz and is at the link segment adjacent to the T. The major stress is on the center conductor where the RF current is 14.3 amperes RMS. The peak voltage at the open is 760 volts, and this occurs at 3.85 MHz. These stresses are readily handled by RG-213 cable in typical amateur service.

Since most amateurs do not accurately know what Q and radiation resistance would exist for their installation, it is desirable to know how sensitive the SWR characteristic is to those parameters. With the aid of the simulation program, a deviation study was made for Q over the range from 10 to 16 and radiation resistance ranging from 50 to 70 Ω. In the analysis, the coax segment lengths were not changed from the values of the design spreadsheet, summarized above. The results, given in Fig 9, show that the coaxial-resonator match dipole is very robust. The SWR is less than 2:1 over virtually the entire 3.5- to 3.85-MHz band for the wide range of Q and radiation-resistance values.

The analysis presented in this paper and the simulation program apply to the topology shown in Fig 5A, where the resonator-transformer match is completely isolated from the dipole antenna. This approach yields only an approximate result for the antenna shown in Fig 5D, where the match and dipole are combined in one collinear structure. However, it is clear from the experimental results previously published[9] that the design equations given here yield an adequate set of dimensions for fabricating a coaxial-resonator match dipole.

Bandwidth-Loss Tradeoff

The bandwidth improvement in this broadband dipole results from two effects—reactance compensation, and the resistive loading of the nonideal coaxial resonator. Hence, lossier coaxial cables yield larger bandwidths, but introduce more matching-network loss. This bandwidth-loss tradeoff is shown graphically in Fig 10, where the maximum bandwidth parameters described in this paper are applied.

The bandwidth improvement factor

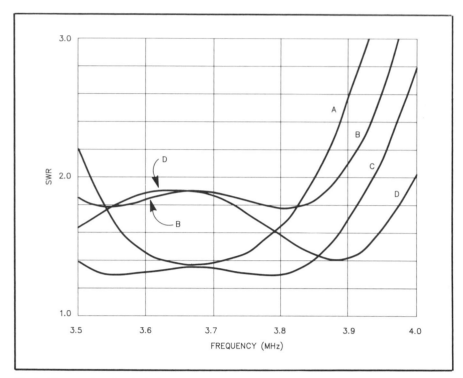

Fig 9—The results of a deviation study reveal the expected performance of the design example for a variety of conditions. The various curves were obtained with these parameters.

Curve A: Q = 16, R = 70 Ω. Curve C: Q = 10, R = 70 Ω.
Curve B: Q = 16, R = 50 Ω. Curve D: Q = 10, R = 50 Ω.

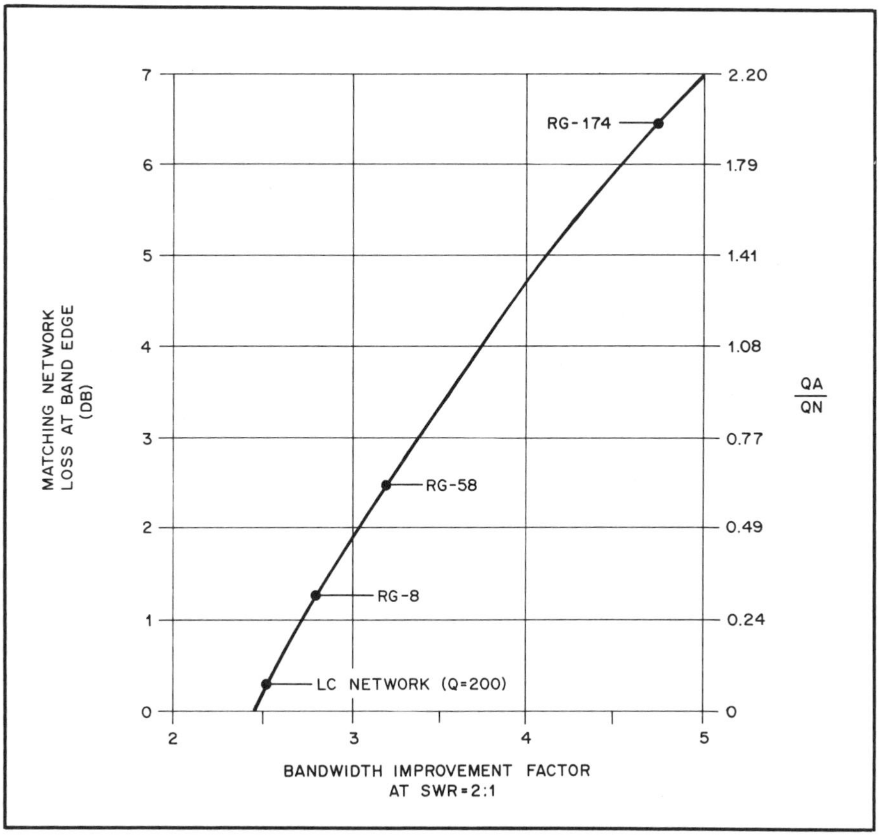

Fig 10—Tradeoff between matching-network loss and bandwidth improvement factor. The points shown for various coaxial cables apply to the case of an 80-meter dipole antenna with a Q of 13.

(BWIF) of Fig 10 is defined as

$$\text{BWIF} = \frac{\text{2:1 SWR bandwidth of broadband antenna}}{\text{2:1 SWR bandwidth of uncompensated } \frac{1}{2}\text{-}\lambda \text{ dipole}}$$

It is clear from Fig 10 that the band-edge matching-network loss depends on the type of coaxial cable used for the resonator. One must keep in mind that higher loss leads to more heating of the coaxial resonator, as well as more signal loss.

Summary

Presented here is a form of matching network for use between the transmission line and the antenna. The coaxial-resonator match, which becomes an integral part of a dipole antenna, serves not only as a matching device, but also has inherent broadbanding properties.

I am indebted to my wife, Barbara, N1DIS, for her encouragement throughout the course of this project. Also, several discussions with John Kenny, W1RR, provided inspiration during the course of the development of coaxial-resonator match, and an example of a broadband dipole shown to me by Reed Fisher, W2CQH, stimulated my analysis which led to the design equations for the optimum matching network impedance level to realize maximum SWR bandwidth.

Notes

[1]F. Witt, "The Coaxial Resonator Match and the Broadband Dipole," QST, Apr 1989, pp 22-27.
[2]J. Hall, "The Search for a Simple, Broadband 80-Meter Dipole," QST, Apr 1983, pp 22-27.
[3]Richard D. Snyder, "Broadband Antennae Employing Coaxial Transmission Line Sections," United States Patent no. 4,479,130, issued Oct 23, 1984.
[4]F. J. Witt, "Broadband Dipoles—Some New Insights," QST, Oct 1986, pp 27-37.
[5]R. C. Hansen, "Evaluation of the Snyder Dipole," IEEE Transactions on Antennas and Propagation, Vol AP-35, no. 2, Feb 1987, pp 207-210.
[6]W. Maxwell, "A Revealing Analysis of the Coaxial Dipole Antenna," Ham Radio, Aug 1976, pp 46-59.
[7]See the article by M. Walt Maxwell, W2DU, "Some Aspects of the Balun Problem," QST, Mar 1983, pp 38-40.
[8]See note 1.
[9]See note 1.

APPENDIX 1

Dipole Parameters

The uncompensated dipole may be characterized by three parameters: (1) the resonant frequency, F0, (2) the radiation resistance at resonance, RA, and (3) the Q, QA. Fig 11 shows the impedance-versus-frequency characteristic for a typical dipole. The resonant frequency is defined as the frequency where the reactance, XA, goes through zero. Notice that the radiation resistance changes with frequency, but its change is much less than the reactance change. In the approximate analysis used in this paper, the radiation resistance is assumed to be constant and equal to the radiation resistance at resonance. In the simulation program, the radiation resistance varies in the way depicted in Fig 11.

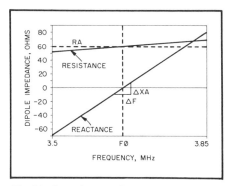

Fig 11—Impedance v frequency characteristic for a typical dipole.

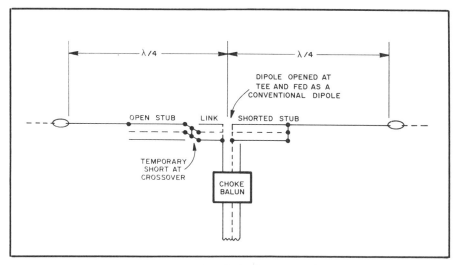

Fig 12—Test configuration for determining dipole Q, resonant frequency, and radiation resistance at resonance. Note that a short circuit is applied at the crossover.

From the information of Fig 11, the dipole Q at resonance may be determined. It depends on the slope of the impedance-versus-frequency characteristic in the following way.

$$QA = \frac{F0}{2\ RA} \times \frac{\triangle XA}{\triangle F}$$

where all of the parameters are evaluated at resonance. Hence, if one can measure the dipole impedance characteristic with an impedance bridge, then QA, RA and F0 may be determined. However, most amateurs do not have sufficiently accurate test equipment to make the measurement. An alternative method of determining the dipole parameters is given below.

Determining Uncompensated Dipole Parameters

Fortunately, one can determine QA, RA and F0 from SWR measurements of the uncompensated dipole. The dipole should be constructed from the materials which will be used in the final compensated version, and it should be placed in the same physical location as the broadbanded antenna. It should be fed in the center in the conventional way as shown in Fig 12. A choke balun should be used or the results will be unreliable. Of course, the SWR method for finding RA and QA depends on the use of an accurate SWR meter.

Refer to Fig 13. By measurements at the transmitter, one can find: (1) F0, the frequency at which the SWR is a minimum, in MHz, (2) SL, the minimum SWR, and (3) BW2, the 2:1 SWR bandwidth, in MHz. It is important that the SWR measurement be corrected for the transmission-line loss between the transmitter and the dipole. The following formula may be used to make that correction.

$$SWR_{ANT}$$
$$= \frac{1}{\tanh\left[\text{arctanh}\left(\frac{1}{SWR_{TRANS}}\right) - 0.1151TL\right]}$$

where TL is the transmission line loss in dB.

There are two cases which may occur, and it is important to recognize which one exists, since the results are different for the two cases.

Case 1:

RA smaller than the transmission line characteristic impedance, ZT

$$RA = \frac{ZT}{SL}$$

$$QA = \frac{F0}{BW2} \sqrt{\frac{5}{2}\ SL - SL^2 - 1}$$

Case 2:

RA larger than ZT

$$RA = SL \times ZT$$

$$QA = \frac{F0}{BW2 \times SL} \sqrt{\frac{5}{2}\ SL - SL^2 - 1}$$

For example, assume measurements establish that F0 = 3.66 MHz, SL = 1.20:1 and BW2 = 0.175 MHz. Since ZT = 50 Ω, there are two possibilities. For Case 1,

$$RA = \frac{50}{1.2} = 41.7\ \Omega$$

$$QA = \frac{3.66}{0.175} \sqrt{\frac{5}{2} \times 1.2 - 1.2^2 - 1} = 15.7$$

For Case 2,

$$RA = 50 \times 1.2 = 60\ \Omega$$

$$QA = \frac{3.66}{0.175 \times 1.2} \times$$
$$\sqrt{\frac{5}{2} \times 1.2 - 1.2^2 - 1} = 13.0$$

One may use experience or RX noise bridge measurements to resolve the ambiguity. The bridge measurement should be made at the frequency F0. Fortunately, not much measure-

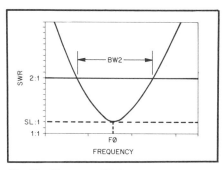

Fig 13—Measured SWR of uncompensated dipole referred to the antenna terminals.

ment accuracy is needed to resolve the ambiguity, so most RX noise bridges will do the job. A simple approach is to balance the bridge with the antenna connected. Then replace the antenna with a 42-Ω resistor and rebalance. Repeat the procedure with a 60-Ω resistor. The right case is the one where the balance settings best match the antenna measurement.

Another way to resolve the ambiguity without making precise antenna impedance measurements or corrections for transmission-line length and loss effects makes use of the Smith Chart. At the transmitter end of the transmission line, use an RX noise bridge to measure the impedance at frequency F0 and at frequencies above and below F0. These latter frequencies are not critical, but frequencies near where the SWR is 2:1 would be ideal. Normalize the impedance data (divide by ZT), and plot the values on a Smith Chart. Sketch the arc of a circle through the three points. If the center of the Smith Chart is inside the circle, you have Case 1. If it is outside the circle, you have Case 2.

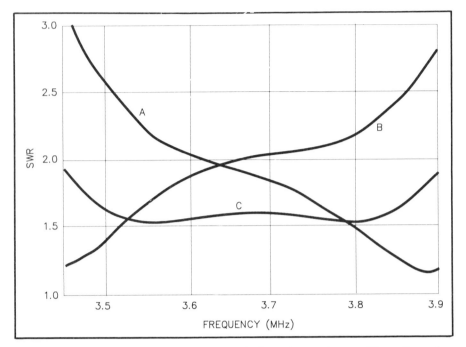

Fig 14—SWR of coaxial-resonator-matched dipole for various tuning conditions. Curve A shows the results when the dipole is resonant 100 kHz too high (about 3.5 feet too short). Curve B shows the effects when resonance is 100 kHz too low (about 3.5 feet too long). Curve C shows a properly tuned antenna. In all cases, the coaxial segment lengths are unchanged.

APPENDIX 2

Tuning the Coaxial-Resonator-Match Dipole

When a coaxial-resonator-match dipole is installed, some tuning will usually be required. The tuning consists of adjusting the overall length of the dipole by adding or subtracting wire at the ends of the antenna. This is analogous to trimming a dipole to resonance.

Fig 14 shows the SWR-versus-frequency characteristic of the design example for various dipole lengths. In all cases, the coaxial segment lengths are unchanged. Note that to improve the SWR at the high end of the band, the dipole must be shortened; to improve the SWR at the low end of the band, the dipole must be lengthened.

A Simple, Broadband 80-Meter Dipole Antenna

By Reed E. Fisher, W2CQH
2 Forum Ct
Morris Plains, NJ 07950

Much has been written about the search for a simple, coaxial-cable fed broadband 80-meter dipole antenna.[1-4] Most modern solid-state transmitters are designed to reduce power if the load SWR exceeds about 2:1—a condition that exists near the band edges with a conventional thin-wire 80-meter dipole. The antenna described in this article maximizes the bandwidth of a wire dipole by exploiting a broadbanding technique based upon modern filter theory.

Construction and Simplified Theory

Fig 1 shows construction details of the antenna. Note that the half-wave flat-top is constructed of sections of RG-58 or RG-59 coaxial cable. These sections of coaxial cable serve as quarter-wave shunt stubs which are essentially connected *in parallel* at the feed point. (Even though the center conductors and shields of the stubs connect to opposite feed-point terminals, the connection can be described as parallel.) At an electrical quarter wavelength (43 ft —inside the coax) from each side of the feed point X-Y, the center conductor is shorted to the braid of the coaxial cable. Except for the feed-point connection, the antenna is similar to the controversial double-bazooka antenna.[5-7]

The parallel stubs provide reactance compensation which has been discussed thoroughly by Maxwell (see Ref 7). Stated briefly, this scheme provides a compensating reactance of opposite sign which tends to cancel the off-resonance antenna reactance. For example, at the band center of 3.75 MHz the antenna/stub combination of Fig 1 looks like a pure resistance of approximately 73 ohms. At the band edges of 3.5 and 4.0 MHz, the reactance provided by the parallel stubs will again make the combination look like a *pure resistance* which now has been transformed to approximately 190 ohms. Suppose the reference resistance (at feed point X-Y) is changed to the *geometric mean value* of the band center and band edge resistances which is $\sqrt{73 \times 190} = 118$ ohms. Then the antenna will exhibit an SWR of 118/73 = 1.61 at 3.75 MHz, and 190/118 = 1.61 at

[1]References appear on page 123.

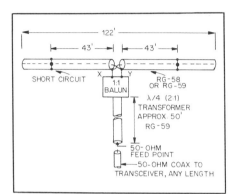

Fig 1—Details of the broadband 80-m dipole.

both 3.5 and 4.0 MHz. In order to achieve this three-frequency compensation, the X-Y feed-point resistance must be near 118 ohms, not 50 ohms. In Fig 1, the quarter-wave transformer, constructed of the 50-ft section of 75-ohm coaxial cable (RG-59) which feeds the balun, provides the required resistance transformation. Such a broadband antenna is never perfectly matched, but the SWR is always less than 2, which satisfies the solid-state transmitter. (It is believed that a similar broadbanding technique is used in a commercially available broadband dipole antenna.[8])

Fig 2 shows that the antenna covers the 80-meter band with an SWR less than 2. The antenna at W2CQH is straight and nearly horizontal with an average height of about 30 feet. The antenna feed point rests over the center of a one-story ranch house.

Adjustment

First, the stubs must be a quarter-wave long at the band center of 3.75 MHz. Good quality RG-58 or RG-59, with a velocity factor of 0.66, should be cut a bit longer than the expected 43 feet. Remember, cheap coax or foam coax may have a different velocity factor. Fig 3 illustrates how the stubs may be resonated by inserting the coil of a dip meter into a small single-turn loop at the shorted end of the stub. Cut the stub until a sharp dip is obtained at 3.75 MHz.

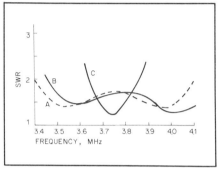

Fig 2—SWR curves for dipoles. Curve A, the theoretical curve with 50-Ω stubs and a λ/4, 75-Ω matching transformer. Curve B, measured response of the same antenna, built with RG-58 stubs and an RG-59 transformer. Curve C, measurements from a dipole without broadbanding. Measurements were made at W2CQH with the dipole horizontal at 30 feet.

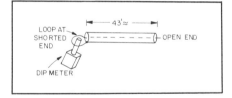

Fig 3—Adjust stub length by coupling a dip meter to a loop at one end of the coax, and trim the other end until a sharp dip is observed at 3.75 MHz.

Next, the balun/quarter-wave transformer combination should be checked. Connect a 120-ohm, 2-watt carbon (noninductive) resistor to the 1:1 balun output terminals. Connect one end of a 55-foot section of RG-59 coax to the balun input. Connect the other end to a sensitive SWR indicator, then drive the indicator with less than 2 watts of transmitter power at 3.75 MHz. Prune the cable length until the lowest SWR is obtained. The shunt (magnetizing) inductance of most commercial 1:1 baluns which employ the 3-winding ferrite transformer circuit, shown in Fig 1A of Ref 9, requires that the cable length be longer than a quarter-wavelength (43 ft) for an input match to be obtained.[9] The shunt inductance will also raise the transformer resistance to about 120 ohms instead of the $75^2/50 = 113$ ohms which would be obtained from the quarter-wave cable alone. This is a desirable condition. The W2AU balun is satisfactory when used with a cable length of about 50 feet.

The balun used with the W2CQH version of this antenna is a modified version of a coreless balun described by DeMaw.[10] The modification consisted of replacing DeMaw's bifilar winding A-D and C-F with a 6-foot section of RG-59 coaxial cable wound *inside* a 6-inch section of 2-3/8-inch OD PVC water pipe. The third (magnetizing) winding (B-E) consists of a coil of no. 18 AWG hookup wire placed between the turns of the coaxial cable. Top and bottom caps are glued to the PVC pipe to make the assembly watertight. The magnetizing inductance of this balun is about 15 μH. Fifty-two feet of RG-59 coax (including the 6 feet within the balun) connects the antenna to the 50-ohm feed point. Beware of baluns with low shunt inductance. The shunt inductance can be measured by connecting a 100-pF capacitor to the output terminals of the balun. The input connector is left open. Insert the coil of a dip meter into the loop formed by the capacitor and balun. If the combination dips below 3.75 MHz, the shunt inductance is satisfactory (greater than 15 μH).

Adjust the center frequency of the antenna flat-top to 3.75 MHz without stubs and quarter-wave transformer. To do this, first disconnect the center conductors of the stubs from feed points X-Y in Fig 1. Leave the coax braids attached. Then connect a length of 50-ohm coax from balun to transmitter and raise the antenna to its final height. Find the frequency where the SWR is lowest, then adjust the outer ends of the antenna (beyond the shorted stub section) until lowest SWR is obtained at 3.75 MHz. At W2CQH the total flat-top length is 122 feet—or 3 feet short of the textbook 468/f value of 125 feet.

Finally, connect the antenna system as shown in Fig 1. An SWR curve similar to curve A in Fig 2 should be observed.

Mathematical Theory

The input impedance of a center-fed dipole, in free space and near its half-wave resonance, is given by Jasik as[11]

$$Z_a = 73 + j[43 - Z_{0a} \cot k\ell] = R_a + jX_a$$
(Eq 1)

where
$k = \dfrac{2\pi}{\lambda}$
ℓ = antenna half-length (approx 20 meters)
λ = wavelength = 80.17 meters at the geometric band center of $f_0 = 3.74$ MHz

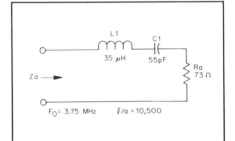

Fig 4—Lumped-constant equivalent circuit of a thin dipole antenna.

The antenna characteristic impedance in ohms is

$$Z_{0a} = 120\left[\ln\frac{2\ell}{a} - 1\right] \text{ ohms}$$
(Eq 2)

where a = antenna radius (*not* its diameter)

Inspection of Eq 1 shows that, near resonance, the antenna behaves like the series-tuned circuit shown in Fig 4. The values of the coil and capacitor may be found by the "slope Q" method.[12,13] This is done by equating the reactance slope (derivative) of the antenna at resonance with the reactance slope of a lumped L-C circuit. By doing this it can be shown that the single-loaded Q of this antenna is

$$Q_a = \frac{\omega_o}{2R_a} \cdot \frac{dX_a}{d\omega}\bigg|_{\omega = \omega_0} = \frac{\pi}{4} \cdot \frac{Z_{0a}}{R_a}$$
(Eq 3)

where
f = frequency
f_0 = center frequency = 3.74 MHz
$\omega = 2\pi f$
$\omega_0 = 2\pi f_0$

For the series-tuned lumped circuit of Fig 4

$$Q_1 = \frac{\omega_0 L1}{R_a}$$
(Eq 4)

To maximize bandwidth, the antenna Q_a

must be made small, and thus Z_{0a} must be made small. Since the antenna length is fixed, the only parameter that can be varied is the dipole radius a, which must be made large. Such a "fat" dipole will yield low Q_a and is the basis of the bulky but broadband "cage" dipole (see Ref 4). Staying with the constraint of a simple antenna, the fattest practical conductor would be RG-58 coaxial cable which has an effective radius (a) of 0.075 inch. Substituting this value into Eqs 2 and 3:

$Z_{0a} = 1074$ ohms
$Q_a = 11.5$

Then, using Eq 4 with $f_0 = 3.74$ MHz, $L_1 = 35$ μH, and $C_1 = 55$ pF

The series-resonant circuit of Fig 4 will yield an SWR versus frequency curve similar to C of Fig 2. Note that the 2:1 SWR bandwidth is only about 220 kHz. The simplest way to broadband the series resonant circuit of Fig 4 is to shunt it with a parallel resonant circuit to form the 2-stage bandpass filter shown in Fig 5. Modern filter theory provides a recipe for finding the optimum shunt reactances for L2 and C2 that will yield the maximum achievable bandwidth, obtained from the Chebyshev or equal-ripple response.[14,15]

Scaling the two-pole, 0.28-dB-ripple Chebyshev filter of Ref 15, pp 9-15, by using the techniques of Ref 14, it is found that L2 = 0.46 μH, C2 = 3950 μF and R_g = 122 ohms. Bloom, doing a computer search, found similar values for L2, C2 and R_g but the antenna was not built.[16]

Because L2 and C2 represent awkward and possibly lossy reactors, the parallel tuned circuit is simulated with a shunt quarter-wave transmission line stub of characteristic impedance Z_{0s}, as shown in Fig 6. Again using the slope Q technique, it can be shown that the single-loaded Q of this shunt stub is

$$Q_{STUB} = \frac{\pi}{4} \cdot \frac{R_g}{Z_{0s}}$$
(Eq 5)

where Z_{0s} is the characteristic impedance of the $\lambda/4$ shunt stub.

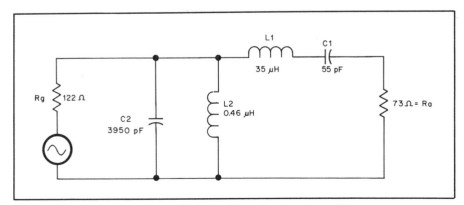

Fig 5—Diagram of a 2-pole, 0.28 dB ripple, Chebyshev band-pass filter.

The single-loaded Q of the L2-C2 tank is

$$Q_{TANK} \frac{R_g}{\omega L2} = 11.3 \qquad \text{(Eq 6)}$$

Combining Eqs 5 and 6,

$$Z_{0s} = \frac{\pi}{4} \cdot \omega_0 L2 = 8.5 \text{ ohms}$$

One way to achieve this low value of Z_{0s} is to connect six 50-ohm stubs in parallel, but this would be a bulky arrangement. Wishing to keep the antenna simple, I set Z_{0s} at 25 ohms, the value for two parallel 50-ohm stubs. To find the consequence of this drastic increase in Z_{0s}, I made a computer study of the circuit of Fig 6 which also includes a λ/4 75-ohm matching section. The study used Suncrest Software's "SNODE" CAD program running on a Hewlett-Packard 9000-216 computer.[17] All transmission-line sections were assigned the textbook-value loss of 0.8 dB/100 feet. The computed values of SWR versus frequency are plotted as curve A, Fig 2. Note that the 2:1 SWR bandwidth exceeds 700 kHz, which is satisfactory.

The study was repeated with Z_{0s} set to the original value of 8.5 ohms. To my surprise, the bandwidth *decreased* about 50 kHz, but the SWR dipped to almost 1 at 3550 and 3950 kHz. Returning to filter theory, the two-pole Chebyshev response ensures that a perfect match is achieved at two frequencies within the band. The "distorted Chebyshev" response obtained by raising Z_{0s} to 25 ohms simply means that a 1:1 SWR is never achieved, but this deficiency is of no importance. A Smith Chart plot of the computed input admittance, as seen looking to the right of points A-B of Fig 6, is shown in Fig 7. Note that the curve neatly encircles (but never touches) the origin or 0.02-siemen point corresponding to 50 ohms.

The impedance characteristics of this broadband antenna system should be virtually independent of antenna height. Fig 8 shows that the antenna resonant input resistance R_a never goes below 60 ohms for a horizontal antenna over a lossy ground.[18] Since, from Eq 3, the antenna Q_a is a function only of Z_{0a} and R_a, and Z_{0a} is a function only of wire length and radius, then Q_a (hence bandwidth) should be independent of antenna height.

However, as the antenna gets very close to the ground, it begins to look like two sections of unterminated quarter-wave transmission line with the earth providing a ground return. The characteristic impedance of this arrangement is given in Ref 15, p 29-20 as

$$Z_0 = 120 \ln \frac{4h}{d} \qquad \text{(Eq 7)}$$

where

h = wire height above ground
d = antenna diameter = 0.15 inch

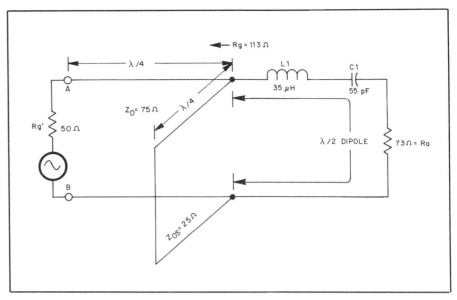

Fig 6—Broadband antenna equivalent circuit.

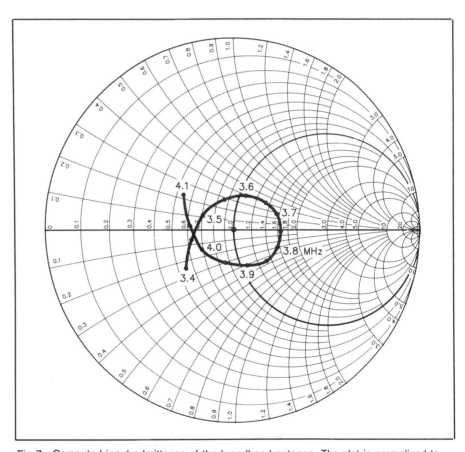

Fig 7—Computed input admittance of the broadband antenna. The plot is normalized to 0.02 siemen, corresponding to 50 ohms.

Eq 7 shows that when $Z_0 = Z_{0a} = 1074$ ohms, then h = 13 feet. Below this height, Eq 7 predominates, Q_a decreases and bandwidth increases. Of course, most of the transmitter power is now dissipated in the earth!

Lower Q Version

The antenna shown in Fig 1 was erected at KE6HU as an inverted-V dipole with the 110° apex at 60 feet and the center over a one-story ranch house. The results were disappointing. Curve A of Fig 9 shows that

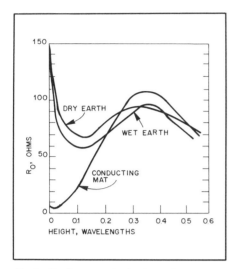

Fig 8—Resistance of a horizontal resonant antenna over earth. (*After Proctor—see Ref 18*)

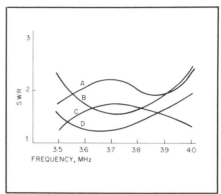

Fig 9—SWR curves for antennas at KE6HU. Curve A, the antenna shown in Fig 1; curve B, an antenna with two 50-Ω stubs and a 1.5:1 RG-59 transformer (see Fig 11). Curve C, an antenna with four 50-Ω stubs (see Fig 12) and a 1.5:1 RG-59 transformer. Curve D, an antenna with four 50-Ω stubs and a direct 50-Ω feed.

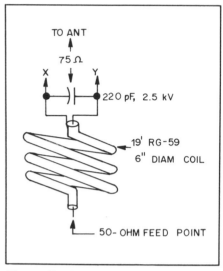

Fig 11—Details of a 1.5:1 transformer and balun for 80 meters.

though the flat frequency response remains, the SWR seldom falls below 2. Measurements made on the dipole alone, fed with 50-ohm coax and stubs disconnected, showed that the SWR reached 1 at 3.75 MHz. This indicated that 50 ohms (not 73 ohms) was the resonant resistance. This low antenna resistance is verified by Lewin whose graph of radiation resistance versus apex angle is shown in Fig 10.[19]

The simplest method of lowering the SWR to 1.5 at band center is to drive this 50-ohm antenna with a 75-ohm (not 118-ohm) source. This was done by building a 1.5:1 matching transformer consisting of a 19-foot section of RG-59 coax shunted by a 220-pF transmitting mica capacitor at the antenna. See Fig 11. A simple balun was built by coiling the coax into a 6-inch diameter, 12-turn roll. The antenna shown in Fig 1 was driven with the 1.5:1 transformer; curve B of Fig 9 shows the SWR results.

Note that the band-center SWR is now 1.5 as expected, but the band-edge SWR exceeds 2. This high band-edge SWR results from the rise in antenna Q as the radiation resistance is lowered. Assuming that the antenna characteristic impedance Z_{0a}, discussed in the theory section, remains constant (with a small change in apex angle), then the lowering of radiation resistance increases the antenna Q_a by a factor of $73/50 = 1.46$.

Johnson and Jasik show that antenna Z_{0a} (hence Q_a) can be lowered by a factor of 0.66 (but not much more) by constructing the antenna with four legs (stubs) as shown in Fig 12.[20] Bill Mumford, W2CU, had been using such a four-stub antenna for several years, in the same inverted-V dipole configuration, and reported broadband performance. The

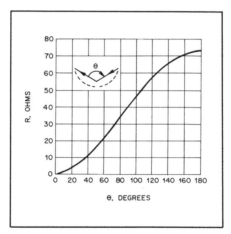

Fig 10—Radiation resistance of inverted-V dipole antenna. (*After Lewin—see Ref 19*)

antenna of Fig 12 was built of sections of RG-58 coax with four cross-connected stubs. The legs were hung as double catenaries with about a 4-foot spacing between the catenary centers.

When this lower-Q four-stub antenna was driven with the 1.5:1 transformer/balun, the response of curve C, Fig 9, was obtained. This antenna, though bulky, easily meets the SWR criterion. When the same four-stub antenna is driven directly with 50-ohm coax and a 1:1 balun (no transformer) the results are those shown in Fig 9, curve D. This is the configuration at W2CU, which also satisfies the SWR requirement. Thus, it appears that for inverted-V or other "bent" configurations, the four-stub antenna is required for acceptable SWR.

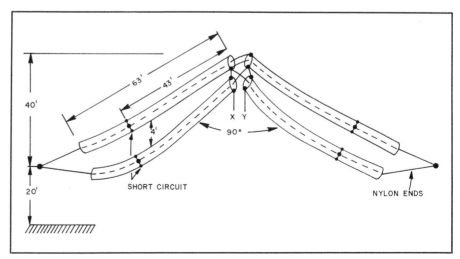

Fig 12—Details of the four-stub antenna.

Additional Topics

The shunt stubs and quarter-wave matching section introduce some *additional* loss into this antenna system which is plotted in Fig 13. Note that a minimum loss of about 1 dB is achieved at 3750 kHz where the shunt stubs are in resonance. Curve C of Fig 13 shows that reduced loss results if the shunt stubs (flat-top) are constructed of RG-8 or RG-213 and the quarter-wave matching section is made of RG-11.

Dual band (80/40-m) operation should be possible if the two shunt stubs are replaced by a single, half-wavelength (68-foot) section of RG-59 which shunts the balun at its feed point. The other end of the half-wavelength stub must be an *open circuit*. This stub would parallel the RG-59 quarter-wave transformer. The 40-meter antenna could run parallel or perpendicular to the 80-meter antenna. A trap dipole will probably perform poorly since the traps greatly reduce the bandwidth of the 80-meter dipole. It may be possible to dispense with the 75-ohm, quarter-wave matching section and feed the balun directly with 50-ohm coax if a 2:1 impedance ratio balun/transformer is constructed using the tapped transformer technique suggested in Fig 1 of Ref 9.

Both the W2CQH and KE6HU versions of this antenna seem to withstand 1 kW PEP in SSB operation with no ill effects. At 1 kW, approximately 300 V RMS exists across the antenna and shunt stubs, well within the voltage rating of solid polyethylene dielectric RG-58. At this power the current flowing in the center conductor of the quarter-wave transformer and the

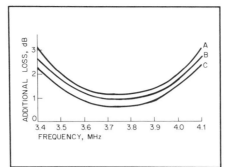

Fig 13—Computed values of additional loss using two stubs and a λ/4 matching transformer. At A, RG-58 stubs and RG-59 transformer; at B, RG-59 stubs and RG-59 transformer; at C, RG-8 stubs and RG-11 transformer.

center conductor of the shunt stubs (near the short circuits) is nearly 6 A, so some heating occurs in these regions. If doubt exists about the power-handling capability, build the flat-top from RG-213 and the transformer from RG-11 coaxial cable.

Acknowledgment

I am indebted to W. W. Mumford, W2CU, former supervisor and mentor, who initiated the quest for this broadband antenna back in 1975. Bill Mumford became a silent key on June 19, 1985. Thanks also to Gil Gray, KE6HU, who eagerly provided his time, labor and encouragement.

References

[1] J. Hall, ''The Search for a Simple, Broadband 80-Meter Dipole,'' *QST*, Apr 1983, pp 22-27.

[2] T. Conboy, ''80-Meter Broadband Antennas,'' *Ham Radio*, May 1979, pp 44-51.

[3] C. Camillo and R. Purinton, ''A Broadband Antenna for 75 Meters,'' *QST*, Jun 1955, p 11.

[4] A. Harbach, ''Broadband 80-Meter Antenna,'' *QST*, Dec 1980, pp 36-37 (presentation of a cage antenna).

[5] C. Whysall, ''The 'Double Bazooka' Antenna,'' *QST*, Jul 1968, pp 38-39.

[6] W. Vissers, ''Build a Double Bazooka,'' *73*, Aug 1977, pp 36-39.

[7] W. Maxwell, ''The Broadband Double Bazooka Antenna—How Broad is It?'' *QST*, Sep 1976, pp 29-30.

[8] R. Snyder, ''The Snyder Antenna,'' *RF Design*, Sept/Oct 1984, pp 49-50.

[9] R. Turrin, ''Application of Broad-Band Balun Transformers,'' *QST*, Apr 1969, pp 42-43.

[10] D. DeMaw, ''Simple Coreless Baluns'', *QST*, Oct 1980, p 47.

[11] H. Jasik, *Antenna Engineering Handbook* (New York: McGraw-Hill, 1961), pp 3-1, 3-3.

[12] R. Fisher, ''Broadbanding Microwave Diode Switches,'' *IEEE Trans on Microwave Theory and Techniques*, Vol MTT-13, Sep 1965, p 706.

[13] W. Joines and J. Griffin, ''On Using the Q of Transmission Lines,'' *IEEE Trans on Microwave Theory and Techniques*, Apr 1968, pp 258-259.

[14] P. Geffe, *Simplified Modern Filter Design* (New York: J. F. Rider, 1963).

[15] E. Jordan, *Reference Data for Engineers*, 7th ed. (Indianapolis: Howard W. Sams and Co, 1985), pp 9-15.

[16] A. Bloom, ''Once More With the 80-Meter Broadband Dipole,'' Technical Correspondence, *QST*, June 1985, p 42.

[17] Suncrest Software, PO Box 40, Veradale, WA 99037.

[18] R. Proctor, ''Input Impedance of Horizontal Dipole Aerials at Low Heights above the Ground,'' *Proc IRE*, Part III, May 1950, pp 188-190.

[19] L. Lewin, ''Mutual Impedance of Wire Aerials,'' *Wireless Engineer*, Dec 1951, pp 352-355.

[20] R. Johnson and H. Jasik, *Antenna Engineering Handbook, 2nd ed.* (New York: McGraw-Hill, 1984), pp 4-16.

Portable, Mobile and Emergency Antennas

Emergency Antenna for ARES/RACES Operation

By Ken Stuart, W3VVN
48 Johnson Rd
Pasadena, MD 21122

T his antenna was designed and built to fulfill the need for a completely portable antenna for emergency conditions. Unlike some which have been described previously in ham publications, this unit has many advantages. It fits into a reasonably small package but can be set up and operational in less than a minute. The antenna provides gain over a quarter-wave antenna but does not have radials. It can be set up in confined spaces, on uneven terrain, or on stairs if need be. There are no detachable parts to drop during assembly and disassembly in the field.

Antenna

The antenna chosen for this system was the familiar "J-pole," or J antenna. Several reasons prompted this choice, the most obvious of which is that radials are unnecessary. This permits its use in emergency offices, shelters and the like, without the worry of having a radial catch in someone's clothing and pulling the antenna over. The antenna can also be placed out of the way near a wall or window.

Another reason for using the J-pole is that it is actually an end-fed half-wave antenna. It has better gain in the horizontal plane than a quarter-wave vertical.

As can be seen from Fig 1 and the photo, this 2-meter J is made from two collapsible whip antennas. One is sold as a replacement for hand-held Citizens Band transceivers, and the other for the familiar TV rabbit ears. These whips are supported by a ¼-inch-thick plastic frame which attaches to the tripod mount. They are connected together at their bases and fed by clamp connections located about 2 inches up from the base of each element. Pieces of string with washers tied to their ends are attached to the top of the plastic support and are used as guides to quickly set the length of each whip.

Antenna Mount

Even though the J is a highly suitable antenna for emergency work, it is not self-supporting. I considered several schemes for a mount, such as a spring clamp or easily assembled fixed base, but none was satisfactory for a wide variety of terrain conditions. Then one day I happened to come upon a discarded camera tripod. The

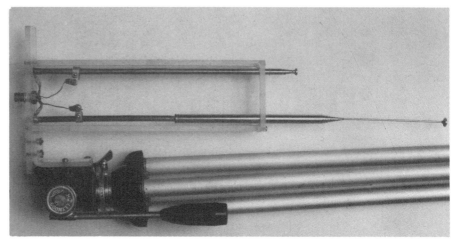

The emergency antenna, ready for transportation. (*Photo courtesy of Dean Alley, NS3V*)

base of the antenna was bolted to the camera mounting plate of the tripod in such a way that the antenna could be tilted down 180° to nest alongside the tripod legs. (Of course, the antenna base could be fitted with a ¼-24 nut so that the antenna could be detached, should the user wish to use the tripod for photographic purposes.) The use of the tripod permitted erection of the antenna on any type of terrain. By adjusting the legs, the height of the antenna can be varied to fit inside a room, or it can be fully extended for better propagation.

Construction

As can be seen in Fig 1, the whip antennas are mounted in a frame made from ¼-inch-thick by 1-inch-wide plastic. Plexiglas® is ideal for this, but other materials can be used as long as they are nonconductive and not affected by moisture. The individual sections of the frame are drilled and tapped to accept no. 2-56 screws for structural strength. The antennas are mounted on the bottom section with their mounting screws and extend through clearance holes drilled in the top section. Solder lugs are mounted under each whip and connected together with tinned copper braid (obtainable from a 2-inch piece of RG-58 coax). The braid must not touch the BNC connector.

The connections to the whips are clamps made from thin strips of copper. These can be easily loosened in order to do the initial tuning of the J antenna. Ordinary stranded hookup wire is used to make the connections to the BNC connector. It makes no difference whether the connector center conductor is connected to the long whip and the outer conductor connected to the short one, or vice versa.

An additional piece of Plexiglas is heated and bent into a right angle, then screwed to the bottom plate of the frame. The upright portion of this section is drilled to accept a screw to mount the antenna to the tripod. The captive screw in the tripod head, which secures the camera, should be removed and a no. 10-32 screw and nut used instead. This additional piece of Plexiglas is necessary in order to provide clearance between the antenna and the tripod when the antenna is rotated down for storage and transport.

Operation

Before initial use, the antenna must be adjusted for minimum SWR at the frequency of interest. This is done by extending the elements to about 19 and 57 inches, and then adjusting the antenna feed points and element lengths to obtain the minimum reading on an SWR meter. When the ele-

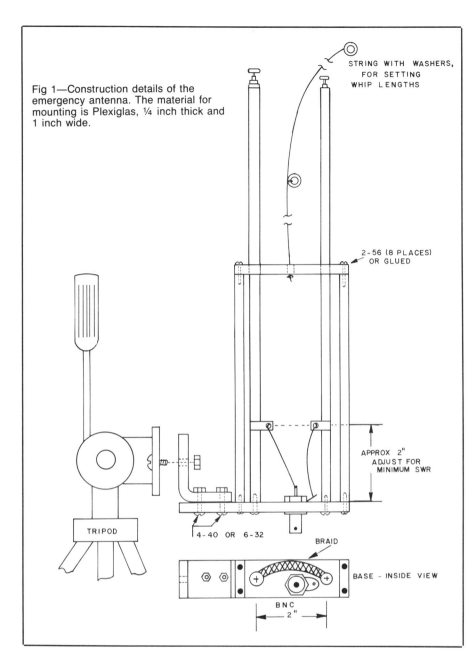

Fig 1—Construction details of the emergency antenna. The material for mounting is Plexiglas, ¼ inch thick and 1 inch wide.

STRING WITH WASHERS, FOR SETTING WHIP LENGTHS

2-56 (8 PLACES) OR GLUED

APPROX 2" ADJUST FOR MINIMUM SWR

TRIPOD

4-40 OR 6-32

BRAID

BASE - INSIDE VIEW

BNC 2"

ment lengths are determined, the string lengths are adjusted so that the washers are located exactly at the tips of the elements. When the antenna is set up again in the future, each element is extended by holding its washer at its tip and extending the element until the string is taut—no measurement is needed!

Erection of the antenna at the site is simplicity itself. The legs of the tripod are extended to the desired length and clamped, and the antenna is then rotated to the upright position. Each of the two elements of the J antenna is extended to its correct operating length as indicated by the two strings and washers, and the coax is connected to the antenna jack. Changing overall antenna height, and leveling on rough terrain, is simply a matter of adjusting the tripod legs.

This antenna has given excellent service for years in many emergency and public service situations. It has been erected on roofs, on the tops of parked cars and vans, on a flight of steps, in emergency operations centers at Red Cross headquarters and hospitals, and even in the sand at one of our local bayside beaches during a walk-a-thon. Also, it is a great attention-getter at public service events, and thereby helps to advertise Amateur Radio.

Table 1

Parts List
The replacement antenna elements and BNC connector are available from Radio Shack® and have the stock numbers indicated.

Element	RS Number
Long whip (72")	270-1408
Short whip (30")	270-1401
BNC connector	278-104

Portable 2-Meter Antenna

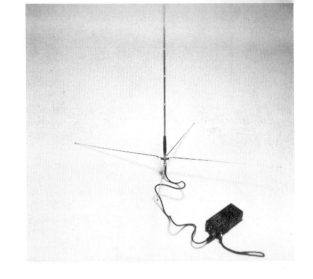

By Michael C. Crowe, VE7MCC

2575 W 20th Ave
Vancouver, BC V6L 1G9

This portable antenna is designed for long-range communications on 2 meters while using small mobile or hand-held transceivers. It is light, simple, rugged and convenient to carry and erect. Although based on common principles and breaking no new technical ground, it has been very effective in the field.

It is best described as a telescoping 5/8-wavelength vertical whip above a ground plane of four folding telescoping quarter-wave radial whips. The theory came from material in older editions of *The ARRL Handbook*; my only contribution is in the simple folding construction to make it portable.

The ubiquitous "rubber duck" is convenient but has severe shortcomings when terrain and distance intervene. In trying different antenna configurations on my hand-held transceiver while searching for a VHF antenna for use in hiking and camping and also for emergency operations, several things became apparent. I noticed that, especially in transmitting, a ground plane gave considerable assistance to most omnidirectional vertical antennas and seemed also to improve the SWR slightly. The height of at least a quarter wave, or even better, half or 5/8 wave, helped transmissions when compared to a compact helical or "rubber duck" antenna. It came down to a decision between loaded 5/8- or 1/2-wave verticals. The 5/8 tested out better, and my efforts then concentrated on building such a ground-plane antenna in a form practical for field work.

After trying various configurations, I came upon the idea to incorporate a ground plane of telescoping whips structurally into the antenna base by affixing them to a sturdy washer mounted there. After that, the actual prototype construction took only an hour, utilizing a second-hand telescoping 5/8-wave whip. The SWR across the 2-meter band is less than 1.5 to 1 (it can be adjusted to no more than 2.0 to 1 on the marine band, so is useful on boats, also). In use, the SWR varies with antenna tilt (see later text), closeness to vegetation, dampness, etc, at any one frequency. So if you are not satisfied with the SWR, try different conditions.

Construction

As Figs 1 and 2 show, construction centers on an ordinary steel washer, about 2½ mm thick and 4 cm in diameter. Four

Fig 1—The ground-plane portion of the portable 2-meter antenna. Construction centers on an ordinary steel washer, about 2½ mm thick and 4 cm in diameter.

equally spaced holes (sufficient to take a 3-mm bolt) are drilled around its circumference. The washer is then mounted on a female-to-female BNC connector (mine is an Amphenol part) designed for panel mounting so that its body is threaded. Tighten the washer around the body of the connector using a large lock washer and the two nuts supplied (if none are, scrounge a couple). It is also possible to find BNC connectors with pre-drilled flanges on their bodies for chassis mounting, but these appear too delicate for portable use.

Several suppliers sell replacement telescoping antennas for portable AM radios which have a flattened mounting portion at one end. Most already have a small hole drilled in the flat portion and an elbow where the whip itself is screwed. Find four of these (they are often sold by Radio Shack, among others, in packages of four for about $10; I wonder who otherwise buys packages of them...). One is mounted on each hole in the large washer, with

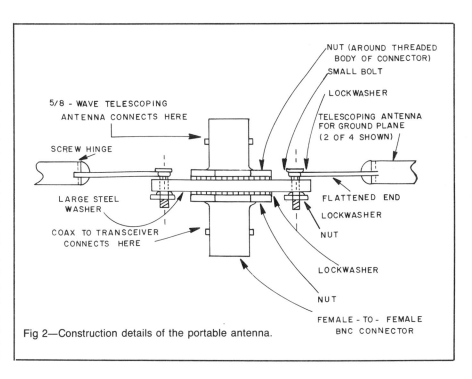

Fig 2—Construction details of the portable antenna.

5/8 - WAVE TELESCOPING
ANTENNA CONNECTS HERE

SCREW HINGE

LARGE STEEL WASHER

COAX TO TRANSCEIVER CONNECTS HERE

NUT (AROUND THREADED BODY OF CONNECTOR)

SMALL BOLT

LOCKWASHER

TELESCOPING ANTENNA FOR GROUND PLANE (2 OF 4 SHOWN)

FLATTENED END

LOCKWASHER

NUT

LOCKWASHER

NUT

FEMALE - TO - FEMALE BNC CONNECTOR

a lock washer on each side and other small washers as needed, to be tightly mounted by the 3-mm bolts and nuts (or such other size as is convenient).

Attach a 5/8-wave base-loaded telescoping whip to the upper BNC connector, and a length of coax cable (mine is 1 meter; somewhat longer would be better) with male BNC connectors at each end to the lower connector. The other coax end then connects to the BNC connector on your hand-held or mobile 2-meter transceiver. That is all there is to it.

Operation

The antenna is basically omnidirectional, and the best results (please do not ask why) are achieved by fully extending all whips and tilting the vertical whip slightly *away* from the desired direction of transmission. Having the horizontal whips bunched on one side further favors transmitting in that direction. For reception, the ground-plane whips can be left telescoped without significant sensitivity loss, but for maximum transmitting efficiency, their extension is essential.

This forms an antenna 130 cm high and 95 cm wide when fully extended, but folds and telescopes to a package only 5 cm in diameter by 24 cm long. It weighs less than half a kilogram, and easily packs in its folded state by wrapping an elastic band around it. Over a rubber duck, the antenna has a theoretical advantage of about 8 dB (I do not have the equipment to measure this, but my results support it). The addition of extra horizontal whips in the ground-plane array seems to have no discernible effect, nor does the use of longer horizontal whips. Do not worry about droop; even if bent at 45 degrees, they are still effective.

Its first use was in the mountains near Vancouver in the fall of 1987. In a clearing (definitely avoid overhead vegetation) in a valley behind a 2000-meter mountain ridge, I was able to make contacts through two repeaters on apartment buildings in Vancouver, about 140 straight-line kilometers away. My signal was adequate to make telephone calls on the autopatch on one of them. Contacts indicated my signal strength to be sufficient for only partial quieting but fully intelligible.

More recently, during an emergency-preparedness exercise, I was able to access repeaters that more powerful mobile transceivers in vehicles with mag-mounted whips could not. Several more of these antennas have been built by local hams in the Vancouver area.

The Half-Wave Handie Antenna

By Ken L. Stuart, W3VVN
48 Johnson Rd
Pasadena, MD 21122

The garden variety 5/8-λ whip is not the ultimate choice for use on a hand-held transceiver. For effective operation, a 5/8-λ antenna must work against a good ground plane, which a hand-held radio certainly is not. Also, the additional 1/8 λ over a half-wave actually produces an out-of-phase component which creates additional lobes and gives a vertical angle to the radiation pattern. (See the treatise on 5/8-λ antennas in *The ARRL Antenna Compendium Volume 1*.[1]) The main advantage of the half-wave is the gain over a 1/4-λ whip, and it is easy to load.

The Half-Wave Handie Antenna is an improvement over the 5/8-λ whip. Actually, the half-wave is the perfect antenna for producing the desired doughnut-shaped radiation pattern. And by end feeding it, the need for a ground system is essentially eliminated. It is also slightly shorter in length than a 5/8-λ, which reduces strain on the hand-held transceiver coax connector. A commercially manufactured version of this antenna has been available for several years and has become very popular (the AEA Hot Rod).

Construction

Being of Scottish descent, I decided that I could build my own version of one of these antennas for less money, so I took several pieces of plastic and wire, a BNC connector, a trimmer capacitor and a Radio Shack replacement whip antenna element and fashioned my own junk-box special. It has operated on my hand-held transceiver for several years, and produces excellent results. In fact, in tests of this antenna, a 1/4-λ whip and a 5/8-λ whip, this antenna "whipped" them all.

This antenna works on the principle that a ½-λ dipole is an excellent radiating element, and by feeding the element at one end where the impedance is highest, we can eliminate the need for a ground plane. One common utilization of this principle is the popular "J-pole" antenna, where the radiating element is actually a ½-λ element fed at the end by a 1/4-λ matching stub. (The J could be used on a hand-held transceiver, and would work very well, but an antenna length of almost 5 feet atop a hand-held unit would be impractical.)

Electrically, the ½-λ antenna consists of a 39-inch collapsible whip with a parallel-tuned matching network connected

[1]D. K. Reynolds, "The 5/8-Wavelength Antenna Mystique," *The ARRL Antenna Compendium, Volume 1*, pp 101-106.

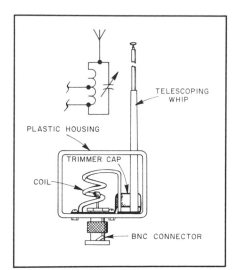

Fig 1—Construction details of the Half-Wave Handie Antenna. The enclosure is formed from a strip of thick plastic.
Whip—Radio Shack 15-232 (page 101 in the 1988 catalog).
Trimmer capacitor—6 to 50 pF, Radio Shack 272-1340 or Arco 404.
BNC connector—UG-260 or UG-88.
Coil—3 turns no. 18 bus wire, ½-inch diameter, ½-inch length. Tap approximately one turn from ground. (Adjust for minimum SWR.)

between it and the hand-held transceiver ground. RF from the radio is applied to a tap on the coil of the matching network. Tuning of the antenna is accomplished by moving the tap feed point and adjusting the trimmer capacitor to achieve a minimum SWR.

Figs 1 and 2 will aid in understanding the following construction steps. Mechanically, I formed a strip of thick plastic into a four-sided box shape with a pair of opposite sides open. (See August 1988 *QST* for detailed information on building enclosures.[2]) A piece of copper was drilled to accept the BNC-connector cable-clamp nut, which was dropped into the hole and its flange soldered to the copper. Four holes were drilled in the copper and the plastic housing for mounting the BNC. Mounting of the whip was done by drilling a snug clearance hole in the top of the housing and a small hole in the bottom to accept the whip mounting screw. The coil and trimmer capacitor were mounted simply by soldering them to the new BNC connector

[2]D. Kennedy, "Build It Yourself—With Plastic," *QST*, August 1988, pp 30-34.

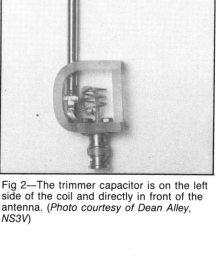

Fig 2—The trimmer capacitor is on the left side of the coil and directly in front of the antenna. (*Photo courtesy of Dean Alley, NS3V*)

copper flange and a solder lug at the bottom of the whip. A piece of bus wire connects the center pin of the BNC to the tap point on the coil. It is soldered to the center pin of the BNC, the pin is inserted in the connector, and the pin and wire potted in place with epoxy cement or hot glue.

Operation

Although the antenna is properly matched only when the whip is extended to full length, I have used it on low power with the whip collapsed with no ill effects to the radio. I don't use high power with the whip collapsed, however. Hand-held transceivers as a breed are pretty bullet-proof, but I prefer not to abuse mine by trying to work into a mismatch at a full power setting.

This antenna has proved itself time and again. It has seen use in numerous RACES and ARES drills and emergencies, and in public service events. It lives on my radio at all times. It is replaced by a duck only when communications will be needed over a very limited geographic area, when it will be more convenient to wear the hand-held transceiver on my belt and use a speaker mike.

Controlled Current Distribution Antennas

The Controlled Current Distribution (CCD) Antenna

By Stanley Kaplan, WB9RQR
11541 N Laguna Dr
Mequon, WI 53092

E. Joseph Bauer, W9WQ
N5415 Crystal Springs Ct
Fredonia, WI 53021

Visit any Field Day site and you know what to expect in the way of antennas silhouetted against the sky. Beam or vertical, long wire or dipole, all will exhibit one characteristic in common —long, uniform metallic conductors made of wire or tubing. On the other hand, visitors to the Ozaukee Radio Club's Field Day Site during any of the last five years who looked closely at the antennas would have noticed something unusual. At least two of the wire dipole antennas at our multiple-transmitter site had little "nodes." These were spaced regularly along the antenna, beginning a short distance from the middle where the open wire feeder was connected to each leg of the dipole. Each node was a small insulator bridged by a capacitor. The dipole itself was not at all a standard dipole. Rather, it was a controlled current distribution (CCD) antenna, designed and built by the authors, based upon articles that have appeared on the subject and on several consulting sessions with Harry Mills, W4FD, the originator of the antenna.[1,2]

Our initial experiments with this unusual design proved that the antenna was an excellent performer. We modeled a computer program to do the calculations necessary to build and test variations of the dipole, as well as vertical and end-fed Zepp versions. This paper presents some of the theory behind the antenna, the advantages and disadvantages, the results of our computer modeling, and some practical hints for building the CCD. We believe that our original series of articles (from which this paper is derived) is the first published anywhere to present accurate values of length and capacitance which can be used to build practical versions of the antenna for several bands.[3]

A conventional dipole is configured so that the two horizontal legs of the antenna are each ¼ λ long, for a total of ½ λ (Fig 1). Our CCD version uses a stretch factor of two, meaning that each leg is ½ λ and the whole dipole is a full wavelength long (Fig 2). We insert capacitors in series with the wire of the dipole, spaced regularly

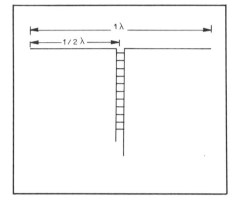

Fig 1—The conventional half-wave dipole. Each leg is ¼ λ, for a total of ½ λ.

along its length, as shown in Fig 3. All of the capacitors in a given CCD antenna are of the same value. However, their value, number and spacing can vary widely, depending upon how the antenna is designed. The band of use will dictate the total length of wire, but the number of capacitors and spacing can be varied depending upon what is in your junk box. In general, the more capacitors, the better.

Why the capacitors? You will recall that a piece of wire suspended in space has inductance, and the value of inductance (which can easily be calculated) depends upon the length of the wire and its diameter. A standard dipole has a value of inductance (and capacitance) which makes that antenna resonant at the chosen frequency. However, in our antenna, which is "stretched" to double the usual length, there is actually more than double the usual inductance. Too much inductance is present to resonate at our chosen frequency. Therefore, we add capacitance. However, we must add only half the capacitance needed to cancel the inductance of the wire; the other half must be left so that the antenna will be resonant at the chosen frequency. The result is an antenna that is physically a full wavelength, but which represents an electrical ½ λ. And there are several important advantages to

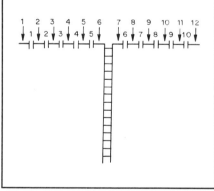

Fig 2—The CCD dipole we describe in this article has a "stretch factor" of two, making it one full wavelength from end to end. For simplicity, the capacitors that are characteristic of the CCD are not shown in this diagram. They are included in Fig 3.

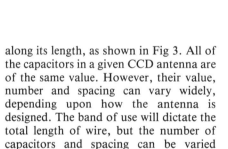

Fig 3—The CCD dipole consists of a total of N sections (12 are shown in this example at the arrows), and N − 2 (10) capacitors. Each *half* of the dipole has N/2 sections and (N/2) − 1 capacitors.

this! Read on, and they will become self-evident.

Fig 4 shows the current distribution of a standard ½-λ dipole. Current is highest in the center, as shown by the graph just

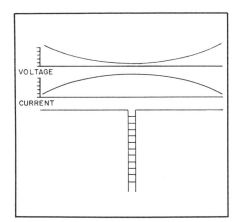

Fig 4—Voltage and current distribution in the conventional dipole. Current (bottom graph) is highest in the center and drops off as the ends are approached. The amount of radiation from small sections of the antenna follows the current distribution—highest at the center and lowest at each end. Voltage is exactly the opposite from current—highest at the ends and lowest at the feed point in the center. High-voltage arcing is most likely at the ends of the antenna, making it necessary to use good end insulators and spacing well away from trees and buildings.

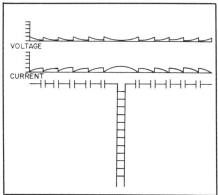

Fig 5—Voltage and current distribution in the CCD dipole. Both voltage and current peaks are much lower than in a conventional dipole, because the capacitors break up the single large current peak and the two large voltage peaks into a series of many smaller ones. Voltage peaks are effectively spread over the entire antenna, and arcing is reduced or totally eliminated. More important, antenna radiation is also spread over the entire wire length, when compared with a conventional dipole. Coupled with the fact that this antenna is a full wavelength long, it is clear that the CCD is a much more effective radiator.

above the antenna. Voltage is low at the center, but high at either end; that is why we must use insulators at the ends, and carefully mount them away from trees and buildings.

On the other hand, the CCD version of the dipole (Fig 5) shows several unique properties with respect to current and voltage. The capacitors change the current and voltage distribution, making variations in both parameters less than in a standard ½-λ dipole fed with the same input power. (This also leads to another plus—a higher feed-point impedance. Our preliminary estimate is that the feed-point impedance is increased by three to four times.)

Since voltage is not high on the CCD antenna, it can be mounted so that it runs through trees or bushes without fear of arcing! That represents a terrific advantage for Field Day, as well as for the home QTH.

Another advantage is based upon how much signal the CCD antenna puts into free space. In any dipole antenna, maximum "pumping" of the signal into space occurs where the current is highest. Thus, Fig 4 shows us that the conventional dipole does most of its radiating at the center, and this radiation drops off as we move toward the ends. In the CCD antenna, there are many current peaks along the wire length (Fig 5), and any given section of wire (a "section" is the wire between two capacitors) is about as effective in pumping out radiation as any other section. Thus, *the entire length of wire is utilized to radiate signal,* as opposed to the drop-off in radi-

ation as the ends are approached, which is seen in the conventional dipole. The capacitors reduce the magnitude of the single current peak seen in a conventional dipole and induce more peaks, thus effectively flattening out the current distribution and

spreading it out. In summary, the CCD antenna utilizes the *whole* wire to radiate your signal, and it is a whole wavelength long. The advantages and disadvantages of the CCD antenna are listed in Table 1.

Calculations

If you have a computer with BASIC, Program 1 will do the dipole calculations for you. If you have an MS-DOS computer, a complied PASCAL version gives a more extensive series of calculations than the BASIC program.[4]

To prepare a second program to calculate long-wire and vertical versions of the CCD, use Program 1 and make the following changes.

```
100 REM This version is for VERTICAL/
    LONG-WIRE CCDs
490 FOR SECTION = 50 TO 1 STEP - 1
510 CAP(I) = CTOT(I)*(SECTION - 1)
```

The rationale behind these changes is given later.

If you have no computer, do not despair. We give you enough data here to get an 80-meter CCD dipole going even if you don't have a computer with BASIC.

The formulas used are

$$L = 0.00508 \, i \, [\ln(4i/d) - 0.75] \text{ or}$$
$$L = 0.00508 \, i \, [2.303 \log_{10}(4i/d) - 0.75]$$

where

L = inductance, μH
i = length of antenna, inches
d = diameter of wire, inches

Note that we suspected, and our correspondence with W4FD confirmed, that garbled typesetting caused his original article to

---

Table 1

Advantages and Disadvantages of CCD Antennas

The CCD has many advantages over more familiar antennas:
It exhibits greater gain than a conventional dipole.
It reduces or eliminates end effects.
It exhibits higher antenna resistance than a standard antenna.
The vertical version of the CCD has a lower radiation angle than a dipole—it is a good DX antenna.
Another Field Day advantage—it works well at only 8 feet above ground. Furthermore, changes in height produce progressively less relative changes in antenna resistance as the number of capacitors and overall length are increased.
No phase-inverting stubs are needed.
You can scale it up or down to fit the available space.
It exhibits very good broadband characteristics.
Cut it for the low end of the band and you should be able to work both phone and CW.
It is less lossy than a standard dipole.
It is effectively adapted to quads, deltas, loops and other antenna forms. Harmonic operation becomes more and more effective as the number of capacitors is increased and the sections are proportionally shortened. Thus, an antenna with many sections is likely to be very effective on several bands.
Both lobes and nulls are very strong, making the CCD much more directional than a conventional dipole. (Too bad that low-band versions are not easily rotated!)

So what are the disadvantages? There are a few, of course:
It costs somewhat more to build, owing to the need for capacitors, as well as twice the wire of a conventional antenna.
Construction takes a longer time than with conventional dipoles, since there are two solder joints to make at each capacitor, and one must use reasonable care in measurement of section length.
Erection space is larger, since the CCD (stretch factor = 2) is a full wavelength long.

present the method for calculation of inductance incorrectly.[5]

The length of the antenna, i, is found by our old standard.

$$i = 39.37 \times 300/f \text{ inches}$$
where f = frequency in MHz

Note that the usual shortening factor (5%) to compensate for end effects is *not* to be used in these calculations.

Capacitance in pF needed to resonate our CCD is found by

$$C = \frac{10^6}{2\pi^2 f^2 L}$$

A typical output from Program 1 is shown in Table 2. The output is calculated for the 80-meter band. Note that the first three lines of output list the frequency (3.5 MHz), the total length of the antenna in inches (round to 3374.6) and in feet (round to 281). Thereafter, the chart lists the number of sections, capacitor values (in picofarads) for no. 20 through no. 12 wire, and the length of each section in inches and in feet.

Thus, if you have 24 (number of sections minus two) 500-pF capacitors in your junk box, you can make a CCD with 26 sections, total length = 281 feet, with each section 129.8 inches long made of no. 14 AWG wire. (Multiply 129.8 × 26 and you should come up with close to 3375.8 inches). Cut your wire and solder in the capacitors so that from the center of one capacitor to the center of the next is 10 feet, 10 inches. Realize that while we often do our calculations to several decimal places, we can rarely measure wire to that accuracy. And even if we could, wire stretches with time. Therefore, close is good enough.

That is how the calculations are done. Given those tools, the next step is to explore your junk box or the upcoming swapfests for a fistful of identical capacitors. Try to find units with a 5% tolerance. Fit your design to those components.

The best of all possible worlds is to feed the CCD with open-wire line originating from a tuner. On the other hand, there is no reason why it cannot be fed with 52-ohm coax, terminated at the antenna with a 4:1 balun (preliminary results of our testing of a 2-meter CCD vertical indicate that a 4:1 balun is about right). In either case, a tuner is desirable, and is a must for multiband use.

Vertical, Long-Wire and Zepp Versions

You will notice in the dipole CCD, Fig 3, that there are a total of N sections (12 are shown in the example), and N − 2 capacitors (10 in this case). The calculations we did for the dipole version resulted in a value of capacitance to cancel the entire inductance of the wire; we doubled this to come up with a value that would cancel only half this inductance so as to resonate at the desired frequency. We then multiplied this value of capacitance by the number of

Table 2

Sample Program Output of Program 1 for 3.5 MHz

Select the capacitor value and wire size you wish to use, then read the number of sections and length of each from this list.

3.500 MHz   DIPOLE CCD
3374.571 inches
281.214 feet

# OF SEC	#20 AWG pF	#18 AWG pF	#16 AWG pF	#14 AWG pF	#12 AWG pF	SECTIONS FEET	SECTIONS INCHES
---	---	---	---	---	---	----	-------
50	949	967	986	1006	1027	5.62	67.49
48	909	927	945	964	984	5.86	70.30
46	870	887	904	922	941	6.11	73.36
44	830	846	863	880	899	6.39	76.69
42	791	806	822	839	856	6.70	80.35
40	751	766	781	797	813	7.03	84.36
38	712	725	740	755	770	7.40	88.80
36	672	685	699	713	727	7.81	93.74
34	633	645	658	671	685	8.27	99.25
32	593	605	616	629	642	8.79	105.46
30	554	564	575	587	599	9.37	112.49
28	514	524	534	545	556	10.04	120.52
26	474	484	493	503	513	10.82	129.79
24	435	443	452	461	471	11.72	140.61
22	395	403	411	419	428	12.78	153.39
20	356	363	370	377	385	14.06	168.73
18	316	322	329	335	342	15.62	187.48
16	277	282	288	293	300	17.58	210.91
14	237	242	247	252	257	20.09	241.04
12	198	202	205	210	214	23.43	281.21
10	158	161	164	168	171	28.12	337.46
8	119	121	123	126	128	35.15	421.82
6	79	81	82	84	86	46.87	562.43
4	40	40	41	42	43	70.30	843.64
2	0	0	0	0	0	140.61	1687.29

capacitors to be used (N − 2). If that confuses you, don't forget that when we place equal-value capacitors in series, as is done in our CCD dipole, the total capacitance is equal to the value of one capacitor divided by the number of capacitors.

In the case of a long-wire CCD, our feed point is not located in the center, but rather at one end. To convert from the dipole configuration, we pull out the feed line and add one more capacitor in its place, as shown

in Fig 6. We then reconnect the feed line at one end of the antenna (note that some sort of impedance matching will invariably be required with solid-state transmitters). If we rotate the antenna 90°, it becomes a vertical (Fig 7). If we feed it with open-wire feed, it becomes a Zepp antenna (so named because it was used as an antenna trailing from Zeppelins, back when they were in style). Fed with open wire, coax, or hooked

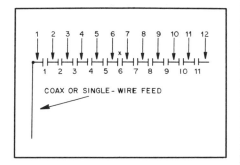

Fig 6—To convert a dipole CCD to a long wire, the center feed line is removed and one more capacitor is added (at X). The feed line is reconnected at one end or the other. The number of sections (N) has not changed; there are still 12, as there were in the dipole version (Fig 3). However, the number of capacitors has increased by one, to N − 1.

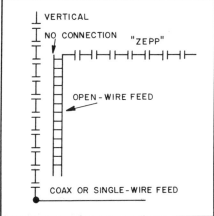

Fig 7—A vertical CCD is nothing more than the long-wire version rotated 90°. The end-fed Zepp is a long wire, fed by open-wire line with one member of the pair unterminated at the antenna end.

directly to the transmitter makes no difference; the basic design is the same. Pull out the center feed line of the dipole, throw in one more capacitor in the center, and we are ready to go. But what about the calculations?

When we calculated the value of the capacitors for the CCD dipole, we used the formula

$$C = C_{tot} (S - 2)$$

where

C = value of each capacitor
C_{tot} = total capacitance needed to resonate the antenna
S = total number of sections

This formula is in line 510 of the BASIC program listing. For long-wire or vertical calculations, the above formula should be changed to

$$C = C_{tot} (S - 1)$$

This change is because the number of capacitors used in long-wire or vertical antennas is one less than the number of sections. The modifications of the program to do these calculations is indicated earlier; only lines 490 and 510, plus the REM statement, need be changed. Line 490 is changed to decrease by 1 for each loop.

Additional Notes and Construction Hints

Here are some points and hints that will make both your design and building tasks easier, and will result in a more durable antenna. In the selection of capacitors for the CCD, precision is more important than accuracy. Good precision means that the capacitors should all be as close to each other in value as possible. For that reason, we suggest you use 5% dipped silver-mica units. On the other hand, the exact value of the capacitors (accuracy) is less critical. For example, if you use 1000-pF 5% capacitors in your CCD, the true value of each is between 950 and 1050 pF. Therefore, if your calculations tell you to use a batch of 981-pF capacitors, 1000 pF 5% will be just fine.

We recommend using supports at each node, especially if your CCD is to be permanent. A no-cost source of really good insulator material for these supports can be found at your local hardware store in the form of scraps of the plastic which must be used (by law in most communities) to replace glass in storm doors. This water-clear plastic is tough, a good dielectric, and best of all, free. You can get all you need by stopping by your favorite hardware emporium from time to time and rummaging in their waste can (with permission). You are sure to find scraps there,

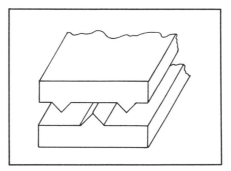

Fig 8—End view of the jaws of "running pliers," used by glass workers to snap glass after scoring. Available for under $10, this tool is especially helpful in snapping apart small pieces of window plastic, suggested for use as node insulators in the text. The opposing fulcrums of the jaws supply the pressure needed to snap this tough plastic.

especially before cold weather. Don't try to cut the stuff with a saw—it is really difficult unless you have a very fine-toothed band saw available. A quick and easy way to cut it is to score it using a special scoring tool for this purpose. Such a tool can be found at the hardware store (and its purchase helps to justify the free scraps). Once scored, break it along the line with pliers. For small pieces such as those you will need for the CCD, running pliers really make the job easy. These are the special pliers used to snap glass after scoring (see Fig 8). Pieces about ½ × 1 inch or slightly larger are about right. Drill holes in them, and use them at each node to take the strain off the capacitor (Fig 9). It is best to "pot" each node after assembly for permanent installations. We have found the plastic goop used to insulate plier handles works well and can be purchased at most hardware stores. You can even get it in different colors to identify different antennas.

When cutting the wire and soldering in the capacitor nodes, it is helpful to do it in a rather systematic way. Solder a hunk of wire at one node, then clamp the plastic insulator in a vise in the center of the node (see Fig 10). Stretch the wire to a chair back or other similar support. Fasten one end of your tape measure to the vise with tape and drape the other end over the chair back. Now measure the wire (120 inches in the case of our 80-meter CCD dipole), but cut the wire at least 4 inches longer. Now thoroughly strip any insulation off the last 6 inches. Place the new node so that its center is 10 feet, 10 inches from the previous node center, and wrap the wire. Solder it thoroughly. When cool, put it in

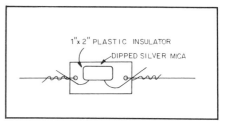

Fig 9—A capacitor can be temporarily taped to the center of a plastic insulator while the bared antenna wire and capacitor leads are assembled and soldered. When twisting the wire and leads together, space them apart, as shown. This permits the solder to bond the wire and leads over the space of several turns. Tightly spaced turns will not permit the solder to penetrate between turns.

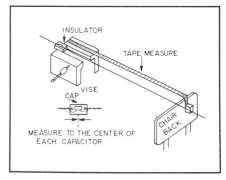

Fig 10—At A, a suggested method for assembly of the CCD antenna. A vise and chair back make it possible for one person to build the sections. Don't forget to cut the wire (at the chair back in this illustration) at least 4 inches longer than the final length of the section. Then strip the enamel or plastic insulation off at least 6 inches, for wrapping and soldering. Section measurements should be made between the center of capacitors, as shown at B.

the vise in place of the previous node. Then start the process over. Only 25 more sections to go!

There you have it. Our experience with this antenna at Field Day has been uniformly excellent, and several of our club members have found similar results at their homes. The proof of the pudding is in the eating; the proof of the CCD is in the sending and receiving. Try it; you'll like it!

Notes

[1]H. A. Mills and G. Brizendine, "Antenna Design: Something New!", 73, Oct 1978, pp 282-289.

[2]D. Atkins, "The High Performance, Capacitively Loaded Dipole," Ham Radio, May 1984, pp 33-35.
[3]This paper is based on a series of articles that appeared in the "ORC Newsletter" (Ozaukee Radio Club, PO Box 13, Port Washington, WI 53074, a Special Service Club located near Milwaukee) during 1984-1985. Commercial versions of the CCD antenna for use from 1.8 to 50 MHz are available from the Ampruss Company, PO Box 551, Aiea, HI 96701-0551. The authors are in no way connected with this commercial source, and provision of this information is not meant to imply endorsement of any commercial product by the authors or the ARRL.
[4]Program 1 of this paper, in BASIC, is available on 5-1/4 inch diskette from the ARRL for the IBM PC and compatibles; see information on an early page of the this book. Also included on that disk is a compiled Pascal version, KAPLAN.EXE, prepared by the authors. The Pascal version offers a more extensive series of calculations than the BASIC program, including for CCD dipoles and CCD verticals. A 3-1/2 inch, 720K disk containing the compiled Pascal version can be obtained from Stan Kaplan, WB9RQR, by sending a blank, formatted disk and a self-addressed disk mailer with return postage.
[5]See note 1.

Program 1

BASIC Listing for CCD Dipole Calculations

The program was written for use with the GW-BASIC interpreter distributed with MS-DOS versions 3.2 or higher, but should work with almost any BASIC with minor modifications. If your BASIC interpreter does not have the "PRINT USING" function, modify each appropriate line by removing the characters: USING "#####.##". [The ARRL-supplied disk filename for this program is KAPLAN.BAS.—Ed.]

```
100  REM This version is for DIPOLE CCD ANTENNAS
110  CLEAR : CLS
120  INPUT"Enter frequency (in MHz): ",FREQ
130  PRINT "Send results to printer as well as screen? [Y/N]: ";
140  P$=INPUT$(1)
150  IF P$="Y" OR P$="y" THEN LPTON=1
160  TOTLINCH=(300/FREQ)*39.37
170  TOTLFEET=TOTLINCH/12
180  DIAM(1)=.032
190  DIAM(2)=.0403
200  DIAM(3)=.0508
210  DIAM(4)=.0641
220  DIAM(5)=.0808
230  PRINT
240  PRINT USING "#####.###";FREQ;
250  PRINT "  MHz          DIPOLE CCD"
260  PRINT USING "#####.###";TOTLINCH;
270  PRINT "  inches"
280  PRINT USING "#####.###";TOTLFEET;
290  PRINT "  feet"
300  IF LPTON=1 THEN LPRINT USING "#####.###";FREQ;
310  IF LPTON=1 THEN LPRINT"  MHz          DIPOLE CCD"
320  IF LPTON=1 THEN LPRINT USING "#####.###";TOTLINCH;
330  IF LPTON=1 THEN LPRINT "  inches"
340  IF LPTON=1 THEN LPRINT USING "#####.###";TOTLFEET;
350  IF LPTON=1 THEN LPRINT "  feet"
360   FOR I=1 TO 5
370   INDUCT(I)=(.00508*TOTLINCH) * ((LOG(4*TOTLINCH/DIAM(I))-.75))
380   CTOT(I)=2*((1/(((6.2832*FREQ)^2)*INDUCT(I))*10^6))
390   NEXT I
400  PRINT
410  PRINT"#      #20     #18     #16     #14     #12"
420  PRINT"OF     AWG     AWG     AWG     AWG     AWG       SECTIONS"
430  PRINT"SEC    pF      pF      pF      pF      pF     FEET    INCHES"
440  PRINT"——     ——      ——      ——      ——      ——      ——      ——"
450  IF LPTON=1 THEN LPRINT"#      #20     #18     #16     #14     #12"
460  IF LPTON=1 THEN LPRINT"OF     AWG     AWG     AWG     AWG     AWG       SECTIONS"
470  IF LPTON=1 THEN LPRINT"SEC    pF      pF      pF      pF      pF     FEET    INCHES"
480  IF LPTON=1 THEN LPRINT"——     ——      ——      ——      ——      ——      ——      ——"
490  FOR SECTION=50 TO 2 STEP-2
500   FOR I=1 TO 5
510   CAP(I)=CTOT(I)*(SECTION-2)
520   NEXT I
530   N=TOTLINCH/SECTION
540   FEET=N/12
550   PRINT USING "###";SECTION;
560   PRINT USING "#####";CAP(1);CAP(2);CAP(3);CAP(4);CAP(5);
570   PRINT USING "####.##";FEET;N
580   IF LPTON=1 THEN LPRINT USING "###";SECTION;
590   IF LPTON=1 THEN LPRINT USING "#####";CAP(1);CAP(2);CAP(3);CAP(4);CAP(5);
600   IF LPTON=1 THEN LPRINT USING "####.##";FEET;N
610   NEXT SECTION
620  PRINT "Do another run? [Y/N]:";
630  X$=INPUT$(1)
640  IF X$="Y" OR X$="y" THEN 100 ELSE 650
650  END
```

The End-Coupled Resonator (ECR) Loop

By Henry S. Keen, W5TRS
Box 11-N
Fox, AR 72051

There is a growing tendency (probably fostered by commercial interests) for the average ham antenna installation to include a rotary beam mounted on a tower, complete with rotator and all the necessary hardware. The station could include a legal-limit amplifier and other expensive accessories. The newcomer to our frequencies may well look at the cost of all this and decide to change his or her hobby to bird watching.

The wire antenna to be described, at a height of 25 feet, with a "barefoot" transceiver and less than 100 watts input, has consistently reached all continents. This was accomplished despite the recent nadir of solar activity and its consequent effect upon DX.

The intended function of the transmitting antenna is to launch the transverse electromagnetic (TEM) mode into space. This mode of propagation may be regarded as using an unrestricted waveguide, without the frequency limitations imposed by transverse dimensions. The TEM mode has its electric and its magnetic fields normal (at right angles) to each other, with the resulting direction of propagation being normal to both. Of these two basic components, the electric field is relatively easy to distort, either by the presence of a conductive body, or something of a dielectric nature. Little, if any attention is usually paid to this phase in the design of the average ham antenna. The ECR loop represents an attempt to correct this oversight.

Dipole Comparisons

Let us begin with a common ham antenna, the horizontal dipole. This is shown in Fig 1. When this antenna is high above ground, it should work well. At practical heights, the electric field at each end will tend to be attracted to ground, with its great size, rather than to the field lines emanating from the other end of the dipole. Ground is such a poor conductor that it must be seen as introducing losses to the system.

A move that is directed toward correcting this problem is to remove much of the high-voltage ends of the dipole. There the voltage distribution tells us that the

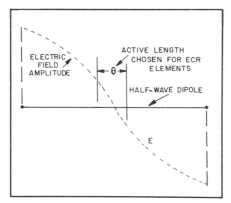

Fig 1—Electric field of a horizontal dipole.

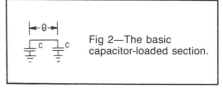

Fig 2—The basic capacitor-loaded section.

electric field is most intense, and likely to be affected by the presence of nearby ground. The removed ends of the dipole will be replaced by appropriate capacitors, as shown in Fig 2, in order to retain resonance.

The wire length as shown in Fig 2 is considered to be a section of transmission line whose characteristic impedance, Z_0, is related to the wire size and its height above ground. This relationship becomes: $Z_0 = 138 \log (4h/d)$, where the height, h, and the wire diameter, d, are both expressed in similar terms. Fig 3 shows the relationship of wire size and height above ground to the resulting characteristic impedance.

The Dipole Center

In developing the basic loaded section used in this antenna, the shortened length of wire is assumed to have been taken out of the center of a dipole. Consequently, a voltage node is still in the center of the wire section. Because identical capacitors are used in the loading process, this can be taken as a valid assumption. An inductive reactance will be presented at each end of the wire section, which must be related to wire dimensions and placement by $X_L = Z_0 \tan \theta/2$, where Z_0 is the previously obtained characteristic impedance of the wire section, and θ is the length of the wire section in electrical degrees, a wavelength being 360 such degrees. X_L is the inductive

reactance of the wire section.

If we are to maintain resonance in the wire section, the capacitive loading reactances, as shown in Fig 2, will have to equal the inductive reactance just obtained. However at this point the computed capacitor assumes that its other terminal is grounded, a most unreasonable situation in an antenna, as we cannot run a ground wire from each capacitor.

This apparently impossible situation is quickly resolved, however, when a number of such sections are connected in series, Fig 4. Experience in the mid 1950's, when I was project engineer in a study of (then new) strip transmission lines, suggested that when connected as a closed loop, a complete phase reversal would be found at each pair of coupling capacitors. Their common terminals were at zero, or ground potential. Thus we have created a "phantom" ground, which serves the purpose without the need of a single ground wire. Carrying the idea a step further, a single capacitor of *half* the

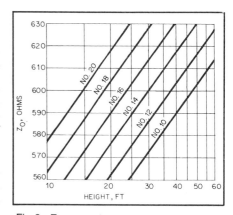

Fig 3—Z_0 versus height and wire size. $Z_0 = 138 \log (4h/d)$

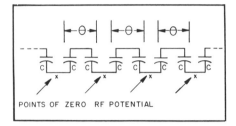

Fig 4—Phantom grounds resulting from series connections of basic loaded sections.

value of that computed for Fig 2 will satisfy all the requirements of the pair.

If now we sketch the voltage distribution along a series of such sections, with a phase reversal across each capacitor, and a voltage null at the center of each wire section, it becomes apparent that the current flow in all wire sections will be in the same direction. There are no current reversals, regardless of the number of sections (Fig 5).

Multiple Radiators

We may therefore deploy the array as a closed loop, with a diameter of ½ wavelength, so that radiation from one side will be reinforced by radiation from the opposite side, thus favoring a low angle of radiation. Originally the loop was deployed as a square, a half wavelength on a side, and supported at the corners. See Fig 6. Although performance was excellent, I felt a closer approximation of a circle might be preferable. Additional supporting lines of monofilament nylon were added to the centers of the sides of the square, and the

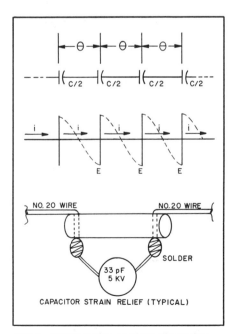

Fig 5—Actual series sections illustrating current agreement and method of strain relief.

shape became an octagon. No direct comparison was possible, but I felt a circular shape might give more uniform coverage.

Some form of strain relief should be provided for the capacitors. At W5TRS they were made from one of those fiberglass poles intended for use on a bicycle to display a flag, making its presence known in traffic.

The 450-Ω open-wire feed line is installed in the center of the most convenient wire section. A total of four supporting points will be needed. With plenty of space to work with, support of the loop at as many as 8 or 12 points should be possible. In this case it would be wise to assemble the array on the ground first, and then raise it, one main line at a time. The nylon lines used can be inexpensive 20- or 30-pound-test monofilament nylon fishing line.

No attempt has been made to measure the actual input impedance of the ECR loop because the input is a balanced line, rather than coaxial. Practically all ham equipment for measuring complex impedances is of the coaxial "persuasion," and measurements would be complicated by whatever form of balun is included. Some early Smith Chart computations that need not be detailed here gave an estimate of somewhere around 200 to 300 Ω as the real component of such quantities. I use a matching network at the input end of the balanced line.

A major problem in developing this loop has been the availability of suitable capacitors. Originally 5% silver-mica units with a 500-V rating were used. However when attempts were made to tune it up on a band for which it had not been designed, one or more of them "objected"; as a result, all were replaced with 10% ceramic capacitors with a 5-kV rating. Because of the estimated high impedance of the loop input, I used no. 20 wire instead of the more familiar no. 14.

Objections will probably be raised by some who feel that the wire sections are too short. One should recall the excellent work several years ago by Jerry Sevick, W2FMI, with top-loaded vertical antennas where he showed that radiation from the high-current portion of a mobile whip was very effective.[1]

In general, the ECR loop was found to be most effective on the band for which it was designed. It should be obvious that when a band change modifies the half-wavelength diameter of the array, it will loose some of its effectiveness.

A further advantage of the ECR loop in a horizontal plane is the low profile. A low 25-foot height reduces the probability of a direct lightning strike when the usual summer thunder storms begin to growl. Furthermore, this low profile may be a factor in keeping peace in the family, when

[1]J Sevick, "The W2FMI Ground-Mounted Short Vertical," *QST*, Mar 1973, pp 13-18, 41.

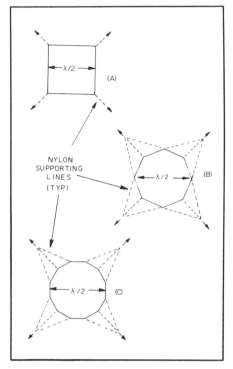

Fig 6—Loop configurations. Capacitor locations are not shown. At A the circumference is 2 λ for a square. The octagon at B has a circumference of 1.657 λ. The circumference for the 12-point figure at C is 1.6077 λ. Place the loop in a horizontal plane for operation.

the XYL or the next-door neighbor objects to a tower, as this antenna is practically invisible! Of course it must not be assumed that one is limited to 25 feet!

Computation of the 20-Meter ECR Loop

The wire size is no. 20, the frequency 14.1 MHz, and the capacitors are 33 pF. From Fig 3, at a height of 25 feet, Z_0 is found to be about 630 Ω. It may be noted that if a value of 600 Ω is assumed, an error of only 5% will be present.

The capacitors of 33 pF may be compared to those of Fig 4, which would show two series capacitors of 66 pF each. At 14.1 MHz they will have a reactance of 171 Ω.

$$X_C = \frac{10^6}{2\pi fC} \qquad \text{(Eq 1)}$$

where

X_C = capacitive reactance, ohms
f = frequency, MHz
C = capacitance, pF

The inductive reactance of half the wire section (from the null in the center to each end) must be equivalent.

$$X_C = X_L = Z_0 \tan \frac{\theta}{2} \qquad \text{(Eq 2)}$$

Solving for θ,

$$\theta = 2 \arctan \frac{X_L}{Z_0} \qquad \text{(Eq 3)}$$

Therefore, 171/630 = tan 15.2 degrees, so θ = 30.4 degrees. This fixes the wire length per section as 5.89 feet, or 70.6 inches, as a wavelength is 69.8 feet. These dimensions are not very critical, as residual reactances are easily compensated by the tuned matching network.

One more decision remains...the pattern in which we will deploy the loop; whether a square, an octagon or a 12 point figure. This will determine the number of sections that must be provided. If a square is used, the circumference is clearly 2 wavelengths, or 720 electrical degrees.

720/30.4 = 23.7 sections (use 24)

If an octagon is chosen, the circumference is 1.657 wavelengths, or 596.5 degrees, so 596.5/30.376 = 19.6 sections (use 20). With the 12-point figure, the circumference is 1.6077 wavelengths, or 578.8 degrees, so 578.8/30.4 = 19.1 sections (use 19). In any case, the nearest whole number of sections will be chosen. Methods of supporting the loop with nylon lines are shown in Fig 6.

All of this may seem complex for a nondirectional antenna. However, when one compares it with all the work necessary to properly install a vertical antenna, with all the dozens of ground wires that would have to be placed, lawns being torn up, and even cement walks, it is obviously a simpler solution.

Appendix

Consider an array of end-coupled half-wavelength resonators (Fig 7). If there are N identical sections, then there are N peaks of resonance before harmonic responses appear. The mechanism responsible for these different resonances is phase reversals that appear across the air gaps coupling the different sections.

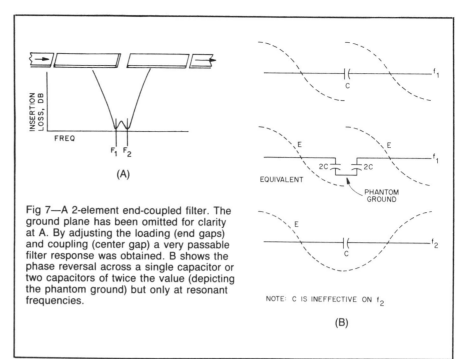

Fig 7—A 2-element end-coupled filter. The ground plane has been omitted for clarity at A. By adjusting the loading (end gaps) and coupling (center gap) a very passable filter response was obtained. B shows the phase reversal across a single capacitor or two capacitors of twice the value (depicting the phantom ground) but only at resonant frequencies.

This is most easily understood when only two sections are considered. At the highest frequency of resonance, the sections may be considered as normal half-wavelength sections. At the lower frequency of resonance, there will be found to be a phase reversal across the center gap. The effect of this action is to place a capacitive load on each of the resonating sections, thus causing a second resonance on a lower frequency than would otherwise take place. This effect was used to arrive at gap capacitance versus spacing relationships to a high degree of accuracy, as frequency measurements (of resonances) are much easier to obtain with accuracy than other parameters.

The fact that such phase reversals actually took place was proved by using a slotted line probe to actually probe the fields of the resonant sections. At the lowest frequency of resonance, there was a phase reversal across every gap in the assembly. It is at this one resonance that the ECR loop functions. While there are 20 such resonances in the possible passband of the loop, the highest such frequency would be when the 6-foot wire length becomes a half wavelength. This would be somewhere about 940/12 or 78 MHz, which suggests that operation in the 20-meter band should be no problem.

Balloon and Kite Supported Antennas

Balloons as Antenna Supports

By Stan Gibilisco, W1GV
871 S Cleveland Ave, Apt P-12
St Paul, MN 55116

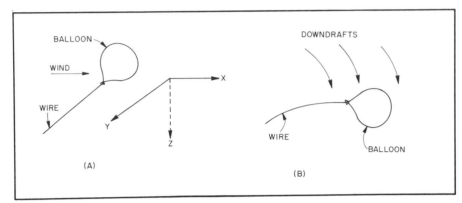

Fig 1—At A, wind tends to drag a tethered balloon downward. The wind vector, X, and the wire pulling back, Y, produce a net downward component Z. At B, downdrafts are another problem when flying a captive balloon; they may blow it down directly.

For operation on the lower frequency amateur bands—160, 80 and occasionally 40 meters—the idea of using a helium-filled balloon as an antenna support has occurred to most of us. This article is concerned with my experiments in using helium-filled balloons as supports for wire antennas on these bands. Ideas for balloon modifications to allow for easy flying in various weather conditions are discussed.

Limitations on Balloon Flying

Balloons will not, in my experience, behave in winds of more than 15 to 20 miles per hour without special stabilizers. Even then, it is doubtful that any balloon will stay up very long in a sustained wind of more than 30 miles per hour. The problem is that the wind tends to drag the balloon down (Fig 1A). This is simply illustrated as a vector diagram in which the sideways wind creates the effect of a downward force on the balloon. Downdrafts are a common problem in irregular terrain (Fig 1B). These air currents literally drive the balloon down. Downdrafts occur most often as wind turbulence but can take place even with a very light wind.

Both drag and downdrafts are countered by using balloons having sufficient lift. The wind resistance of a balloon increases with the square of the radius for a spherical balloon, while the volume, and consequently the lifting power, increase with the cube of the radius. A large spherical balloon is the best. I have used 40-inch and 54-inch diameter spherical balloons.

Precipitation will present a problem. Frozen precipitation causes the most trouble. Dry snow may not stick to a balloon, but wet snow, sleet and freezing rain will almost always bring it down. A heavy downpour will do the same.

The wire must be light so that it does not weigh down the system. I have used fine aluminum welding wire, 0.030 inch in diameter, which has about 1200 feet per pound. This has allowed me to fly antennas up to about 1.6 wavelengths at 1.8 MHz with no trouble.

We will not deal with the theory of antennas here, except to say that any length of wire can be flown, as long as the balloon will lift it. Generally, a practical maximum is about 2 wavelengths at 1.8 MHz, flown as a sloping long wire. Balloon antennas often fly at an angle because of the wind.

Safety

Of primary importance when flying a balloon antenna is that no one gets hurt and no property gets damaged. The rules are simple enough but are plainly stated here:

1) *Never* fly a balloon antenna where it can come down on a power line. This means that the nearest above-ground power line should be farther away than the length of the antenna wire.

2) *Always* tether the balloon with a backup cord such as 20-pound-test fishing line, so that the balloon will not be likely to carry away the antenna wire. The wire could come down on a power line.

3) *Never* fly balloon antennas near or in thundershowers.

4) Don't try to fly a balloon antenna if the weather is, or is expected to be windy, rainy, snowy, or severe in any way except for very cold or hot.

5) Use helium, not hydrogen, to inflate the balloon.

6) Keep an eye on the balloon when it is up, and reel the antenna in when it is not being used. Never leave the balloon antenna up unattended.

Attaching the Wire

The first time I flew a balloon antenna I attached the antenna to the neck of the balloon using twist ties, the kind that are supplied with garbage bags. This worked alright for a while, but one balloon got away in a gust of wind and another had its neck mutilated by incessant pulling and stretching; this balloon popped when I took the twist tie off.

I decided that it would be necessary to use some sort of stopper that could be pulled out of the balloon neck for refilling, to which the wire could easily be attached. I came up with the scheme shown at Fig 2. The stopper was an empty 35-millimeter camera-film container. A screw eye was inserted and the end sealed

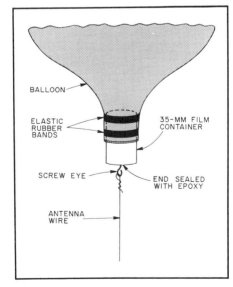

Fig 2—Method of attaching wire to avoid damage to the balloon.

with five-minute epoxy. Elastic rubber bands secured the neck of the balloon around the tumbler-shaped plastic stopper. When it became necessary to recharge the balloon, it was quite easy to pull the stopper out. Some care was needed in peeling off the neck of the balloon since breakage was a high risk.

For balloons with larger necks, I settled on the idea shown in Fig 3. The stopper is actually left in the neck of the balloon all the time. The gas is kept in by putting the top (actually the bottom) on the shampoo bottle. A hole is punched in the bottom of the bottle, about 1/8 inch in diameter, to allow for slow filling and also to allow some time for placing the bottle top on after the balloon is taken from the tank nozzle. The screw eye is placed in the bottle top, using five-minute epoxy for filler. This arrangement works quite well and is physically rugged.

One problem to be watched for is that the wire will try to slip off the screw eye. The wire should be twisted about eight times, at least, and a closed screw eye is preferable to a screw hook. Kinks in the wire must be avoided since hard-drawn aluminum (alloy 5356) wire will break if it kinks.

Winding and Unwinding

The antenna wire should be wound on a spool that will allow easy retrieval and will prevent kinking. A spool for electrical cord is very good for this purpose. It has a turning circumference of about 18 inches

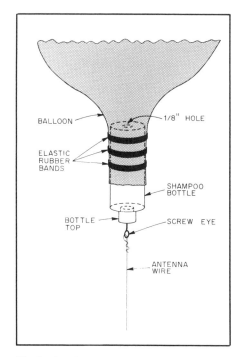

Fig 3—Another method of tying the wire to the balloon. This method does not require that the stopper be removed to add helium. The balloon is recharged by unscrewing the bottle top and replacing it after inflation is completed.

and a built-in crank. The aluminum wire from the original spool should be fitted with a rubber band to keep it in place, then wound from the original spool onto the electrical cord spool. The tether line, such as 20-pound-test monofilament fishing line, may be wound at the same time. The fishing line comes in lengths such as 110 yards, and this is very close to 5/8 wavelength at 1.8 MHz. Simply wind the wire and fishing line together onto the spool until there is no more fishing line. Care should be exercised to ensure that both the line and the wire are securely fastened to the spool prior to winding. The two should be securely fastened together at the balloon end, as well.

The antenna is unwound by simply letting the balloon go. This should be done slowly, so the wire doesn't become too slack, but it should be done fast enough to get the balloon up and in the clear before it has a chance to rub against something and pop. (Perhaps Murphy's Laws have an explanation for why a balloon seems to "lay for" objects like eaves corners and sharp tree branches when it's on a short leash!) A rate of about six feet per second is reasonable.

The wire length will depend on the band and antenna configuration you want to use. It is a good idea to measure the number of crank turns of the spool that are needed to obtain, say, a quarter wavelength at 7 MHz (33 feet), then keep track of the approximate amount of wire you let out by counting the turns of the spool as the balloon is launched. An antenna tuner can then be used to get the desired 1:1 SWR for the transmitter. The point on the wire can be marked with masking tape for future reference. If you want to be precise about having the correct length, you must test for resonance at an odd number of quarter wavelengths, bypassing the antenna tuner and adjusting for a dip in SWR. Once you know the length, then you can add or subtract any length up to 1 wavelength by physical measurement. Remember to include the length of the lead-in when estimating initial antenna length.

It is important that the spool be properly anchored once the balloon is at the desired altitude. The spool must be tethered so that no more wire can unwind, to ensure the balloon will not carry the entire spool away. (If a 40-inch helium-filled balloon can carry off a medium-sized hammer, then it can carry more than you might expect.) I secured the spool to a balcony railing by passing an old belt through it and securing the belt to the tightest notch. Turning was prevented by making a looped end on a piece of stiff coat-hanger wire, putting the loop around the spool crank, and securing the other end of the stiff wire to the railing (Fig 4). Even then, the stiff wire had to be pulled tight or the spool would break free and all of the wire would be let out.

Reeling the wire in is quite easy unless there is a wind and you've let out a lot of

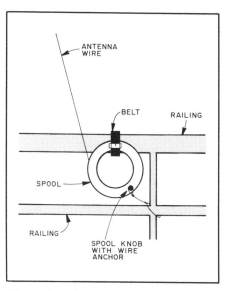

Fig 4—Method of anchoring wire spool to balcony railing. In any case the anchoring must be very secure, since balloons can pull with surprising force.

wire. In any case, don't hurry, and make sure you do not lose the spool! When the balloon is down with about 66 feet of wire left (a quarter wavelength at 3.5 MHz), the antenna can be left alone for the next use. Leaving some wire up will keep the balloon from being batted around and possibly jerking loose or popping. Of course if the wind gets too strong the whole assembly will have to be taken into the garage.

Connecting the lead-in during actual use is done via an alligator clip from the lead-in to the antenna wire. When the antenna wire is long, static electricity will build up on it, and it is common for this to result in sparking when the clip is first attached. It is wise to avoid touching the lead-in and the antenna simultaneously before they are electrically connected.

On the Air

It is not very practical to constantly watch the antenna while on the air. If the antenna comes down, there will be a large change in the SWR and in the loading. Glancing at the SWR meter with the switch in the REFLECTED position is a good idea from time to time. The SWR will not change much as the antenna blows around, unless it becomes almost parallel to the ground. When there is so much wind that the antenna is constantly threatening to come down, you are better off to reel it in and delay using it until conditions are better.

For receiving, a balloon-supported antenna will be susceptible to noise. This is partially offset by the higher signal levels in some cases, but when the noise level exceeds S9, it is worthwhile to consider other antennas for receiving. Low long wires, loops and Beverages are excellent for this.

Signal reports at night for stateside contacts are generally S8 to S9-plus, with 100 watts or more RF output on 160 and 80 meters. It was rare to get anything but S9 reports when running 500 watts. This was not during contests (when most reports are routinely given as S9); many times I had reports of S9 plus 20 or even 30 dB. Remarks such as, "You've got the strongest signal I've heard in a while on this band," were commonplace.

The most universal thing was the enthusiasm that other hams seemed to show when I told them what was holding up my antenna. They often wanted to know details, such as the type of wire used, where the balloons were obtained, how many radials there were, and such things. Often I was told that such a project had been considered but never done.

Stabilization

The problem of wind was reduced somewhat by adding a disk at the base of the balloon in the hope of countering the natural tendency of air resistance to pull the balloon downward. I thought that if something were to deflect air downward, the reaction would push the balloon upward. One device that seemed to help is illustrated in Fig 5.

A cardboard disk was traced from a 33.3-rpm record. The disk was then sprayed with acrylic for rigidity and water resistance, and a hole punched in the center. This disk was cemented with epoxy to the base of the stopper for the balloon. Thus, if the balloon were to fly vertically, with the wire straight up and down, the disk would be horizontal. If a wind blew the balloon aside, the disk would remain

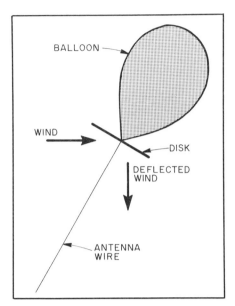

Fig 5—Addition of deflecting disk to stabilize balloon in winds up to approximately 20 miles per hour. A typical flying angle is 40 to 60 degrees above the horizon, unless there is no wind. Perfectly vertical flight is very rare.

perpendicular to the antenna wire, deflecting air downward and tending to counteract the force pushing down on the balloon.

The idea proved fruitful and the stability of the antenna was virtually guaranteed in sustained winds of up to 20 miles per hour. Gusts over 25 miles per hour still caused trouble; under these conditions even kites usually become unstable, and more down-to-earth antennas are necessary (figur-

atively and literally).

Other ideas for stabilization have been considered. The most interesting is flying a kite with the balloon several feet beyond, and the whole assembly held together with twine. The ideal kite would have no tail, such as the Jalbert parafoil or Hargrave box kite. The kite would have to be light enough, or the balloon big enough, to allow flight under conditions of little or no wind.

Choice of Antenna Length

The best antenna lengths for good all-around signal propagation are between 1/2 and 5/8 wavelength. The impedance at the feed point is very high and contains no reactance when the length is 1/2 wavelength. This length is best if a good RF ground system, consisting of numerous radials, cannot be installed. At 1.8 MHz, 1/2 wavelength is about 260 feet.

If several radials, each at least 1/4 wavelength, can be laid down at the station, then a 5/8-wavelength wire is a better choice. This is about 325 feet at 1.8 MHz. An antenna tuner will sometimes match a 5/8-wavelength antenna better than it will a 1/2-wavelength antenna, perhaps because the resistive part of the impedance is lower with the 5/8-wavelength wire. The voltage is also much lower for a given power level.

Various kinds of antennas can be flown using balloons when there is wind and thus a slope to the antenna. I have used slopers of up to 1.6 wavelengths at 160 meters—a whopping 830 feet—to take advantage of the directional properties of long wires. When this wire was at a slant of 45 degrees, the far end was up about 590 feet. I wonder what the record height for a 160-meter antenna is.

Kite-Supported Long Wires

By Stan Gibilisco, W1GV

871 S Cleveland Ave, P-12
St Paul, MN 55116

This article discusses an antenna that is elementary, and also as old as the wireless art, although the theory is somewhat complex. Any shortwave listener who has used a kite or balloon to support a long wire will agree that such an antenna is very effective at medium and high frequencies. Its value for amateur transmitting has been somewhat ignored in recent years, because of the availability of other, more complicated types of antennas.

The Concept of the Very Long Wire

A very long wire (VLW) is exactly what its name suggests: a wire that is not just long, but very long. Ideally a VLW is of such a length that adding more wire will not change its characteristics significantly. From a practical standpoint, we may consider for amateur use that a VLW is at least one full wavelength at 1.8 MHz, so that it has multiple lobes. This length is represented by 533 feet. At 14 MHz, such an antenna is about eight wavelengths; at 28 MHz, 16 wavelengths.

The VLW tends to exhibit diversity in both receive and transmit, as well as a more or less omnidirectional radiation pattern when supported by a high-angle kite or balloon. The gain normally associated with a long wire will occur, but is of little importance in most situations. The numerous minor lobes account for most of the useful radiation from a kite- or balloon-supported VLW. These lobes tend to blend together because of wire curvature or sag. The feed-point impedance is generally quite high. The use of a kite ensures that the wire will be high above the ground, maximizing low-angle radiation and minimizing ground losses to practically zero. Helium-filled balloons may also be used when there is not enough wind to support a kite.

Radiation and Response from End-Fed Wires

The radiation patterns from end-fed wires are fairly well known, and are discussed in detail in *The ARRL Antenna Book* and other publications. The shortest commonly used end-fed wire is ¼ λ long and often is erected as a vertical, operated against ground and fed with coaxial cable (Fig 1). The wire may be shorter than ¼ λ and tuned to resonance by means of a series inductor. Some verticals are up to 5/8 λ in height. The reactance may be tuned out and the remaining resistance matched to 50 or 75 ohms in a variety of ways. All of these

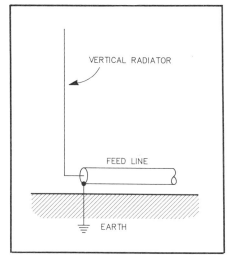

Fig 1—Typical quarter-wave vertical operated against ground. This is one of the simplest types of end-fed antennas.

vertical antennas have simple donut-shaped radiation patterns with the greatest radiation perpendicular to the wire.

When the length exceeds about 5/8 λ, maximum radiation is no longer perpendicular to the wire, but is in the form of lobes at various different angles. As the wire is made longer, the major lobe develops at an ever-decreasing angle with respect to the line of the wire. These lobes become more numerous as the length of the

wire increases. When the wire is very long —several wavelengths—the lobes are so numerous that the radiation pattern may be considered omnidirectional for all practical purposes. Eventually the wire may become so long that adding more wire does not make any practical difference. When this condition is reached, we have a true VLW. A wire that is at least one full wavelength at 1.8 MHz will act as a true VLW at frequencies above about 14 MHz (Fig 2), and will be a fairly good approximation of a true VLW at 3.5 and 7 MHz. It may be considered as a simple long wire at 1.8 MHz, but will operate on all bands with excellent efficiency.

Directional effects will be noticed with such a wire, although they will not be especially pronounced. The best radiation will be generally in a direction "downwind" from the feed point; the same is true for response.

The minor lobes take the form of double cones with greater apex angles than that of the main lobe, but having the same axis, namely, the line containing the wire. For a wire that is M quarter waves long, M being an integer (1, 2, 3,...), there are exactly M radiation lobes.

In theory the radiation pattern from a VLW that is eight wavelengths long is that shown in Fig 2. The maxima result from addition of the radiation from each current maximum in the wire. The minima result from cancellation of the radiation from the current maxima. In practice, with an end-

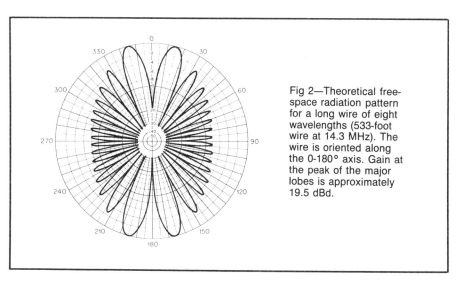

Fig 2—Theoretical free-space radiation pattern for a long wire of eight wavelengths (533-foot wire at 14.3 MHz). The wire is oriented along the 0-180° axis. Gain at the peak of the major lobes is approximately 19.5 dBd.

fed VLW, the maxima in the direction away from the feed point (toward the far end of the wire) are of greater magnitude than those in the opposite direction, because some of the electromagnetic field is radiated as it travels along the wire. Also, the nulls become less and less well defined as the length gets greater, primarily for this same reason, and also because of surrounding physical objects that tend to absorb and reradiate energy from the wire. A practical radiation pattern for an eight-wave VLW is illustrated in Fig 3.

With a true VLW antenna the minor lobes tend to merge, so the antenna can be considered an all-around fair radiator. The antenna radiates just about equally well in directions generally broadside to the wire. In the case of a wire running straight upward from the feed point, the antenna approaches the behavior of an isotropic radiator in the half space above the horizon. Energy directed downward toward the ground will be reflected back upward anyway. The high gain of the main lobes is directed almost straight up, but in a very narrow cone that does not contain much of the energy in the transmitted signal. This situation is shown in Fig 4.

Input Impedance of Very Long Wires

For practical reasons, all of the antennas I have tested are end-fed very long wires. The antennas are fed as unbalanced systems against ground. For simplicity the wires are brought directly to the station window, where a Transmatch is employed. The RF ground consists of a connection to a cold-water pipe, the utility ground at a 120-volt socket, and one ¼-wave "radial" wire for each amateur band 160 through 10 meters, calculated for length ℓ according to the formula

$$\ell_{ft} = \frac{240}{f_{MHz}}$$

Whenever possible, the wire is such that it is near an integral multiple of ½ λ at the operating frequency. The formula for determining the length of a wire N wavelengths long, where N is some integral multiple of ½, is generally agreed to be

$$\ell_{ft} = 984 \frac{N - 0.025}{f_{MHz}}$$

where ℓ is the length of the wire and f is the frequency of operation.

The complex impedance of an end-fed wire varies according to the graph of Fig 5. Lengths shorter than ¼ λ are not considered for the purposes of this discussion. Note that the reactance alternates between capacitive (negative) and inductive (positive), being zero at each integral multiple of ¼ λ. The resistive component is rather low at odd multiples of ¼ λ, but high at even multiples of ¼ λ. With each succeeding increase of ¼ λ

in the length of the wire, the resistance becomes closer and closer to a certain limiting value, approaching this value from below at odd multiples of ¼ λ and from above at even multiples. We are concerned only with the even multiples of ¼ λ.

For low values of N, where N is the number of half waves in the wire, the resistance R is on the order of several thousand ohms. As N becomes large, however, R decreases to perhaps one or two hundred ohms for, say, N greater than 30. I have not actually measured the change in the value of R as the length increases, but I have found that the standing-wave ratio (SWR) without a Transmatch seems to approach a value of about 3, suggesting that the wire feed-point impedance converges on some value near R + jX = 150 + j0.

We can expect that the exact length of the wire, and whether or not it is some precise integral multiple of ½ λ, becomes relatively unimportant when the length of the antenna is very great, say, 10 wavelengths (N = 20). The fluctuations in reactance, as well as in resistance, will become smaller and smaller although the resonant bandwidth, as a percentage of the operating frequency, gets narrower and narrower.

I decided to use a length of 555 feet in practical tests. This is exactly two wavelengths at 3.5 MHz as calculated using the formula above. The length is found to be four wavelengths at 7.048 MHz and

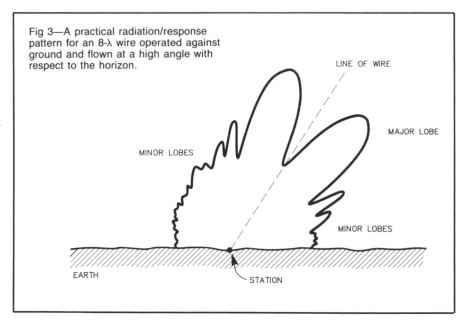

Fig 3—A practical radiation/response pattern for an 8-λ wire operated against ground and flown at a high angle with respect to the horizon.

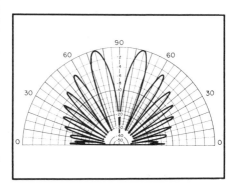

Fig 4—Theoretical radiation pattern for a 533-foot VLW at 14.3 MHz, oriented vertically. Relatively little energy is concentrated in the main lobes, despite their theoretical high gain (8.4 dBi). Pattern calculated with MININEC, a method of moments procedure, for a flat earth having average soil conditions (a conductivity of 5 mS/m and a dielectric constant of 13).

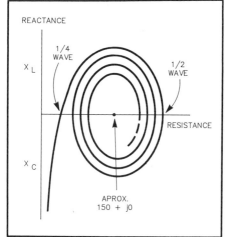

Fig 5—Variation in complex impedance of an end-fed wire as it becomes longer. The curve intersects the resistance axis (j = 0) at each integral multiple of ¼ electrical wavelength. The curve converges on a point of zero reactance with a resistance of about 150 ohms (experimentally found).

eight wavelengths at 14.139 MHz. These are all CW frequencies.[1] The resistive part of the input impedance will be high at these frequencies, minimizing ground losses. At the 21- and 28-MHz bands, the wire is probably so long that length is not of much importance in determining the resistive portion of the input impedance. The antenna is also close to one wavelength long at 1.8 MHz.

Expectations were that, with this VLW attached to a kite and flown at a high angle above the horizon, results would be good for DX as well as for stations at close-in distances. There would be plenty of high-angle radiation, good for local work at the lower frequencies. This prediction was verified dramatically in actual operation. The more interesting tests were on 14, 21 and 28 MHz. Although this antenna does not have much low-angle radiation in theory, the horizon distance at an elevation of several hundred feet might allow for useful radiation even at angles of less than 1° with respect to the horizon. I have yet to perform extensive tests to compare the VLW with other antennas at these frequencies.

In any event there can be little doubt that the VLW is a good antenna for Stateside contesting, especially on the lower frequency bands at 1.8, 3.5 and 7 MHz. Its suitability for Field Day is obvious.

The general radiation pattern in the vertical plane for a VLW antenna 555 feet long, flown with a kite at an angle of 55° (average) is shown in Fig 6 for 1.8, 3.5 and 7 MHz. Maxima are indicated by arrows with lengths denoting approximate relative field strength or response. At bands above 7 MHz, the pattern becomes similar to that of an isotropic radiator, somewhat favoring the "downwind" directions. The wire sag tends to fill in the nulls and slightly attenuate the maxima of the actual radiation/response pattern.

Advantages

We would expect the VLW to have certain advantages, as follows.

Simplicity: The VLW is easy to put up and take down.

Moderate to low cost: The whole setup, not including tuner, can be had for less than $100.

Wide-band coverage: The efficiency is excellent on all bands, 160 meters and down, with a Transmatch and single ground lead for each band.

High efficiency: There is minimal ground loss and zero feed-line loss.

Portability: The entire antenna fits in the back seat or trunk of a small car.

Diversity: Because of its great length, the VLW will act as a diversity system for receive, and presumably for transmit also.

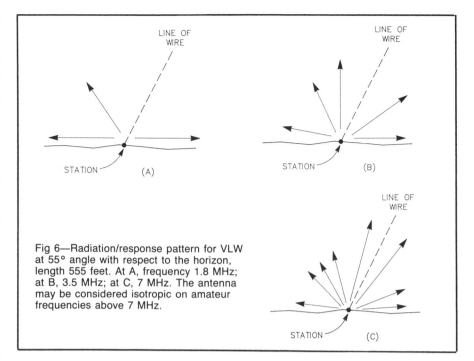

Fig 6—Radiation/response pattern for VLW at 55° angle with respect to the horizon, length 555 feet. At A, frequency 1.8 MHz; at B, 3.5 MHz; at C, 7 MHz. The antenna may be considered isotropic on amateur frequencies above 7 MHz.

Each half-wave section may be considered as a separate antenna, all connected together at the feed point and spanning a length of one to several wavelengths.

Practical Considerations

Now we will examine the mechanics of kite flying. Though "electric" kites are over a century old, they still are useful. Having chosen a 555-foot very-long wire to be supported by a kite, the problem now becomes a mechanical hurdle. A certain amount of wind is necessary, of course; in practice this wind must be over 5 to 7 miles per hour, but less than gale force. We must have a wire that is very lightweight yet is a good conductor. The whole apparatus must not be unduly expensive.

I use two different kite arrangements, one for light winds and one for heavier winds. A 7-foot delta wing "ultralight" suffices for winds in the range of about 5 to 25 mi/h; a Ferrari ram kite is used for winds in excess of 25 mi/h up to perhaps 35-40 mi/h. Aluminum welding wire of diameter 0.035 inch is employed. The alloy, called "5356," is a hard-drawn wire having approximately 900 feet per pound. It comes in rolls of 1 pound. Table 1 shows the parts needed for flying a VLW antenna, not including an excellent location and an antenna tuner with a fairly wide-range matching capability from 1.8 through 29.7 MHz.

Table 1

Parts List for the Very Long Wire (VLW) Antenna

Not included: choice of location and antenna tuning network.

Description	Quantity	Total Cost
Kite, delta, 7 ft	1	$35.00
Kite, Ferrari ram, 4 ft²*	1	29.00
Welco Alloy 5356 aluminum welding wire, 1-pound roll, 0.035 in.†	1	5.50
Kite swivel attachments, 50 pound*	12	1.00
Tethering twine	1	2.50
Spool, electric cordΔ	1	7.50
Alligator clip	1	.29
Lead-in wire (min length possible)	1	<1.50
Belt for securing spool	1	—
Grounding wire	1	<1.50

*Into The Wind, 1408 Pearl St, Boulder, CO 80302. Kites I use are part numbers 367 (Sunburn delta) and 505 (Ferrari ram).

†Available at welding supply shops.

ΔAvailable at most hardware stores. The spool has a hand-held sliding attachment and crank for rapid turning and a minimum diameter of about 5½ inches.

[1]Notes appear on page 149.

Safety

There are several important precautions to be taken when flying an "electric" kite. First, the wire must *never* be able to fall on a power line. Second, you have to be aware that substantial electrostatic voltage will build up on the wire under certain conditions, and this can be quite dangerous even in clear weather. Third, *the wire must be tethered to prevent it from flying away if the wire should break.* Kite line, with a strength of the rated amount, should work well. I am fortunate to live in a neighborhood where all of the power lines are underground. The nearest above-ground lines are about 600 feet directly west of the house. Another set is at about 800 feet north-northeast; another is at approximately 1200 feet south of the house. With a prevailing west wind, it is safe to fly a kite-supported antenna to more than a quarter of a mile.

It is essential that the area near the station be surveyed and the nearest above-ground power lines found. Then, do not fly kite-supported wires that are likely to fall on any of these power lines. Close watch must be kept on the kite so that changes in wind direction will be noticed before they can cause a hazard.

Atmospheric static is a different problem. The antenna must not, of course, be flown in or near thundershowers, in rainy weather, or if the weather is threatening. Also, a kite antenna must not be left unattended for long periods of time, even if it will stay up. Once the kite is at a distance of about 200 feet from the station, it should be grounded by means of a wire loop that the antenna wire passes through (Fig 7). This looped wire is connected to the station ground, and this station ground must be good at direct current. I have seen sparks jump more than ½ inch from ungrounded, kite-supported wires. I have a friend who has seen sparks up to 3 inches in sunny weather with a 600-foot kite-supported wire for shortwave listening. These voltages can knock you down if they are allowed to discharge through your chest cavity.

Once the kite antenna is at the operating altitude, a spark gap can be connected between the output of the antenna tuner and ground. This will keep electrostatic voltages from building up on the antenna. For high power levels, a larger gap must be used. The protective wire-loop ground should be left around the antenna wire until after the lead-in to the station has been connected, and should be reconnected prior to disconnecting the station lead-in. In other words, the antenna should always be protected for static charges. Voltages build up very quickly on a free wire.

As long as the kite and line together weigh less than 5 pounds no Federal Aviation Administration (FAA) restrictions apply, with one exception: The kite and line must not present a hazard to people,

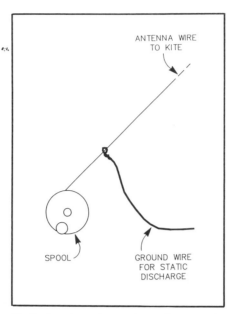

Fig 7—Looping a heavy, bare wire around the kite-supported wire as it is let out protects against electrostatic buildup. The heavy wire should be grounded.

property or aircraft. The kites I have used weigh less than ½ pound by themselves. Theoretically, I could fly up to 4½ pounds of wire—about a mile—if the wind were strong enough. I haven't tried such a daring adventure yet.

Favorable Winds

Kite flying is, itself, an art that requires practice. It makes an interesting biathlon with ham radio operating. A knowledge of the sorts of winds that are good for a particular kite, and the kinds of conditions in which kite flying is a pleasure versus those that make it a nightmare, comes with experience. This experience can be gained only by playing with kites. I am not especially concerned with what people think of a 34-year-old man playing with kites in the streets, but I have become rather popular with children in the neighborhood.

I have used just these two kites for my antennas, because they do not pull extremely hard and do not require sophisticated lines. Larger kites, with Dacron flying line along with the hard-drawn aluminum wire, would probably provide increased stability and reliability. I have in mind the possibility of obtaining a large delta-Conyne kite having a wingspan of 13 feet, 4 inches for use with VLW antennas even longer than the present standard of 555 feet. A better winding spool would probably also be necessary for use with such a kite. A winch would be almost mandatory.

Table 2 is a Beaufort table showing various wind speeds and the types of kites best suited for work at these speeds. The kite configurations may include others besides those shown here.

In my experience, winds of less than 5 to 7 mi/h will not support any kite reliably enough to allow comfortable operation of a VLW. Winds of more than 40 mi/h cause all kites to become unstable; they are apt to dive and get tangled up in trees and other nearby objects.

For no-wind conditions, helium balloons may be used. A balloon of 40 inches diameter will support a considerable amount of wire. (See the accompanying article in this chapter.) The problem with balloons is that they tend to blow down in even a light breeze, and mishaps with birds

Table 2

Beaufort Scale of Wind Speeds, and Kite Types Best Suited for Work at Various Speeds

Beaufort Number	Speed, mi/h	Description of Wind Action	Best Kite Types
0, 1	0-3	Smoke rises vertically or drifts; wind cannot be felt	None
2	4-7	Light breeze. Wind felt on face. Leaves rustle. Wind vanes move.	Delta
3	8-12	Gentle breeze. Leaves and twigs in motion. Light flags extended.	Delta Delta-Conyne
4	13-18	Moderate breeze. Wind raises dust and paper. Small branches in motion. Flags flap.	Delta Delta-Conyne
5	19-24	Fresh breeze. Small trees in leaf sway. Wavelets form on lakes and ponds.	Delta-Conyne Airfoil Ram
6	25-31	Strong breeze. Large branches move. Telephone lines whistle.	Airfoil Ram
7	32-38	Strong wind. Whole trees in motion. Walking impeded.	Ram
8	39 +	Gale. Progressively greater effects.	Not recommended

are almost invariably fatal for the balloon—and perhaps for the bird as well.

Again, experience is necessary in order to get the best results with kite-supported antennas. So play around a while first with ordinary kite line before attempting to fly a VLW antenna. You might even want to read some books about kites and kite flying. Some interesting historical information can be found, such as the experiments of Marconi, and this is good reading in any case.

Measuring the Wire

The length of 555 feet is not especially critical and need not be directly measured. In fact a very close measurement can be made by resonating the antenna as a 2.25-λ radiator at a frequency of 3.944 MHz. This frequency is found by using the formula for determining length ℓ_{ft} for a wire of N = 2.25 λ, and setting ℓ_{ft} = 555.

$$555 = 984 \, \frac{2.25 - 0.025}{f_{MHz}} = \frac{2189}{f_{MHz}}$$

$$f_{MHz} = \frac{2189}{555} = 3.944$$

We simply reel out wire until we reach the fifth SWR minimum without a Transmatch at 3.944 MHz, indicating a length of 9/4 or 2.25 λ (SWR minima will occur when the length is 1, 3, 5, 7, 9,...quarter wavelengths). This SWR will not be exactly 1:1, but the minimum will indicate near-resonance and in this way the length can be found quite accurately.

It is best to have a tunable antenna-matching network. This allows continuous frequency coverage, whereas a simple transformer will not. The settings of the antenna tuner will be critical, especially at the higher frequencies where reactance fluctuations occur, and it can be expected that changes in adjustment will be quite rapid even for relatively small changes in operating frequency. This is because, for an antenna several wavelengths long, a slight change in wavelength moves the current nodes a fairly large distance at the feed point. The difference adds up for each wavelength along the wire, beginning at the kite end.

It should be mentioned again that the nearest above-ground power lines must be farther from the station, in the general direction of flight, than the length of the antenna wire. If there is any doubt about this, the wire should be shortened by reeling it in until you are certain it cannot fall on a power line.

Comparison Tests

I have compared this antenna with a "backward inverted L" on all bands, and have found that the kite-supported VLW is superior almost without exception at 1.8, 3.5 and 7 MHz.[2] There is more testing still to be done, as of this writing, at 14, 21 and 28 MHz. I do not have 10-MHz capability and cannot perform tests on this band.

For receiving, gain of 5 to 15 dB is typical for the VLW over the backward inverted L. The difference seems more pronounced for relatively nearby stations (out to perhaps 1000 miles) than for DX at 7 MHz. The gain is dramatic for all signals at 1.8 and 3.5 MHz. There is less frequently a difference at 14, 21 and 28 MHz, although rarely are signals louder on the backward inverted L than on the VLW.

The VLW does receive more noise than the backward inverted L on all bands. Generally the signals are proportionately louder on the VLW along with the noise at 1.8, 3.5 and 7 MHz; however, on 14, 21 and 28 MHz, signals are occasionally more readable on the backward inverted L because of this noise difference. Contacted stations agree that the VLW is superior at 160 and 80 meters. This is generally true at 40 meters also, especially when the contacted station is "downwind."

Tests on the higher bands (20, 15 and 10 meters) so far indicate that the VLW performs at least as well as the backward inverted L in almost every case. When the station is "downwind" the VLW has definite gain over the other, up to around 10 dB by observation. This is probably the result of the general directional effect of an arbitrarily long wire. There is a noticeable diversity effect on receive, so that there is less fading on the VLW than on the backward inverted L. It would be expected that this diversity characteristic would hold true for the VLW over other smaller antennas such as the Yagi or quad, for both transmitting and receiving.

The VLW appears to have certain advantages and disadvantages, as follows.

Advantages: (1) simplicity, (2) moderate to low cost, (3) wide-band coverage, (4) high efficiency even with marginal ground, (5) portability, and (6) diversity on receive (and presumably on transmit, too).

Disadvantages: (1) large space requirement, (2) somewhat greater susceptibility to noise as compared with narrow-band antennas, (3) power-line and electrostatic hazards, (4) need for steady winds, and (5) need for supervision.

Future Plans—The Ultimate VLW

A true very long wire may be defined as one that behaves essentially as an infinitely long wire; that is, such that adding more wire would not significantly change its behavior. When I can afford it, I will test such a very long wire system. I prefer the lower frequency bands and have the room needed to use this kind of antenna regularly. The system will include a 160-inch delta-Conyne kite, a large Ferrari ram kite (probably 10 square feet), Dacron flying line for the delta-Conyne, and, most probably, a heavier aluminum wire such as 0.040-inch diameter.

For windless conditions, helium-filled balloons can be used, although windless conditions are rare in my part of the country, especially in the winter when the low bands are best. For those with the money and the space, an arrangement of this kind might make a good starting setup, although I recommend getting some kiting experience before working with a kite capable of overpowering a person in a moderate to strong wind.

The VLW is an antenna that may have escaped notice because of its utter simplicity, and because of our universal tendency to dismiss such an idea as grandiose or too simple to work. For those with a touch of madness as well as an appreciation that an antenna can have artistic value as well as practicality, and for those with the necessary physical space, the VLW might well prove a worthwhile antenna to try.

Notes

[1] This simply reflects the writer's operating preference.

[2] This is an inverted L with a top section about 100 feet long and a vertical section about 30 feet long, with the vertical section at the far end and grounded at the base. It works like a typical inverted L.

Antenna Potpourri

Antenna Selection Guide

By Eugene C. Sternke, K6AH

117 Ridgetop Pl
Sequim, WA 98382

Shortly after a radio amateur passes the stage where the only requirement of an antenna is to "get on the air," he or she begins to set up a more complex set of specifications. This "dream antenna" plan first starts out with a number of features to be included, but to make the design more complicated many constraints also appear. Cost is one important factor, but so are other considerations. The plans may have to allow for the size of your property, the covenants, conditions, restrictions, and wishes of the XYL or OM, so the ultimate design is a collection of many factors and compromises. Through the use of the tables and discussions presented here, you can select an antenna to provide for the type of communication desired on the HF bands. The data in the tables provides the basis for selecting the best approach based on performance and cost. When deciding between antenna improvements or more power, remember that the antenna equally benefits the receiver.

If one of the important features of the antenna system is to have the ability to work DX, the designer soon finds out that low angles of radiation are important. In the past, the general rule of thumb has been to make the antenna as high as possible if using a horizontal antenna or to use a vertical mounted at ground level. If the horizontal design is required to get above local obstructions or for some other reason, the question is, "What is the best antenna system that will provide the desired results?" This sometimes has resulted in the examination of vertical radiation patterns of horizontal antennas at various heights above a perfectly conducting ground. The type of information available left much to be desired, for in most cases it was hard to determine decibel levels versus angles of elevation, and at best required a considerable amount of interpolation. Furthermore, most graphs do not show the effects of an imperfect ground.

One of the first questions to be answered as specifically as possible is, "What angles of radiation does my antenna need for working DX?" The results of one study conducted to answer this question are presented in *The ARRL Antenna Book*.[1] From this table it can be seen that when propagation permitted, 99% of the time for reception between England and New Jer-

sey on 7 MHz, the signals arrived at angles from 10 to 35° above the horizon; for 10 MHz, from 8 to 26°; for 14 MHz, from 6 to 17°; for 18 MHz, from 5 to 12°; for 21 MHz, from 4 to 11° and for 28 MHz, from 3 to 9°. The ideal antenna system should provide maximum radiation and reasonably uniform gain between the angles specified. An antenna with a pencil-sharp beam within the angles specified may give a super signal at times, but for consistent DX it would be best to cover as equally as possible the entire range of angles provided in *The Antenna Book*.

With the availability of low-cost and relatively simple computer software, it is possible to determine how important the antenna-system design variables are, and how they affect the vertical radiation pattern. The antenna analysis program called ANNIE was made available in 1984 by James C. Rautio, AJ3K.[2] Using this program, I examined the information for conducting tradeoffs in the selection of antenna parameters in respect to antenna height, ground characteristics and antenna directivity in the vertical plane. Unfortunately, ANNIE does not automatically take into consideration the change of antenna radiation resistance as a function of antenna height. Instead, it requires that an adjustment (element weight) be made to keep the antenna power constant for all heights above ground. Fortunately, at antenna heights above ½ λ, the errors introduced are not more than 0.2 dB with the element weight held constant. To simplify the analyses once an element weight

was selected, it was held constant in the data presented here, but this should not reduce the validity of the conclusions.

Comparative Analysis of Antenna Types

I started the vertical directive analysis by examining the most popular low-cost antennas used at many amateur stations. The abbreviations in the "Ant Type" column of Table 1 should be interpreted as follows: Vert 0.25 is a vertical monopole ¼ λ long with the base at ground level; Vert 0.5 is a vertical ½ λ long; Hor 0.25B is a ½-λ horizontal dipole mounted ¼ λ above ground and the gain is measured broadside to the antenna; an "E" indicates gain off the end of the dipole; the "V" antenna is the typical inverted-V dipole with an apex angle of 90°, antenna center 0.225 λ high and each leg ¼ λ long. In Table 2, the V-S 0.18B antenna is a ¼-λ vertical monopole. Its base is 0.18 λ above ground level. It uses a ¼-λ monopole sloping at a 45° angle from horizontal at the base of the vertical section but not grounded at that point. The antenna is fed at the center or base of the vertical section with a balanced transmission line. The "S" antenna is a ½-λ antenna sloping 40° from vertical.

Reference to the quality of the ground beneath an antenna system includes ground below the antenna and up to 20 or more wavelengths away. The quality of the ground immediately below the antenna becomes very important when the antenna is vertical and ground becomes part of the antenna circuit, as with a ground-mounted ¼-λ vertical antenna. Then the quality of

---

Table 1

Gain of Popular 3.5-MHz Antennas Over Average Ground

Decibel Levels at Angles Above Horizontal

Ant Type	Max Ang	Max dB	40°	30°	20°	10°
Vert 0.25	25°	0.8	0.1	0.8	0.6	− 1.8
Vert 0.5	20°	− 0.1	− 4.1	− 1.3	− 0.1	− 1.6
Hor 0.25B	90°	4.7	4.1	2.8	0.3	− 5.1
Hor 0.25E	90°	4.7	− 1.6	− 5.3	− 10.2	− 15.9
Hor 0.125B	90°	2.6	− 0.1	− 1.9	− 4.9	− 10.4
Hor 0.125E	90°	2.6	− 4.8	− 8.2	− 11.9	− 16.3
V 0.225B	90°	3.5	1.4	− 0.3	− 3.1	− 8.5
V 0.225E	90°	3.5	− 0.2	− 1.8	− 3.7	− 6.9
S 0.0B	30°	− 0.9	− 1.2	− 0.9	− 1.4	− 4.0
S 0.0E	30°	0.7	0.5	0.7	0.0	− 2.6
S 0.0 − E	25°	− 4.0	− 4.8	− 4.0	− 4.3	− 6.6

---

[1]Notes appear on page 156.

152

Table 2

Gain of Popular 7-MHz Antennas Over Average Ground

			Decibel Levels at Angles Above Horizontal			
Ant Type		Max Ang	Max dB	40°	30°	20°	10°
Vert	0.25	25°	0.9	0.1	0.9	0.7	-1.5
Vert	0.5	25°	-0.1	-5.4	-2.3	-0.8	-2.0
Hor	0.25B	90°	4.3	3.8	2.5	0.1	-5.2
Hor	0.25E	90°	4.3	-2.2	-5.9	-10.6	-15.7
Hor	0.5B	30°	5.1	3.8	5.1	4.4	0.2
Hor	0.5E	45°	-2.8	-2.8	-4.6	-9.2	-15.6
V	0.25B	90°	3.6	2.0	0.5	-2.2	-7.5
V	0.25E	90°	3.6	-0.5	-2.2	-4.5	-8.1
V	0.5B	30°	5.0	4.5	5.0	3.7	-0.8
V	0.5E	45°	-1.7	-1.9	-3.0	-5.2	-9.1
V-S	0.06B	20°	-0.9	-2.6	-1.3	-0.9	-2.7
V-S	0.06E	25°	0.4	-0.8	0.2	0.3	-1.7
V-S	0.13B	25°	0.8	-0.6	0.6	0.8	-1.2
V-S	0.13E	25°	1.7	0.7	1.6	1.6	-0.4
V-S	0.18E	20°	2.3	0.7	1.9	2.3	0.5
V-S	0.18B	20°	1.6	-0.6	1.0	1.6	-0.2
V-S	0.22E	20°	2.7	0.1	1.9	2.7	1.1
V-S	0.22B	20°	2.0	-1.2	1.0	2.0	0.5
S	0.3B	35°	0.6	0.6	0.6	-0.0	-2.7
S	0.3-E	90°	-1.4	-8.1	-9.8	-7.8	-8.0
S	0.3E	60°	-0.1	-0.6	-0.8	-1.1	-3.0

Table 3

Horizontal Pattern of the V-S Type Antenna on 7 MHz Over Average Ground

Azimuth Angle	Decibel Levels at Angles Above Horizontal		
	10°	22°	35°
-90°	-0.6	0.5	-2.6
0°	0.5	1.8	0.0
30°	0.9	2.4	0.7
60°	1.0	2.6	1.0
90°	1.1	2.6	1.1

the ground near the antenna base determines the ratio of the power radiated to the power used to heat the earth in close proximity to the antenna base. For example, with a perfect ground (zero loss) 100% of the power is radiated. If the ground loss is equal to power in the radiated signal, half or 3 dB is lost to heat.

The values used in this analysis for the ground conductivity (in millisiemens/meter) and dielectric constant are, respectively: desert 1, 7; average 5, 15; good 20, 30; and sea 1000, 81.

Tables 1 through 4 show the dB gain of the antenna as a function of height over average soil and are relative to the broadside gain of a dipole in free space. Also shown is the gain as a function of other soil conditions. In some cases these tables contain values carried to three significant figures, which is somewhat of an overkill as far as any one value is concerned. However, this accuracy was included to show that in many cases the differences between some antennas is rather small. Fractions of a dB difference between different types of antennas may be open to some question because the gain shown is directly related to element weight as discussed above. However, when antennas are similar, the element weight is about the same so that the results are not far off. Furthermore, when listening to the beacon stations transmitting on 14.1 MHz, one can realize how little change there is in signal readability with their 10-dB changes in output power. To observe how trivial a few dB are, go to a Hi-Fi radio shop and try out one of their sets that has a dB read-out of the volume level. Check to see if you can detect the difference in 2 dB. Therefore, small

differences in dB levels are unimportant unless you are communicating under marginal conditions and that small difference is just enough to push the signal above or below the minimum signal-to-noise ratio for copying. Another justification for small improvements is if you can collect a fair number of small gains, their sum may make a significant change.

Discussion of Antennas Typically Used

The selection of an antenna becomes the greatest challenge when required for the lower frequency bands, namely 1.8 and 3.5 MHz. Antennas for these bands are normally physically large compared to a typical city lot, and can place a strain on a small budget. The design of the support for these antennas may require the help of a good civil engineer when restrictions and esthetics are included in the problem. Examples of popular antennas for the

3.5-MHz band are given in Table 1. Antennas for the 1.8-MHz band are similar, but remember that a $\frac{1}{2}$-λ radiator for that band is about 260 feet long, so some form of a compromise probably will be necessary.

One form of compromise that is not employed frequently enough is the use of a center-fed radiator. The feed should be in the form of an open-wire balanced transmission line. For low power, TV twin-lead can be used. If radiation broadside to the antenna is desirable, a center-fed antenna makes a very versatile radiator. This form of antenna can be operated on almost any frequency. At the length of 0.64 λ each side of center, the antenna is known as an extended double Zepp, which provides about 3 dB gain over a $\frac{1}{2}$-λ dipole and has a beamwidth of approximately 35°. On lower frequencies where the length is less in terms of wavelengths, the pattern broadens and the gain drops until the antenna approaches the characteristics of an isotropic radiator. Because of resistive losses in the antenna, feeder system and matching network, the efficiency of the antenna will drop when the radiator is less than $\frac{1}{2}$ λ long. These losses are typically small when compared to losses in an antenna depending on a good ground system (as in a vertical working against ground). With good construction techniques, satisfactory operation may

Table 4

Gain of Popular 14-MHz Antennas Over Average Ground

			Decibel Levels at Angles Above Horizontal			
Ant	Type	Max Ang.	Max dB	40°	30°	20°	10°
Vert	0.25	26°	0.8	-0.1	0.7	0.5	-1.8
Vert	0.5	20°	-0.6	-4.9	-1.9	-0.6	-2.0
Hor	0.5B	30°	5.0	0.9	5.0	4.3	0.1
Hor	0.5E	45°	-2.7	-2.9	-5.0	-9.7	-15.9
Hor	0.75B	20°	5.3	-10.4	1.8	5.3	3.0
Hor	0.75E	90°	4.1	-10.4	-7.8	-9.8	-16.5
Hor	1.0B	14°	5.5	2.9	-13.0	3.7	4.7
Hor	1.0E	50°	0.4	-3.1	-11.3	-11.1	-17.2
V	0.5B	30°	4.8	4.4	4.8	3.5	-1.0
V	0.5E	50°	-1.6	-1.9	-3.1	-5.4	-9.5

be found with a center-fed horizontal antenna 20 to 25 feet long on each side of center. Personal use of such an antenna on 80 meters has convinced me of the fact.

The vertical antenna gives good omnidirectional coverage and may be the best for you of the 3.5-MHz antennas listed in Table 1. A sloping ¼-λ monopole (noted as S 0.0E) may be a very convenient substitute for the vertical. This antenna could be an insulated guy for a mast with a height of 55 feet or greater. By selecting which guy is fed, a front-to-back ratio of 4 dB or more can be realized. Not shown in the tables is the fact that maximum gain is realized at 20° off broadside. The few dB superiority of the horizontal antenna at high angles and broadside to the antenna is useful mainly for local communication. Another type of antenna that could be used on 3.5 MHz or even 1.8 MHz is covered in Table 2. In that group for 7 MHz is the vertical sloper (V-S), which shows up rather favorably. Such an antenna could be operated on 80 or 160 meters if a good low-loss matching network and feeder system are provided.

The antennas considered as candidates for use on the 7-MHz band are shown in Table 2. On this band the size and cost of an antenna are much more reasonable. Here the possibility of two horizontal ½-λ antennas oriented at right angles to each other and mounted at a height of ½ λ is hard to beat except at radiation angles of 10° and below. If these low angles are not important, the inverted-V dipole with its center at ½ λ is a very close competitor.

A compromise antenna is the one labeled V-S, vertical and sloping. This antenna consists of a ¼-λ monopole with its base mounted at 0.22 λ (31 feet) above ground. The other half of the antenna is a monopole which runs from the base of the vertical element to ground level at an angle of 30° from vertical. An open-wire feeder is attached at the junction of the two elements. The desirable features of this antenna are its good low-angle performance and moderately omnidirectional characteristics. If four sloping elements are provided (which also can serve as guys if properly insulated from the mast and ground) and if a means for switching to different sloping elements is used, some small amount of directivity can be obtained; see Table 3. Once again, remember that plus or minus a few dB on a good signal will be hard to detect. The height of the vertical element can be reduced to make for easier construction if the angle of the sloping monopole is increased toward 90°. See V-S 0.06B (or E), but note that some low-angle gain is sacrificed.

The directional characteristics of the V-S antenna are shown in Table 3. The −90° angle is off the back side away from the direction of the sloping element. Zero degrees is off the side and 90° is in the direction of slope. The dB levels are tabu-

Table 5

Gain at Critical Angles of a 14-MHz Ground-Mounted ¼-λ Vertical Monopole Versus Ground Conditions

Ground	17°	6°	Max dB	Angle of Max dB
Desert	− 1.34	− 6.47	− 0.34	28°
Average	0.13	− 4.50	0.77	26°
Good	1.59	− 2.15	1.91	24°
Sea	4.76	4.21	4.82	14°
Perfect	5.86	6.0	6.0	1°

Table 6

Gain at Critical Angles of a 28-MHz Ground-Mounted ¼-λ Vertical Monopole Versus Ground Conditions

Ground	9°	3°	Max dB	Angle of Max dB
Desert	− 4.56	− 11.6	− 0.30	29°
Average	− 2.40	− 9.2	0.78	27°
Good	− 0.68	− 6.8	1.73	25°
Sea	4.15	1.58	4.48	16°
Perfect	5.86	6.00	6.00	1°

lated for 10° (the lower angle of DX reception on 40 meters), 35° (the upper limit), and 22° (the median angle).

At 14 MHz and above, the size of an antenna becomes even more manageable for the average ham. With ½ λ at 14 MHz being only 33 feet, it is possible to either rotate the antenna or have several fixed arrays to cover the desired directions. As can be seen from Table 4, the competition is between the heights of the horizontal ½-λ antenna. Later in this paper, I will show the optimum height for DX. *No*, it's not always the higher the better!

Should you desire omnidirectional coverage, minimum space requirements and/or to disguise your antenna, you may choose a vertical antenna. Table 5 provides the decibel-response levels at the upper and lower angles for optimum DX coverage when the antenna is mounted over different grounds. It is obvious that the better the conductivity and the larger the dielectric constant of the ground, the better. This table will tell you whether to move to the seashore or buy an amplifier. Remember, that ground must extend for many wavelengths from the base of the antenna. This is because the data presented here is based on the signal reflected from the ground, which adds to or subtracts from the signal radiated skyward depending upon the phase and amplitude. Note that losses in the ground for the return of current in your antenna system are not included here.

Table 6 provides the same type of data as above except it is for operation on 28 MHz. These data indicate that for a ¼-λ radiator, the type of ground has greater effect on the radiated field strength at the

low angles required for this frequency. This provides another good argument for using horizontal polarization for the bands at 14 MHz and above. The advantage of using horizontal polarization on 14 MHz and above is shown in Tables 7 and 8. With horizontal polarization it can be seen that the quality of the ground has less effect on the field strength over the range of angles for optimum DX. The program ANNIE provided the data shown in Table 9. This table shows essentially that, irrespective of the antenna type, there is just one best height to maximize the signal in the range of angles best for DX. This height is slightly over 1.2 λ. Less height reduces the signal strength at the low angles while raising the signal level at the upper angle. Heights over the 1.2 λ do the reverse, and will place a very deep null in the range for optimum communication. This 1.2 λ at 14 MHz equates to an antenna height of approximately 85 feet. An interesting fact is that 85 feet turns out to be about the optimum height for all frequencies from 14 MHz to 28 MHz.

The question about vertical steering of the antenna lobe in the 14 to 28 MHz range was answered by the program ANNIE. The answer is that the best configuration has all of the elements horizontal and in the same plane. Any deviation will result in less signal strength. The way to steer the antenna lobe in the vertical plane is to change its height above the ground.

Conclusions

The information presented here will serve as a quantitative guide for antenna selection and for conducting tradeoff studies for improving your ability to hear and work DX. It will give you a good idea where the best place is to spend your effort and money to maximize the gain. This information evaluates the antennas in the real world, not hypothetically with an infinitely conductive ground or in free space. The table values permit a numerical evaluation so you can determine whether you should buy a new beam, a new tower, more power, or move to the sea coast.

Remember that 1 or 2 dB is a trivial matter unless your typical communication involves digging signals out of the noise. In the choosing of an antenna the important area to check is if the antenna will put the signal where you want it. The large loss typically off the end of a dipole at low radiation angles is something worth considering and avoiding. The fact that a vertical antenna provides a few dB less in a favored direction may be acceptable considering that it is omnidirectional and has low-angle radiation capability. The vertical sloping antenna (type V-S in the tables) may be worth a second look if you want to favor some direction or need a slight front-to-back ratio and the ability to roughly choose the direction (within 90°).

If you really want to maximize your signal in some favored direction, a beam will help, but the beam must be at the correct height if you want to place the vertical lobe correctly for optimum communication. The vertical position of the antenna lobe is just as important as rotating your beam to place your antenna lobe at the correct azimuth angle.

ANNIE provides interesting conclusions regarding vertical antennas where it shows the gain versus length of antenna as a function of operating frequency. The quality of the ground determines if gain is increased as the antenna height is increased from ¼ to ½ λ (both antennas mounted at ground level). When the quality of the ground is good or better, the longer antenna is better and the greatest improvement for the longer antenna is at the lower angles of radiation.

Although the vertical antennas presented here were limited to ¼ and ½ λ, ANNIE does shown that several dB of gain can be realized by increasing the length to 0.64 λ.

After reviewing the above data, the next antenna I erect will be a version of the 8JK type. It will have two parallel horizontal radiating elements spaced about 10 feet apart and each driven at its center with a balanced transmission line to provide equal power with a 180° phase difference. The length of the radiating elements will be 18 to 22 feet on each side of center. The ideal height for this array is about 85 feet above ground level. Less height is acceptable at a cost of decreased gain at low angles of radiation.

Note that my antenna differs from the

Table 7

Gain at Critical Angles of a 14-MHz ½-λ Horizontal Dipole Versus Ground Conditions

Height = 0.5 λ; indicated responses are broadside to the antenna.

Ground	17°	6°	Max dB	Angle of Max dB
Desert	3.18	− 3.99	4.52	27°
Average	3.53	− 3.86	5.03	27°
Good	3.76	− 3.78	5.41	29°
Sea	4.01	− 3.78	5.93	30°
Perfect	4.02	− 3.81	6.01	30°

Table 8

Gain at Critical Angles of a 28-MHz ½-λ Horizontal Dipole Versus Ground Conditions

Height = 1 λ; indicated responses are broadside to the antenna.

Ground	9°	3°	Max dB	Angle of Max dB
Desert	3.92	− 5.44	5.20	14°
Average	4.11	− 3.87	5.48	14°
Good	4.23	− 3.28	5.66	14°

Table 9

Beam Coverage of 14-MHz Reception Band With Average Ground Versus Antenna Type Comparison

Antenna Type	17°	6°	Max dB	Angle of Max dB	Antenna Height*
Hor Dipole	3.02	3.02	5.57	11°	1.235
8JK Type	7.10	7.11	9.49	10°	1.215
Yagi 3 El	10.02	10.04	12.54	11°	1.210

*Height in wavelengths

8JK antenna, which specifies ½-λ radiators on each side of the feed point. As long as the radiators on each side of the feed point are of equal length, the antenna will work.[3] The only time length becomes critical is when each side of center exceeds 5/8 λ where multiple lobes start to form; the other condition is when the length is much less than ¼ λ where losses must be considered in the feed line and matching network. If these losses are minimized, operation on 80 meters should be acceptable. In fact, on-the-air tests have proved this fact. I will probably raise a few eyebrows when I tell my 80-meter contacts to wait a few seconds while I rotate my beam.

This antenna will operate with good efficiency on any frequency from 10 to 30 MHz, and between these frequencies it will provide within 3 dB of the maximum gain of a 3-element monoband Yagi. As

operation approaches 30 MHz, its gain will almost equal that of the Yagi. No dimension is crucial as long as each radiating element is the same length. The antenna can be light weight and offer low wind load, which will make it simple to rotate. If I want to equal or exceed the gain of a 3-element Yagi, I will add another identical array about 0.4 λ (at the lowest operating frequency) above. The vertical height of the antenna will be about 85 feet to the center of the array. The two sets of arrays will be fed in phase. A big feature is that all adjustments are made at ground level. No tower climbing is necessary to make adjustments for a lower SWR.

What will be an inconvenience is the fact that the feed to the elements must be via a balanced transmission line. The feed line must be tuned to transform the antenna load to a 50-ohm unbalanced line to match the modern transceiver. Also, the radiation pattern is bidirectional, which may or may not be a disadvantage. I certainly won't have to be concerned whether the short path or the long path is best.

If the inefficiency of this antenna turns out to be a problem at the lower frequencies, a second antenna will be the vertical sloper noted as V-S in the tables. By selecting which of four sloping elements is driven, a preferred direction can be selected to provide a small gain with a small front-to-back ratio. This antenna will have to be fed with a tuned balanced transmission line but will have the advantage of

operating with good efficiency on any frequency where the element lengths are more than $1/10$ λ long. The maximum frequency should be limited to the point where the element lengths exceed $5/8$ λ unless a multiple-lobe vertical pattern is acceptable.

Perhaps the small gain figures of the V-S antenna may not be very impressive, but the 10 dB or so improvement over the radiation off the end of a fixed horizontal antenna is highly significant. Also the V-S antenna compares favorably with the gain of a horizontal dipole mounted less than $1/4$ λ above ground.

In my opinion, the single-band Yagi for the HF bands is about as out of date as a

crystal-controlled transmitter of pre-World War II vintage.

Notes

[1] G. L. Hall, Ed., *The ARRL Antenna Book*, 15th edition (Newington: ARRL, 1988), p 23-17, Table 2.

[2] J. C. Rautio, "The Effect of Real Ground on Antennas," *QST*, Nov 1984, p 35.

[3] A true W8JK array consists of four $1/2$-λ elements operated in a collinear/end-fire arrangement. If the radiators are shortened to only $1/4$ λ each side of center, the antenna becomes simply a two-element phased array and approximately 3 dB of gain will be sacrificed.—*Ed.*

A Ham's Guide to Antenna Modeling

By Steve Trapp, N4DG

24768 Mango St
Hayward, CA 94545

Computer modeling of an antenna system is a practical approach, now that many hams have computers at home or work. If you are putting up a standard, well-matched antenna, you may need no more analysis than referring to the tables in *The ARRL Antenna Book*. But if your antenna is a short dipole or loop, or if it needs a matching network, you may want to use a computer to estimate the impedance at the transmitter. This paper presents equations and suggestions on using a computer to calculate antenna system efficiency and impedance.

My particular dilemma was having to live with a moderately invisible 33-foot-long apartment wire to an adjacent building and the desire to work all the HF bands. My shack location puts my rig at the other end of 80 feet of RG-8X coax. I don't mind going out to the balcony to flip a band switch if there is no decent way to tune the antenna from my matching network at the rig. My goal was to make rough predictions on the various matching networks I wanted to try at the antenna base and at the rig. This forced me to know the effects of the antenna, coax and matching network.

One rule came up several times in the model and design iterations: Without good measurements, your results will be rough. This is particularly true with an antenna, where impedance is hard to predict. It is also generally easier to refer to a Smith Chart (or other faster analysis) to guide yourself in the rough design. However, computer analysis helps you to understand the compromises in your design, and allows you to see the antenna system response at many frequencies, through the line and matching network.

Component Modeling

To perform network calculations you need to find the impedance of all the components. Eqs 1 and 2 are nothing new, just the reactance of capacitors and inductors, scaled to use more convenient units. My analysis program (on CP/M Turbo Pascal)

overflowed until I scaled the frequency, capacitance and inductance.

$$X_C = \frac{-j}{2\pi\, f_{MHz}\, C_{\mu F}} \text{ohms} \qquad \text{(Eq 1)}$$

$$X_L = +j\, 2\, \pi\, f_{MHz}\, L_{\mu H} \text{ ohms} \qquad \text{(Eq 2)}$$

As RF current is confined to a thin layer near the surface of the wire, the wire loss is important to consider, especially in low-impedance (loaded) antennas.[1] Since the current through an antenna decreases toward the tip, you should use around half the formula resistance value. Conversely, the resistance value should be increased for inductors, because the magnetic field causes increased concentration of the current. The optimum wire size for air coils should give a 70% to 45% wire diameter to pitch ratio, decreasing with the coil length-to-diameter ratio.[2] With the precautions mentioned, the wire resistance in ohms is

$$R_{skin} = 0.0010 \text{ length}_{ft}\, \frac{\sqrt{f_{MHz}}}{\text{diam}_{in.}} \text{ohms}$$
$$\text{(Eq 3)}$$

It is a good idea to make rough estimates of the series resistance of all coils and capacitors. Doing so helps avoid the computer giving you impossibly good results or overflows at resonance. To model the effect of core or dielectric loss in an inductor or capacitor, you could add a resistance with a value of $R_{par} = QX$.

Transformers can be modeled to include the effects of imperfect coupling, capacitance, and finite reactance-to-impedance ratios. The following formula will give the impedance that is seen at the primary terminals of a transformer with the secondary loaded[3]

$$Z_{pri} = X_{Lpri} - \frac{k\, X_{Lpri}\, X_{Lsec}}{X_{Lsec} + Z_{load}} \text{ohms}$$
$$\text{(Eq 4)}$$

[1]Notes appear on page 161.

where

X_{Lpri} and X_{Lsec} = primary and secondary impedance of the windings

Z_{load} = impedance connected to the secondary

Z_{pri} = impedance that could replace the transformer and its load

k = coefficient of coupling (approaches 1 in a tightly coupled transformer)

This formula reduces to the familiar N_{pri}^2/N_{sec}^2 step-up of impedance when k approaches 1 and X_{Lsec} is much greater than Z_{load}.

Network Analysis

The fastest and easiest way to calculate the response of a matching network at many frequencies is as a ladder network. The computer figures the effective Thevenin source voltage and series impedance as each part of the network is added. The calculation is repeated for every frequency to be analyzed.

To analyze starting from the antenna, assume you have a one-watt signal existing at the antenna impedance. The calculations "walk through" the rest of the circuit, modifying the effective source voltage and impedance to reflect the influence of each part. Finally, you note the voltage at your 50-Ω rig. You can calculate the loss of your antenna system as well as the load impedance along the way. Remember, in calculating the output power, even a perfect match would have *half* the Thevenin voltage across the load.

The effective source voltage and impedance are calculated for each successive part. For series components, use Fig 1 along with Eqs 5 and 6. For parallel components, use Fig 2 along with Eqs 7 through 9.

$$V_{new} = V_{old} \qquad \text{(Eq 5)}$$

$$Z_{new} = Z_{old} + Z_{series} \qquad \text{(Eq 6)}$$

$$V_{new} = V_{old}\, \frac{Z_{parallel}}{Z_{old} + Z_{parallel}} \qquad \text{(Eq 7)}$$

157

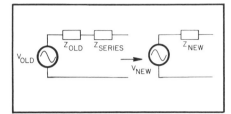

Fig 1—Series component circuit reduction.

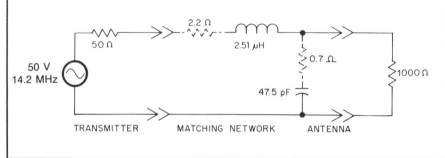

Fig 3—Antenna matching network. Resistors shown with broken lines represent loss resistance in the coil, capacitor and associated connections.

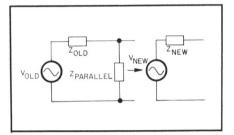

Fig 2—Parallel component circuit reduction.

$$z_{new} = \cfrac{1}{\cfrac{1}{Z_{old}} + \cfrac{1}{Z_{parallel}}} \quad \text{(Eq 8)}$$

$$\frac{1}{Z} = \frac{Z_{real}}{Z_{real}^2 + Z_{imag}^2} - j \frac{Z_{imag}}{Z_{real}^2 + Z_{imag}^2}$$

(Eq 9)

The impedance of a transformer primary with a secondary connected to a known impedance was given earlier in Eq 4. If you are simulating a circuit and know the Thevenin equivalent source connected to the primary, the source can be "transformed" to a new effective source from[3]

$$V_{Thev} = \frac{X_{LM} V_s}{Z_{Thev} + X_{Lpri}} \quad \text{(Eq 10)}$$

$$Z_{Thev} = X_{Lpri} - \frac{X_{LM}^2}{Z_s + Z_{Lpri}} \quad \text{(Eq 11)}$$

where

V_s, Z_s form the source connected to the transformer primary

V_{Thev}, Z_{Thev} form the resultant source that includes the transformer

$X_{LM} = \sqrt{k \ X_{Lpri} \ X_{Lsec}}$, the mutual reactance

Example of Thevenin Calculation

Suppose the network of Fig 3 is at the base of an antenna, fed by a 50-Ω coax line from a transmitter supplying 50 watts at 14.2 MHz.

Step 1. Start with the source impedance and voltage.

$Z1 = 50 + j0 \ \Omega$
$V1 = 50 \ V \ rms$

Step 2. Add the series inductor by Eqs 5 and 6; assume the coil has a Q of 100.

$Z2 = (50 + R_{loss}) + j(0 + X_L)$
$\quad = 52.2 + j224 \ \Omega$
$V2 = V1 = 50 \ V \ rms$

Step 3. Add the parallel capacitor using Eqs 7 and 8. Assume the capacitor and its connecting hardware have 0.7 Ω resistance.

$$Z3 = \cfrac{1}{\cfrac{1}{Z2} + \cfrac{1}{0.7 - jX_C}}$$

$$= \cfrac{1}{\cfrac{1}{52.2 + j224} + \cfrac{1}{0.7 - j236}}$$

Apply Eq 9 to each 1/Z value in the denominator and combine terms. Applying Eq 9 to the result yields $Z3 = 1000 - j3 \ \Omega$.

$$V3 = V2 \times \frac{0.7 - j236}{Z2 + (0.7 - j236)}$$

$$= 50 \ \frac{0.7 - j236}{52.9 - j12}$$

$$= 50 \ (0.7 - j \ 236) \ (0.0180 + j \ 0.00408)$$

$$= 50 \ (0.975 - j \ 4.24)$$

$$= 48.8 - j \ 212 \ V = 218 \ V \ \underline{/-77°}$$

[When multiplying two complex terms such as here, remember that $j^2 = -1$.—*Ed.*]

Not a bad match for a 1000-Ω 1-λ dipole! To see what the voltage would be under a 1-kΩ load, we proceed to the last step.

Step 4. Apply the load impedance in parallel (across the source), and use V4 to get the power. Since this is the last step, we don't need to know the Thevenin source impedance. The resistance used in the power calculation should be the resistive portion of the load.

$$V4 = V3 \times \frac{Z_{ant}}{Z_{ant} + Z3} = (48.8 - j \ 212)$$

$$\times \ \frac{1000 + j0}{(1000 + j0) + (1000 - j3)}$$

$$= 24.6 - j106 \ V = 109 \ V \ \underline{/-77°}$$

$$Power = 4 \ \frac{V4_{real}^2 + V4_{imag}^2}{Z_{load \ (real)}}$$

$$= 4 \ \frac{605.2 + 11236}{1000} = 47.4 \ watts$$

The constant 4 is required, as mentioned previously, because even a perfect load has only half the Thevenin voltage across it. Thevenin models assume the transmitter can be represented as a voltage, twice the value under load, in series with a source impedance. The load inefficiency was already considered in the transmitter design.

I used this circuit at the base of my 40-m antenna to allow operation on 20 m, where the antenna is resonant, but at a high impedance.

Suggestions for Calculation

1) Store your impedances in an array so you can easily modify the component values and add components.

2) Break up long equations into common pieces. It is easier to understand and often faster for the computer.

3) It will make analysis easier if you can write complex addition, multiplication and reciprocal subroutines. In BASIC, or languages without subroutines with arguments, you could use the position of the real and imaginary elements in an array to refer to the complex quantities.

4) Decide what values will probably be changed often during your design iteration. The frequently changed values can be typed in at run time and the more constant values can be stored in the program text. Use of a quick edit-compile-run environment such

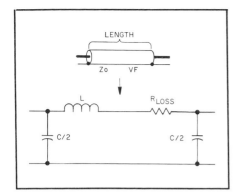

Fig 4—Model of transmission line.

as QuickBASIC or Turbo Pascal will help blur the distinction and make program changes easier.

5) Print out intermediate results to check your program and help guide yourself in "tweaking" component values. You can eliminate the printouts later, or have a variable that enables these printouts.

6) The Thevenin source voltage is *not* the voltage measured across that portion of the circuit. It is the voltage at that point in the circuit without any further loading. It is generally greater, but in the middle of a high-Q circuit can actually be much less than the actual voltage.

7) All simple circuit combination calculations can be performed without the use of any trig functions to preserve accuracy and speed, as shown in Eq 9 above.

8) For easy design iteration, it is good to have both the ability to do simple impedance calculations and a full network simulation. I frequently found my program too restrictive in not being able to jump into smaller subcircuits to quickly see what would resonate away small mismatches. My HP-15C calculator can perform complex calculations, which I use while the program runs to modify component values. A more versatile program interface, where you can save and combine networks, would help.

Modeling Coax Cable

The impedance Z_i, looking into a transmission line of impedance Z_0 and terminated by an impedance Z_{ant}, is

$$Z_i = Z_0 \frac{Zr \cosh vx + \sinh vx}{\cosh vx + Zr \sinh vx}$$

$$(Eq\ 12)[4]$$

where

$$Zr = \frac{Z_{ant}}{Z_0}$$

$v = a + jB$
a = attenuation constant in nepers (8.686 dB) per unit length
B = phase constant in radians per unit length ($2\pi \times$ length/λ)
x = length of line

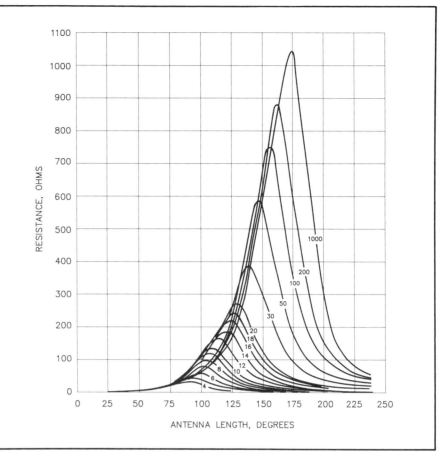

Fig 5—Radiation resistance of a dipole antenna versus its length for various length/diam ratios. Note that for a fixed conductor diameter, the length/diam ratio changes when the length is adjusted. (*After Johnson and Jasik*)

$$\cosh vx =$$

$$\frac{(e^{ax} + e^{-ax}) \cos Bx + j (e^{ax} - e^{-ax}) \sin Bx}{2}$$

$$\sinh vx =$$

$$\frac{(e^{ax} - e^{-ax}) \cos Bx + j (e^{ax} + e^{-ax}) \sin Bx}{2}$$

Simulation Method of Solving Transmission Line

The above equation assumes a uniform dB loss per length. I was interested in checking the loss of the cable under high SWR conditions, where the loss would probably be worst at the impedance extremes. It is possible to break up the line into a series of small L-C pi networks that simulate the cable transformation of load impedances, Fig 4. For accurate modeling, the pieces should be no bigger than one percent of the wavelength. This method of analysis is slowed by the need for looping the calculation down the transmission line, but does not require any trig or log functions. The resistance R_{loss} assumes all the loss is due to skin resistance and none due to the dielectric. A better model would have

a portion of the loss as a parallel resistance representing leakage loss.

$$C_{\mu F} = \frac{length}{Z_0\ VF\ c} \qquad (Eq\ 13)[5]$$

$$L_{\mu H} = \frac{length}{Z_0\ VF\ c} \qquad (Eq\ 14)[6]$$

$$R_{loss} = Z_0 \frac{2.3\ dB_{loss} \times length}{20\ \ell_{db}} \sqrt{\frac{f}{f_{dB}}}$$

$$(Eq\ 15)[7]$$

where

Z_0 = characteristic impedance of line
VF = velocity factor of line
c = velocity of light, scaled as are L, C and length (983.6 ft/μs or 299.8 m/μs)
length = length of each iterated piece, less than 1% of a wavelength
dB_{loss}, f_{DB} and ℓ_{DB} = set of loss, frequency and length data from coax loss tables. For accuracy, use loss data near the frequencies of interest.

Modeling the Antenna

Having an equation for something seems to give a result some legitimacy and exactness. In modeling the real world, that isn't always true. This is certainly the case with antennas. But we are only looking for a result good enough to preserve your signal and transmitter. An impedance error of 10% would be a good prediction, and would affect your SWR only marginally. The exception is with high-Q loaded antennas or networks, where a small change in value will throw the SWR off.

An antenna impedance will depend on height, ground conductivity, orientation, wire diameter, length, frequency, the effect of surrounding objects and the connection hardware.[8,9] Because of the great variability on antenna impedance, it is best to either build the antenna and make measurements, or to accept a rough estimate in advance.

The antenna impedance equation I used included the effect of conductor diameter. The best bandwidth comes with wide elements; my no. 18 wire was a necessary compromise that I wanted to investigate. I included a rough multiplying factor in my antenna model to account for the V shape and the ground proximity.[10]

$$Z_i = R(kl) -$$

$$j \left[120 \left(\ln \frac{\ell}{a} - 1 \right) \cot (kl) - X(kl) \right]$$

$$(Eq\ 16)[11]$$

where

Z_i = impedance of a center driven dipole antenna of total length 2ℓ and radius a

(kl) = 2$\pi \ell/\lambda$ = electrical length of one element in radians

ℓ and a are the length and radius of each element

$R(kl) = 19.94 (kl)^2 + 2.83 (kl)^4 + 0.19 (kl)^6 + 0.10 (kl)^8$

$X(kl) = 7.98 (kl) + 23.96 (kl)^2 - 32.99 (kl)^3 + 22.73 (kl)^4 - 4.17 (kl)^5$

Eq 16 is good for antenna elements shorter than ¼ λ. Above that length, it is best to refer to graphic data. To avoid the need for storing and interpolating the table data for R(kl) and X(kl), I fit the data to polynomials, accurate to 0.1 Ω.

When the length of each element is longer than ¼ λ, the real impedance will rise and the reactance becomes inductive. The change in impedance and reactance will be much smaller if you can use wide elements, as shown in Figs 5 and 6.[12]

Summary

My antenna system, simulation program and calculation methods have evolved over the last few years. My Pascal program creates a data type, Thevenin, which is a combination of two complex data types. I wrote subroutines to create, add, multiply and invert complex numbers. Thevenin sources can be created and placed in series

or parallel. The routines would be faster if they combined sources with impedances and ignored the phase of the voltage, but I made all impedances Thevenin networks for generality. My program shows the impedance and power at several places in the circuit numerically, and finally in a graphic form.

Ladder-network simulations are good for problems with one source and one load, which includes most antenna systems without multiple radiators. Full N × N matrix solutions require more computation and are not required in these cases. Solving op-amp filters, Wheatstone bridges, etc cannot be described by a simple ladder network and requires a more sophisticated solution.

The L-C network coax model does a good job of simulating test cases (10:1 im-

pedance mismatch, etc) and transforms my simulated antenna impedance to match the antenna tuner settings.

I began the antenna simulation by interpolating graphic data.[13] The wider conductor used in the graph seemed to compensate for the proximity of my antenna to the ground, giving good agreement with my measurements. When I began using formulas, I had to modify the impedance values and effective antenna length slightly to account for the effects of the L-shaped antenna, ground and metal proximity.

I wanted to find a "canned" equation for antenna impedance in terms of length, diameter, shape, ground, etc. I couldn't find anything that reduces all the complex interactions into a formula. There are really too many more effects to be considered.

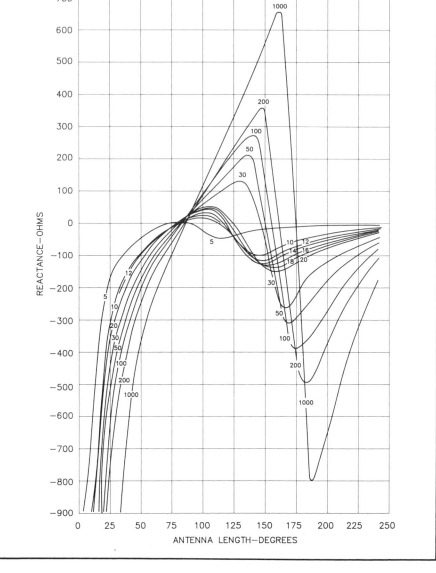

Fig 6—Reactance of dipole antenna versus its length for various length/diam ratios. See note in Fig 5 caption. (*After Johnson and Jasik*)

This is why measurements, graphs and experimentation are necessary.

I initially avoided putting a second matching network at the antenna base, but the high impedance of the 33-foot wire on the 80, 20 and 10 m bands makes the system lossy and critical to tune. However, I was able to operate on 20 m for over a year by using a Transmatch at the rig, and the matching settings could be verified by the computer.

On 10 and 15 m the coax is a few wavelengths long, the matching network had a high Q and unknown stray reactances, and the computer predictions did not do as well. This demonstrates that your computer predictions will never be any better than your data, verifying the computer proverb, "Garbage in, garbage out."

I designed an antenna base-matching network where the antenna is plugged into one socket for 80 m with inductive loading, one for 20 and 10 m that reduces the high impedance to near 50 Ω, and one straight to the coax for the naturally resonant 40- and 15-m bands. The trickiest part in modeling the performance was the 80-m loading coil.

The reactance of the antenna is about ten times its resistance, so my 20% deviation in actual reactance made the initial coax SWR high.

One other lesson I learned is to think as much about an unbalanced "grounded" radiator as the hot one. Unless the antenna is a ground plane, the voltage and current at the junction of the grounded legs will be as high as the hot leads. If the antenna is asymmetrical, the ground leg will have an independent contribution to the impedance. For example, my loaded 80-m radiator has a radiation impedance of 10-15 Ω, but the $\frac{1}{4}$-λ element connected to the shield raises the total impedance to about 50 Ω. Another thing, your computer won't tell you if you have a nonresonant ground-side radiator. Both elements should be loaded independently or your "ground" may become very hot, but the loading impedances can be combined for the ladder network model.

I hope this information will let you start making computer estimates of what is happening with your antenna, feed line and tuning network. Computer modeling allows you to experiment with a concept and estimate the outcome. Your result will be as good as your model.

Notes

[1] G. Hall, Ed., *The ARRL Antenna Book,* 15th ed. (Newington, CT: ARRL, 1988), p 5-14.
[2] C. J. Michaels, "Optimum Wire Size for RF Coils," *QEX*, Aug 1987, pp 6-7.
[3] Author's calculation from 2-port description of a transformer in a circuit.
[4] See p 174 of R. K. Ghormley, "Mr. Smith's 'Other' Chart and Broadband Rigs," *The ARRL Antenna Compendium, Volume 1*, pp 170-175.
[5] R. C. Johnson and H. Jasik, *Antenna Engineering Handbook*, 2nd ed. (New York: McGraw-Hill, 1984), p 42-6.
[6] See note 5.
[7] Author's calculation, linearized for R_{loss} much less than Z_0.
[8] R. B. Dome, "Impedance of Short Vertical Antennas," *QST*, Jan 1976, pp 32-33.
[9] R. B. Sandell, "The Horizontal Dipole Over Lossy Ground," *The ARRL Antenna Compendium, Volume 1*, pp 148-151.
[10] See Johnson and Jasik (note 5), pp 4-4, 4-7 and 4-8.
[11] See Johnson and Jasik (note 5), p 4-5.
[12] See Johnson and Jasik (note 5), pp 4-7 and 4-8.
[13] See note 12.

A Window Slot Antenna for Apartment Dwellers

By Ermi Roos, WA4EDV

7400 Miami Lakes Dr, Apt D-208
Miami Lakes, FL 33014

Hams who live in apartments or condominiums face the severest restrictions imaginable to their Amateur Radio activities. Often, no outdoor antenna of any kind is permitted. This usually limits the ham to using a short indoor antenna. Short antennas are acceptable if operation is restricted to the VHF and UHF range, but if the antenna must be indoors, performance is still likely to be poor. This is because the steel and concrete used in the construction of large buildings shields and absorbs radio waves.

The Window Slot antenna described here is an indoor antenna that functions nearly as well as a comparably sized outside antenna. The concrete and steel building materials actually help, rather than hinder, the performance of this kind of antenna. Not only will the apartment building not obstruct the radiation of RF energy, but the building itself will radiate some of the energy. By converting the building from a shield to a radiator, it is possible to obtain much of the performance of an outdoor antenna from an indoor antenna.

Slot v Wire Antennas

Most antennas used by amateurs are made of wires or metal rods. A transmitter causes RF current to flow along the lengths of the wires or rods, and the current generates RF radiation. Energy from a radiating wire inside an apartment building is attenuated by the walls of the building.

A slot antenna is made by cutting a long narrow slot in a metal sheet. RF power is applied to the slot, and RF current flows all over the area of the sheet. Fig 1 illustrates the essential differences between the two kinds of antennas. The current in the sheet causes the radiation of RF energy. By placing the slot antenna on a window, the walls of the building become an extension of the metal sheet. Some RF current flows through the walls, causing them to radiate power. Therefore, a slot antenna mounted on a window converts the building from a shield to a radiator. Of course, the walls are not as effective at radiating power as a metal sheet, but fortunately, most of the current flows in the immediate vicinity of the slot. Thus, the poor efficiency of the

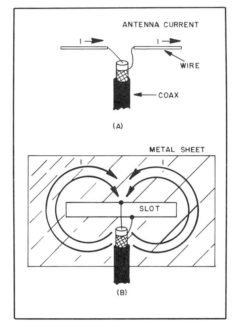

Fig 1—A comparison between the wire antenna and the slot antenna. At A, RF current of a wire antenna is carried only along the wire. At B, RF current of a slot antenna is spread out over the metal sheet surrounding the slot.

walls as radiators will not greatly affect the antenna performance.

If the slot is cut in the horizontal direction, the radiation will be vertically polarized. Unlike a slot, a wire or rod antenna must be oriented vertically in order to obtain vertical polarization. This characteristic of slot antennas is very useful because the windows of most modern apartment buildings are longer horizontally than vertically. It is therefore usually easier to mount a horizontal slot in a window than a vertical one. Because of limitations in the size of windows, the Window Slot is most useful in the VHF-UHF spectrum, and vertical polarization is most often used on these frequencies.

Construction

Each Window Slot antenna is a custom construction project because it is necessary to design an antenna for the particular window being used. Some general construction hints will be given here. The severest restriction in the use of the Window Slot is that the window should face in the direction of desired operation. Usually there will be little choice in this matter. In most apartments the windows face in one, or at most, two directions. In your apartment, if you are lucky, the window you wish to use faces a favored direction. Even if this is not the case, the antenna will radiate in other directions. The best operation will be in the direction outside your window. A power amplifier will help make the antenna useful in unfavorable directions.

A second construction constraint is that the antenna should not be conspicuous when viewed from the outside, and the light shining through the window should not be blocked. This means that the antenna should be made with metal screen instead of solid metal sheet. Copper is the best material to use, but today copper screen is very expensive and difficult to find. It is far more economical to use aluminum screen. The main disadvantage in using aluminum is that it is not as easy to solder as copper. Although soldering of aluminum is difficult, it is by no means impossible, and the necessary techniques are described later in this paper.

The size of the window used is another important constraint. To accommodate different window sizes, three different configurations of the Window Slot are described here. The largest is the full-wave antenna, Fig 2, which is useful for very large windows or extremely high frequencies. This would be the best slot size to use if window dimensions and the operating frequencies permit it. The half-wave antenna, Fig 3, is a little trickier to use, but will function as well as the full-wave antenna once it is properly set up. The short antenna, Fig 4, is used if the window is not large enough to permit the use of a half-wave antenna at the desired operating frequency. This antenna is the most difficult to tune, but will give very good performance if the antenna length is not a great deal shorter than a half wavelength.

The wavelength, λ, may be calculated for a particular frequency, f, by the use of the

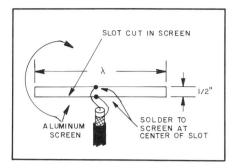

Fig 2—A full-wave slot antenna for vertical polarization.

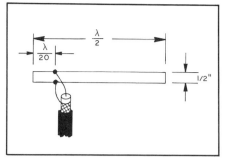

Fig 3—A half-wave slot antenna.

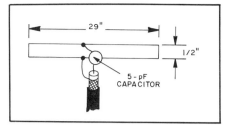

Fig 4—A short slot antenna intended for use in the 144-148 MHz band. The capacitor cancels inductive reactance at the feed point.

following formula:

$$\lambda = \frac{11,800}{f_{MHz}}$$

For example, at 147.0 MHz, λ = 80.27 inches. For a half-wavelength slot operating at this frequency, the total length of the antenna is 40.14 inches, and the coax connector is soldered 4.0 inches from one end of the slot. (See Fig 3.)

To solder the coax cable to the aluminum screen, it is necessary to rub the screen with the soldering iron tip while performing the soldering. The difficulty with soldering aluminum is that a very thin oxide coating rapidly forms on the metal surface, and it is impossible to solder to the oxide. The rubbing action removes the coating from the aluminum, and permits the solder to stick to the bare metal. No special aluminum solder or flux is necessary. The same solder used for ordinary copper wire may be used. Aluminum screen is very fragile, and care should be taken not to separate the strands while performing the soldering. To reduce the mechanical stress on the solder joints, lightweight RG-174 coax cable should be used. In case RG-174 cable is not available, ordinary RG-58 cable will function very well. RG-174 cable has very high losses (0.13 dB/ft at 147 MHz), and the length should be kept as short as possible. If RG-58 cable is used, the losses will not be as great (0.07 dB/ft at 147 MHz), but extra care should be taken when mounting the antenna to avoid breaking the solder connections.

The screen may be mounted to the window with duct tape, but this is not a very secure mounting method. The duct tape will tend to become loose with time. It would be best to have the screen mounted in a frame of the proper size for your window. You will probably be able to find several firms in your area that make window screens with any size frame.

In the full-wave antenna of Fig 2, the impedance at the center of the antenna is 50 Ω, making it ideal for use with 50-Ω coaxial cable. The impedance at the center of the half-wave antenna of Fig 3 is 500 Ω, but at approximately 1/20 wavelength from one end of the antenna, the impedance is 50 Ω. You may have to experiment with the position of the feed point to find the location that gives the minimum SWR.

Short Slot Antennas

Fig 4 shows a short nonresonant slot antenna that is tuned with a capacitor in series with the coax cable. The dimensions and capacitor value shown in Fig 4 are applicable only to the 144-148 MHz band. For other frequencies, the dimensions and capacitor values would have to be determined by experiment.

Unlike short wire antennas, which have capacitive reactance, short slot antennas have inductive reactance. Because of the capacitive reactance of short wire antennas, they are resonated with loading coils. On the other hand, because short slot antennas are inductive, they are resonated with loading capacitors (such as in Fig 4). Because inductors are usually very lossy compared to capacitors, short slot antennas should be more efficient than short wire antennas of the same size. The antenna of Fig 4 has a radiation resistance of approximately 50 Ω at the operating frequency, allowing coupling to 50-Ω coax cable. If the slot is significantly shorter compared to a wavelength than the antenna of Fig 4 (0.36 λ), however, the radiation resistance will become very low. An antenna tuner would be needed to step up the antenna

impedance to obtain proper coupling to a coax cable. To increase the radiation resistance of a very short antenna, the slot can be made wider. This will make it easier to match the antenna to the coaxial cable.

As with most indoor antenna projects, the builder must use a great deal of imagination to make a good slot antenna. The information in this paper should be sufficient for making VHF and UHF antennas, but the real challenge for a serious experimenter would be in making an effective short slot for an HF band. This can be accomplished only with a great deal of trial and error, but the references at the end of this article can provide some additional useful information.

The most attractive feature of slot antennas is that they have no protruding structure, and can therefore be used in many situations where an antenna should not be conspicuous. A slot could even be placed over a ditch in the ground, resulting in an antenna that has no height whatsoever. The earth surrounding the slot would actually become a radiator of RF energy. Admittedly, such an "underground" antenna is more useful for underground organizations than radio amateurs, but the example illustrates the extreme concealability that can be achieved.

We hams are very proud of our hobby, and usually like to show off our antennas and equipment. But the slot antenna should be kept in our bag of tricks for those situations where we would rather not attract too much attention, and simply blend into the scenery.

Bibliography

H. Jasik, *Antenna Engineering Handbook* (New York: McGraw-Hill, 1961).
J. D. Kraus, *Antennas* (New York: McGraw-Hill, 1950). Also 2nd ed. (McGraw-Hill, 1988).

Polar Pattern for the C64

By Steve Cerwin, WA5FRF
3911 Pipers Ct
San Antonio, TX 78251

This paper describes two BASIC programs for the Commodore 64 which graphically plot polar coordinate data. One program outputs to an 80-column printer (MPS801), the other to the 40-column TV screen. The major elements of the programs specific to the C64 are the I/O statements, and translation to other computer types will involve I/O and screen format control.

The programs were inspired by the call of Henry Elwell, N4UH, for software to graphically display polar MINIMUF predictions.[1] However, the programs are not limited to this application and are useful for the general-purpose display of polar relationships. In fact, the mathematical relationship loaded in the programs listed here defines the radiation pattern of a two-element phased array of specified element spacing and phase delay—a useful program for predicting antenna patterns of custom or multiband arrays.

Program Output

Examples of the graphic output from the programs are given in the figures. Fig 1 shows the familiar cardioid pattern obtained when two verticals are spaced ¼ λ apart ($\pi/2$) and fed 90° out of phase (also $\pi/2$). Fig 2 shows the broadside and endfire patterns obtained with two antennas spaced ½ λ apart and fed either in phase (0) or 180° out of phase (π). The double plot was obtained by running the same piece of paper through the printer twice, with the delay variables and plot symbols entered differently for each run. The delay-variable differences are indicated in the figure caption. Antenna orientation is indicated by the dark dots on the Y axis, and phase delay is positive when feeding the upper antenna and negative when feeding the lower.

Figs 1 and 2 are examples of the printer output; photographs of the TV screen output are given in Fig 3. Shown are the endfire versus broadside patterns for the case of ½-λ spacing, as in Fig 2. The fourfold reduction in resolution in going from the 80-column to 40-column format is evident when comparing the two. However, a substantial penalty in process time is paid for

the hardcopy output: five minutes plot time versus 35 seconds for the screen output. Computation time adds an additional 30 seconds or so. For experimentation, the screen output yields quick results, which can be followed with higher resolution hardcopy if desired.

All of the patterns available for a two-element array with spacings from 1/8 to 1 λ and delays from 0° to 180° are given in the *The ARRL Antenna Book*.[2] With the exception of linear versus log output (which can be included in the polar equation in the program, if desired), the outputs of these programs essentially mimic *The Antenna Book* plots. Obviously, these programs offer an element of flexibility for spacings and delays which are beyond those shown in *The Antenna Book,* or are at in-between values. This becomes valuable when analyzing the patterns of fixed arrays operated on multiple bands. For example, if the array with ½-λ spacing depicted in

Figs 2 and 3 were cut for 40 meters but operated at the third harmonic on 15 meters, then the multilobed radiation pattern shown in Fig 4 would result. Here the pattern for in-phase feed is plotted.

The Programs

The BASIC program listings for the screen and printer output programs are given in Programs 1 and 2, respectively. The first line of the screen program sets the cursor color (Commodore key and the number 6 key depressed), clears the screen (shift-CLR), and centers the cursor (multiple down arrow) for the print statements in lines 20-40. (These statements give the programmer something to look at while the computer is off doing trig calculations). Line 50 dimensions the rectilinear array in which the polar pattern, axes, and antenna symbols will be plotted. For both the screen and printer output programs, the largest XY array of X columns by Y rows which

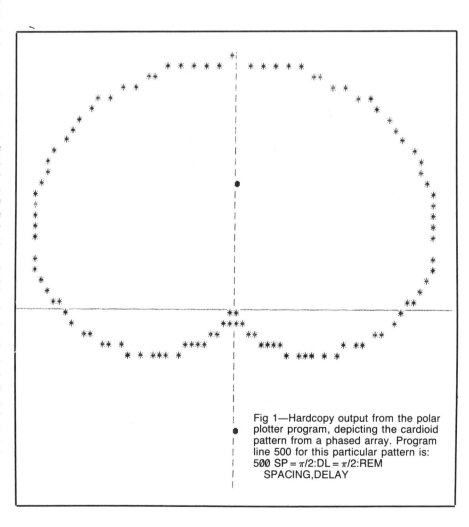

Fig 1—Hardcopy output from the polar plotter program, depicting the cardioid pattern from a phased array. Program line 500 for this particular pattern is:
500 SP = $\pi/2$:DL = $\pi/2$:REM SPACING,DELAY

[1]Notes appear on page 167.

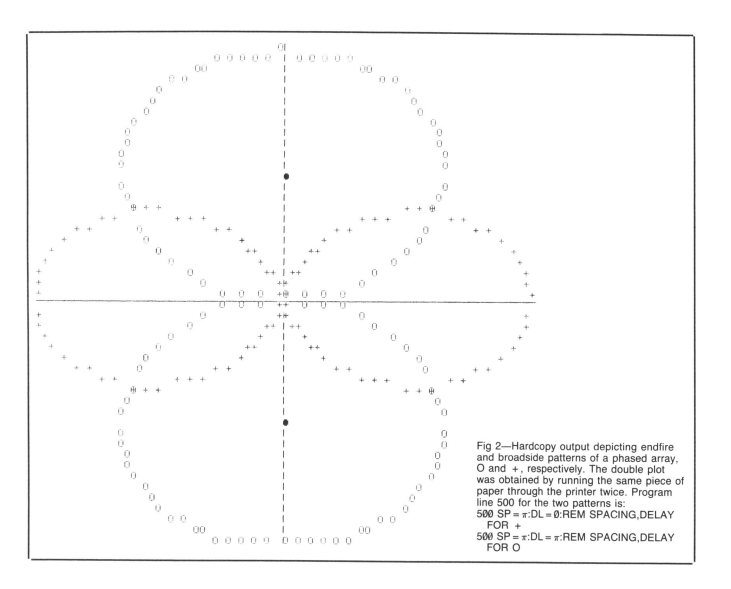

Fig 2—Hardcopy output depicting endfire and broadside patterns of a phased array, O and +, respectively. The double plot was obtained by running the same piece of paper through the printer twice. Program line 500 for the two patterns is:

500 SP = π:DL = Ø:REM SPACING,DELAY FOR +
500 SP = π:DL = π:REM SPACING,DELAY FOR O

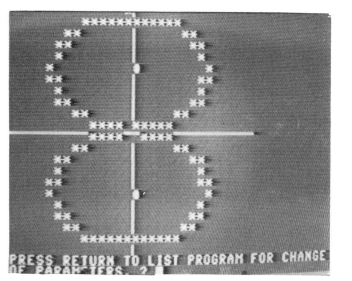
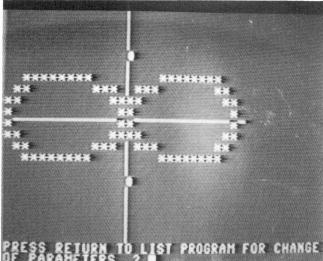

Fig 3—Screen output of the polar plotter program. At A, endfire pattern, and at B, broadside pattern of two elements spaced ½ λ apart. There is a reduction in resolution from hardcopy printouts because of the 40-column screen format, but process time for screen displays is significantly shorter.

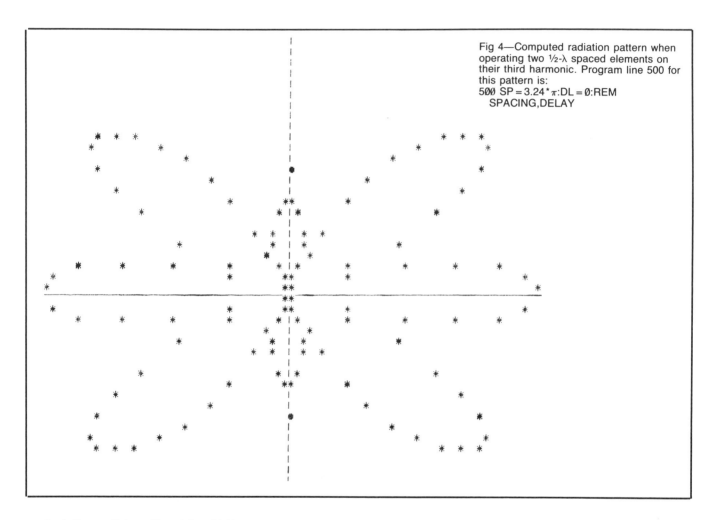

Fig 4—Computed radiation pattern when operating two ½-λ spaced elements on their third harmonic. Program line 500 for this pattern is:

500 SP = 3.24*π:DL = 0:REM SPACING,DELAY

was both "square" (equal length by width) and would fit in the output area was determined. This turned out to be 29 columns by 22 rows for the screen program, and 79 columns by 47 rows for the printer program. The idea behind the plotting program is to use the individual cursor locations on the screen or printer as possible data point locations and to fill only the computed data locations with a data symbol, leaving all others blank. Symbols depicting the axes and antenna orientation are added at the appropriate locations.

Line 60 of Program 1 starts the 120-step loop which calculates polar data at 3-degree increments—a resolution which pretty well matches that available in the output area. Line 70 converts degree data to radians and line 80 calls the subroutine which calculates the length of the polar radius given by the trigonometric formula in line 510. The equation shown gives the radiation pattern of two antennas spaced distance SP radians apart and fed with a phase difference of DL radians. The desired values of SP and DL are entered into the program by editing line 500 directly. At the end of each execution of the program, the program is LISTed (lines 260-270) which allows access to line 500.

The relationship defining the polar function is wholly contained in the subroutine beginning at line 500. Therefore, if data on

something other than the two-element phased array (for example, the 360-degree MINIMUF propagation prediction) is desired, then the appropriate function can be loaded in the subroutine and will be invoked by the call in line 80. The angle TH (radians) is passed to the subroutine, and the amplitude of the polar radius is returned in PR. PR must be a positive number, normalized to the range from 0 to 1.

Lines 90 and 100 perform a polar-to-rectangular conversion using the computed length of the polar radius and the polar angle. These lines determine the X and Y locations in the matrix which are to contain a data point. The coordinates are loaded into the matrix in line 110.

The nested loops for output begin at line 140 and end at line 250. Each column of every row corresponds to an element in the matrix and is tested for a data point. If it contains one, then the data symbol defined in line 190 is printed. The Y and X axes are printed in lines 170 and 180, with the data point having priority in the event of conflict. Lines 200 and 210 plot the symbols indicating antenna orientation, with the symbols having priority over axis marks or data points.

After the program is through calculating and plotting the data (be patient—there are a lot of numbers to crunch!) the polar pattern is displayed until a RETURN is entered

in response to the INPUT statement in line 260. RETURN causes the program to LIST (line 270) returning the computer to the BASIC mode with the last part of the program listed on the screen. The programmer then can modify the spacing and delay parameters in line 500 or the polar equation in line 510 and RUN the program again.

The listing for the hardcopy program is given in Program 2. It operates in much the same manner as the screen version except that there are no messages sent to the screen while the program is in the trig calculation loop, and the computer output remains assigned to the line printer after execution is completed. If the hardcopy program is LISTed to the screen before it is RUN, successive iterations can be accomplished by entering the desired alterations to the program (which remains listed on the screen), placing the cursor over the previous RUN line (which appears at the end of the program listing), and pressing RETURN. Line 260 of the hardcopy program causes lines 500 and 510 to be printed under the polar graph. To return output control to the screen after running the hardcopy program, PRINT#1 and CLOSE1 commands must be executed (or you can just do a "Commodore 64 Reset" by cycling power off and on).

Remember that the spacing and delay variables must be entered in radians. One wavelength = 360° = 2π radians = 6.28. A half wavelength will be half these amounts, and so on. The C64 will accept the π key as a numeric input (SHIFT-↑).

The programs have been fun to use in playing around with the different antenna patterns and combinations of patterns. They have even proved useful in developing new attempts to turn my city lot into a directional black hole for forty meters. The alternate subroutine listed in Program 3 predicts the radiation pattern for a phased array of four verticals arranged in a diamond, spaced ½ λ across a diagonal. The N-S pair are driven endfire and phased with the E-W pair, which are driven broadside. It is a relatively simple task to electronically rotate the pattern to eight points around the compass by switching phasing lines. To see what the pattern looks like you'll have to load up the code and run the program!

Notes

[1]H. G. Elwell, Jr., "360-Degree MINIMUF Propagation Prediction," *Ham Radio*, Feb 1987, pp 25-31, 33.
[2]G. L. Hall, Ed., *The ARRL Antenna Book*, 14th or 15th ed. (Newington, CT: ARRL, 1982 or 1988), pp 6-5 and 6-6 (14th ed.) or pp 8-6 and 8-7 (15th ed.).

Program 1

BASIC Listing for Screen Output Program for C64 Computer

Boldface type in lines 10, 30 and 40 indicates inverse Commodore characters, ie, dark characters on illuminated background.

[This program is *not* included with those available on disk from the ARRL.—*Ed.*]

```
10 PRINT "(space)(heart)QQQQQQQ"
20 PRINT SPC(9)"WA5FRF POLAR PLOTTER"
30 PRINT SPC(7)"QQQ30 SECONDS COMPUTE TIME"
40 PRINT SPC(9)"QQQ25 SECONDS PLOT TIME"
50 DIM XY(29,22):REM DEFINES MATRIX
60 FOR I = 3 TO 360 STEP 3:REM 3 DEG DATA
70 TH = I*π/180:REM DEG TO RADIANS
80 GOSUB 500:REM CALCULATE POLAR RADIUS
90 X = INT(14*PR*SIN(TH) + 15):REM POLAR TO RECTANGULAR CONVERSION
100 Y = INT(10*PR*COS(TH) + 12)
110 XY(X,Y) = 1:REM LOADS CARTESIAN MATRIX WITH POLAR DATA
120 NEXT I
130 PRINT"POLAR PLOTTER OUTPUT"
140 FOR I = 22 TO 1 STEP −1:REM PLOT
150 FOR J = 1TO 29
160 M$ = " "
170 IF J = 15 THEN M$ = "|":REM PLOTS AXES
180 IF I = 11 THEN M$ = "—"
190 IF XY(J,I) = 1 THEN M$ = "*":REM MATRIX SYMBOL SELECTION
200 IF I = 17 AND J = 15 THEN M$ = "•":REM PLOTS ANTENNA ORIENTATION
210 IF I = 6 AND J = 15 THEN M$ = "•"
220 PRINT M$;
230 NEXT J
240 PRINT CHR$(13);
250 NEXT I
260 INPUT"PRESS RETURN TO LIST PROGRAM FOR CHANGE OF
    PARAMETERS. ";R$
270 LIST
500 SP = π:DL = π:REM SPACING,DELAY
510 PR = SQR(2 − 2*COS(π − (SP*COS(TH) − DL)))/2:REM POLAR RADIUS EQUATION
520 RETURN
```

Program 2

BASIC Listing for Printer Output Program for C64 Computer

[This program is *not* included with those available on disk from the ARRL.—*Ed.*]

```
10 REM WA5FRF POLAR PLOTTER
20 DIM XY(79,47):REM DEFINES MATRIX
30 FOR I = 3 TO 360 STEP 3:REM 3 DEG DATA
40 TH = I*π/180:REM DEG TO RADIANS
50 GOSUB 500:REM CALCULATE POLAR RADIUS
60 X = INT(39*PR*SIN(TH) + 40):REM POLAR TO RECTANGULAR CONVERSION
70 Y = INT(23*PR*COS(TH) + 24)
80 XY(X,Y) = 1:REM LOADS CARTESIAN MATRIX WITH POLAR DATA
90 NEXT I
100 OPEN 1,4
110 PRINT#1,"POLAR PLOTTER OUTPUT"
120 FOR I = 47 TO 1 STEP  – 1:REM PLOT POLAR PATTERN
130 FOR J = 1TO 79
140 M$ = " "
150 IF J = 40 THEN M$ = "|":REM PLOTS AXES
160 IF I = 23 THEN M$ = "—"
170 IF XY(J,I) = 1 THEN M$ = "*":REM MATRIX SYMBOL SELECTION
180 IF I = 35 AND J = 40 THEN M$ = "•":REM SHOWS ANTENNA ORIENTATION
190 IF I = 12 AND J = 40 THEN M$ = "•"
200 PRINT#1,M$;
210 NEXT J
220 PRINT#1,CHR$(13);
230 NEXT I
240 PRINT#1
250 CMD 1
260 LIST 500 – 510:REM PRINTS CONSTANTS AND EQUATION
500 SP = π:DL = π:REM SPACING,DELAY
510 PR = SQR(2 – 2*COS(π – (SP*COS(TH) – DL)))/2:REM POLAR RADIUS EQUATION
520 RETURN
```

Program 3

Subroutine to Predict the Radiation Pattern of a Four-Element Phased Array in a Diamond Configuration.

[This subroutine is *not* included with programs available on disk from the ARRL.—*Ed.*]

```
500 SP = π:DL = π:REM SPACING,DELAY
510 PA = .813*SQR(2 – 2*COS(π – (SP*COS(TH) – DL)))/2
520 IF TH > = π/2 AND TH < 3*π/2 THEN PA = PA*( – 1)
530 PB = SQR(2 – 2*COS(π – (SP*COS(TH – π/2) – DL + π)))/2
540 PR = 1.1*ABS((PA + PB)/2)
550 RETURN
```

A VHF RF Sniffer

By Don Norman, AF8B

41991 Emerson Court
Elyria, OH 44035

The VHF RF Sniffer is one of the simplest, yet one of the handiest pieces of antenna test equipment that the VHF antenna experimenter can own. Technically, the device is an absorption wave meter, and its use and electrical design date back into the early days of Amateur Radio. The circuit is nothing more than a tuned circuit with a pilot lamp in series. It is sensitive enough to detect the presence of RF on antenna elements and feed lines, and the relative brightness enables the user to compare current loops on an antenna with the theoretical current loops in *The ARRL Antenna Book*.

The sniffer is shown schematically in Fig 1. A sketch of the physical layout of the sniffer is shown in Fig 2.

Construction

The sniffer is easiest to build on a piece of nonconducting material, such as phenolic perforated board or plastic sheet. Perforated board is easiest to use, as the conductor loop may be threaded through the holes in the board to keep the conductor in place. As an alternative, the loop may be glued to the board with epoxy cement. The conductor loop should be made of enameled magnet wire, no. 16 or larger, with one side of the loop within 1/8 inch of the edge of the board. The insulated board should be large enough to minimize the effects of hand capacitance. A nonconducting handle may be attached. Sensitivity of the sniffer is governed by the spacing of a side of the loop to the conductor being tested. The pilot lamp should use a socket; you *will* blow it out. A no. 48 lamp may be used as readily as a no. 49. A no. 48 lamp has a screw base, while a no. 49 has a bayonet base. Use what you have.

The sniffer can be tuned to cover the 2-meter or the 220-MHz band. Place the sniffer near a source of RF in the band of interest and tune the variable capacitor for the greatest brilliance on the pilot lamp. A hand-held transceiver is an excellent signal generator for tune-up (Fig 3). Once the sniffer is tuned you are ready to examine your feed lines for unwanted RF.

Uses

I use the sniffer extensively in the development of VHF antennas. Other testing devices are available, but one of these sniffers fastened to a wooden yardstick shows immediately and unmistakably the presence or absence of RF in the various parts of an antenna and its feed line. The sequence of steps in checking an antenna is to mount the antenna six or so feet above ground. (A two-meter antenna six feet above the ground is roughly equivalent to

a 40-meter antenna 120 feet above the ground.) Use the sniffer to check for the presence of RF on the radiator. Next, check for RF on the radials. Then check for the presence of RF on the outside of the feed line. The radials should have RF present, while the feed line should not.

One commercial VHF antenna exhibited current loops for the full 50 feet of a coaxial feed line when running 10 watts output. A popular published home-built design showed no detectable RF on the radials, while current loops were present on the coax all the way to the transmitter. Such radials serve no electrical function and are relegated to the roles of humming in the wind and serving as bird perches.

This article falls short of depicting all the possible applications of the RF sniffer. The basic unit has been equipped with clips and hung on feed lines that were going to be out of reach. It has also been attached to radiators with masking tape, and used in a number of other applications. However,

no method has been found to use it efficiently in bright sunlight.

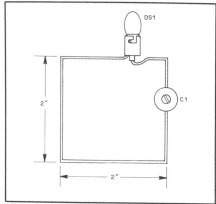

Fig 2—The physical layout of the RF Sniffer can be just like the schematic. Also see the photo.

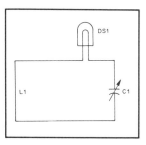

Fig 1—The RF Sniffer is a tuned circuit with a small incandescent lamp used as a current indicator.
C1—5-15-pF variable capacitor.
L1—Wire loop; see text.
DS1—No. 48 or 49 pilot light.

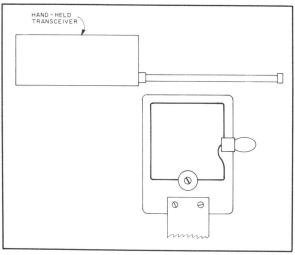

Fig 3—A hand-held transceiver is useful for tuning and checking operation of the RF Sniffer.

Baluns and
Matching Networks

Some Additional Aspects of the Balun Problem

By Albert A. Roehm, W2OBJ
22 Brookdale Road
Cranford, NJ 07016

A simple change in balun style and location can improve the operation of your multiband antenna system. Walt Maxwell, W2DU, wrote a very informative article about baluns for the March 1983 issue of *QST*.[1] Some of the mystery surrounding the connection between a coaxial feed line and a balanced antenna was clarified for the first time in ham radio literature. The balun as a matching device was de-emphasized while trying to promote the idea that "its *primary* function is to provide proper current paths between balanced and unbalanced configurations." The construction of a choke-type balun using ferrite beads on the outside of coaxial cable was also described. If you haven't already read this excellent article, you should. And if you did, perhaps you should reread it just to refresh your understanding of the principles mentioned in it.

Another balun article worthy of your attention appeared in *The ARRL Antenna Compendium, Volume 1,* and was written by Roy Lewallen, W7EL.[2] This paper describes a series of experiments designed to study the performance of a center-fed dipole using no balun, a voltage-type balun, and finally, a choke-type balun. The superiority of the choke-type balun becomes obvious from the data presented.

While both of the above referenced articles address the problem of feeding a dipole antenna with coaxial cable, a more universal approach needs to be investigated. Specifically, what can we expect when using a single, random-length wire antenna and tuner for multiband operation? Note that we are ruling out multiple resonant dipoles fed by a common transmission line (see Fig 1A) or the use of traps (see Fig 1B) because they are specific designs. Let us also assume that our antenna of interest is fed at the center with an open-wire line as shown in Fig 2. The dilemma, of course, is how to connect a balanced circuit to an unbalanced circuit. The answer up until now has been to use a balun at the juncture. This is still the correct solution, but hopefully this paper will help you understand more about it.

¹References and notes appear on page 174.

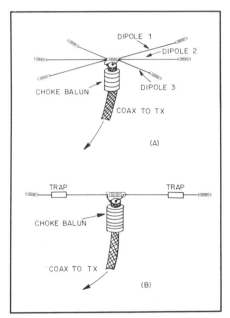

Fig 1—Examples of resonant dipole antennas that do not need tuners.

The Basic Requirements

To begin with, we are faced with two requirements: (1) to resonate the antenna system at the desired operating frequency, ie, obtain a conjugate match, and (2) to go from an unbalanced circuit at the output of the transmitter to a balanced circuit at the input end of the transmission line (points A and B of Fig 2). The first requirement is easily met by using an antenna tuner or stub matching. Since the tuner is so much more flexible than the stub, it is almost universally used by amateurs at HF. Among the most popular tuner circuits we find the L, pi, T, Ultimate and SPC types.[3-6] They have all been described and their virtues outlined any number of times. However, they invariably have one thing in common, eg, for simplicity's sake, they are unbalanced. Although the T circuit is shown in Fig 2 it could be replaced by any of the others.

There are also a number of ways of accomplishing the second requirement. Most often, a balun of the 1:1 or 1:4 variety is used at the output of the tuner. A review of the information in Ref 2 will show that these baluns fall into the voltage-type designs because when they are properly constructed they produce output voltages that are equal in magnitude and opposite in phase. If points A and B of Fig 2 have somewhat unequal impedances, Ohm's Law predicts that either balun will cause unequal currents in the transmission lines. This could lead to a skewed radiation pattern. On the other hand, if the impedances are equal, not only will the transmission-line currents be equal, but no current will flow in the third or magnetizing winding of the 1:1 balun. Paradoxically, if this winding carries no current it is not needed (see Ref 2).

There are other disadvantages to conventionally wound transformer-style voltage baluns. For example, they are generally restricted in bandwidth if constructed with

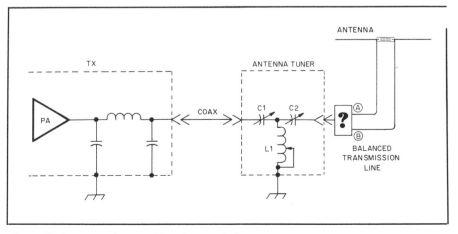

Fig 2—Typical setup for a multiband center-fed antenna.

an air core. If enough inductance is used to cover the 160- or 80-meter bands then there is the danger of having too much interwinding capacitance. Although a solid core can increase the operating bandwidth of the voltage balun, the core is also required to support all the magnetic flux associated with the load current. This principle is exactly the same as the iron core in a power transformer. Conversely, Refs 1 and 2 show that a choke-type balun is used to oppose (choke) only the imbalance current, not the entire load current. In addition, a choke balun constructed of coaxial cable and ferrite beads is not limited by bandwidth, power handling at amateur levels, or SWR considerations.

The Current-Path Approach

The amateur fraternity seems to have overly complicated its understanding of balanced versus unbalanced circuitry. In many applications it doesn't matter if a circuit is balanced or not, as long as the current paths are controlled. For example, in the driven element of a Yagi-type beam or the simple center-fed dipole, the predicted radiation pattern is obtained with equal RF currents in both halves of the antenna. Note that it is not the voltages nor the impedances that need to be equal—only the currents.

Until now, amateurs have been led to think of a balanced circuit as symmetrical about its (sometimes imaginary) center. This is, no doubt, a carry-over from push-pull vacuum tube circuitry as well as audio and telephone technology. Stated another way, a balanced circuit was thought to be an unbalanced circuit plus its mirror image. This definition is true for voltage or impedance considerations between various parts of the circuit and ground. Fig 3 shows an example of a balanced antenna tuner.

It has been recognized for a long time that installing the balun after the tuner subjects it to the same SWR as at the beginning of the transmission line. The impedance at this point can vary by many thousands of ohms. Placing the balun on the input side of the tuner (generally at 50 ohms) allows the tuner to buffer the balun from these transmission-line variations.[7] Until now, this meant using a balanced tuner configuration with its inherent high cost, complexity, and bulk.

To resonate the antenna system shown in Fig 2, all that is needed is an *unbalanced* tuner to obtain the conjugate match and a *choke* balun on the input side to control the current paths. This concept is shown in Fig 4. The explanation which follows is based on the operation of the T-match tuner shown in Fig 4, but can be adapted to any of the other tuner circuits. According to Kirchhoff's Law, the phasor sum of all currents into and out of any branch point in a circuit must equal zero. This idea can be expanded, for example, so we can say that the currents going in and coming

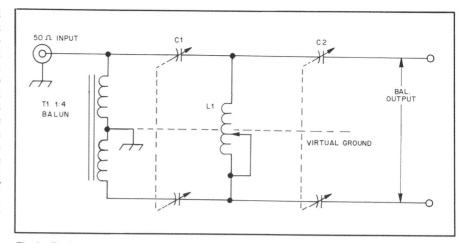

Fig 3—Typical balanced tuner. Note that the dual-section capacitors require insulated shafts.

out of the antenna tuner must also (vectorially) add to zero. This is true provided the dimensions of the tuner are a fractional part of a wavelength and radiation losses are negligible. For operation at HF both of these conditions are met.

For convenience, we can label the various current paths as follows:

I1—the (outside of the) inner conductor of the feeding coax.

I2—the inside of the outer conductor (shield) of the feeding coax.

I3—the outside of the outer conductor of the feeding coax.

I4—capacitive coupling from the tuner to ground.

I5 and I6—the two antenna transmission line conductors.

In a coaxial cable, I1 and I2 are inherently equal in magnitude and 180° out of phase. This can be proved mathematically using Ampere's second law, but a simple experiment suggested by Roy Lewallen, W7EL, also illustrates the idea.[8] This is shown in Fig 5. Here a short length of coax is connected to two unequal resistors

labeled R and 2R. If a high enough impedance is added to the outside of the shield, the cable will drive the equal inner currents through the load resistors *even if the resistors are unequal*. Measurements taken at 400 kHz and 4 MHz indicate that the voltage drop across resistor 2R is twice the voltage developed across resistor R, where R equals 27 ohms. This test proves that the currents are equal, and leads to the substitution of an unbalanced tuner in place of the resistors. Incidentally, the resistors can be interchanged with the same results.

It is desirable, in fact crucial, to reduce the current in path I3 (Fig 4) to nearly zero. As stated above, this is accomplished by increasing the impedance through ferrite loading of the shield. This isolation of the outer surface of the shield results in the transmitter end of the modified coaxial cable being *unbalanced* while the tuner end is *balanced*.

Likewise, path I4 should be reduced to nearly zero. First and foremost, remove any ground connection to the tuner circuit. Second, place the tuner at least 4 inches

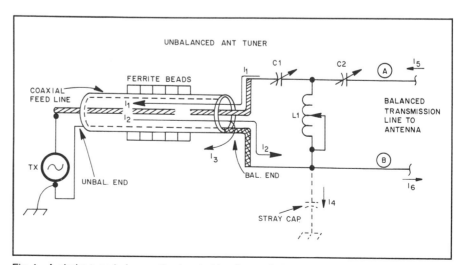

Fig 4—A choke-type balun on the input side of an unbalanced tuner controls current paths and permits the tuner to operate in a balanced circuit.

away from grounded metallic objects and wiring to reduce capacitive coupling.

Another look at Fig 4 will aid in understanding the following key statements. With both I3 and I4 reduced to negligible amounts, and with I1 equal and opposite to I2, then I5 must also be equal and opposite I6. After all, paraphrasing Kirchhoff, "what goes in must come out." Also, I2 cannot equal I6 because of the circulating current through inductor L1. However, the impedance for these two current paths remains close to 25 ohms at the tuner when the antenna system is matched for the operating frequency. This quantity represents the coax-shield half of the tuner's input impedance. While it is nice to have the appropriate test equipment on hand for various measurements, a simple check using only a small neon lamp and a 100-watt transmitter will confirm that the impedance for I5 is usually higher than for I6. For most antenna systems on most bands the lamp will light when coupled to point A but remain dark for point B. This indicates that the voltage developed at point A exceeds the ignition voltage of the lamp. Since I5 is equal in magnitude to I6, the only way V5 can exceed V6 is to have Z5 greater than Z6.

There is another important reason for locating the choke-type balun on the input side of the tuner. Using the recommended design of the balun employing only type 77 bead material, the estimated impedance on the outside of the shield varies from 1560 ohms at 160 meters to 2640 ohms at 10 meters, peaking at about 3480 ohms at 20 meters. You can see that the ratio of the impedance controlling I3 compared to the tuner's low input impedance forces practically all of the load current to enter the tuner. This is not true when the same choke balun is used on the output side of the tuner. Here, for some frequencies, the transmission-line impedance could easily equal or exceed the balun impedance, forcing the current on the outside of the shield to increase dramatically.

Measurements I took with an RF ammeter on my antenna system show excellent current balance on the transmission lines from 160 through 15 meters (10 meters was not tried). Current imbalance was generally less than 5% on all bands and never exceeded 8% after a conjugate match was achieved.

Balun Construction

Depending on the power and operating

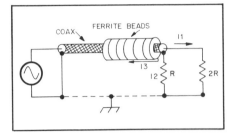

Fig 5—Test setup to show that I1 equals I2 when I3 is reduced to zero.

frequency, there are several ways of making a good ferrite-loaded choke balun. First, select the type of coaxial cable needed to meet the impedance and RF power requirements of the station. Since most amateur installations use 50-ohm cable, RG-58A with an OD of 0.195 inch is a good choice for power levels up to about 600 watts. RG-8/M (Radio Shack no. 278-1328) cable has an OD of 0.242 inch and handles well over a kilowatt of power. An economical balun can be constructed by slipping 12 Amidon type FB-77-5621 ferrite beads ($1.05 each) over an 18-inch length of either of these coaxial cables.[9] The beads measure 0.562 inch OD by 0.250 inch ID by 1.125 inches in length, and can be strung on the cable either touching each other or separated. This balun design can handle about 100 watts of power on 160 meters, 250 watts on 80 meters, and over 600 watts on the other HF bands.

Type 77 bead material is semi-conductive and should be insulated or dressed away from nearby bare wires or other metallic surfaces. The beads are also lossy at HF and under some conditions will get extremely hot at flux densities well below saturation. This material has a permeability of 2000 (which is considered high) and the beads closest to the tuner tend to operate at higher flux densities compared to the beads farther down the cable. This effect is somewhat similar to uneven voltage drops across resistors connected in series. As expected, the beads doing the most work have the greatest rise in temperature. One method of controlling heat build-up is to use shorter beads (presently unavailable in the required ID) or, alter-

natively, to use a lower permeability material, such as type 43 with a mu of 850. An excellent 80- through 10-meter high-power balun consists of 12 type FB-77-5621 beads plus 6 type FB-43-5621 beads ($0.95 each) on a 28-inch length of RG-8/M cable. Be sure to place the type 43 beads on the tuner end of the balun. To extend high-power operation to the 160-meter band, use a 36-inch length of cable and add 6 more FB-43-5621 beads on the input end of the balun.

If RG-8 or RG-213 cable is used, a minimum of 20 type FB-77-1024 beads ($1.50 each) are recommended. These beads measure 1 inch OD by 0.5 inch ID by 0.825 inch long. Again, if the balun is to be supplemented with some lower permeability material, add 7 or 8 type FB-43-1020 beads ($1.50 each) on one or both ends of the balun.

The information presented in this paper should help you understand the operation of tuners and baluns in multiband antenna systems. Three new concepts were introduced: (1) the use of a choke-type balun at the tuner input, (2) a balun (of any type) ahead of an unbalanced tuner, and (3) construction of a balun using a mixture of ferrite materials. Now you can put the right kind of balun in the right place and never worry about it failing.

Special thanks are extended to Walt, W2DU, and Roy, W7EL, for their generous assistance, patience, and encouragement. Without their help I would still be using a toroidal balun on the output side of my tuner.

References and Notes

[1]M. W. Maxwell, "Some Aspects of the Balun Problem," *QST*, Mar 1983, pp 38-40.
[2]R. W. Lewallen, "Baluns: What They Do And How They Do It," *The ARRL Antenna Compendium, Vol 1* (Newington, CT: ARRL, 1985), pp 157-164.
[3]G. Grammer, "Simplified Design of Impedance-Matching Networks," in three parts, *QST*, Mar, Apr and May 1957.
[4]D. K. Belcher, "RF Matching Techniques, Design and Example,"*QST*, Oct 1972, p 24.
[5]L. G. McCoy, "The Ultimate Transmatch," *QST*, Jul 1970, pp 24-27, 58.
[6]W. Bruene, "Introducing the Series-Parallel Network," *QST*, Jun 1986, p 21.
[7]J. S. Belrose, "Tuning and Constructing Balanced Transmission Lines," Technical Correspondence, *QST*, May 1981, p 43.
[8]R. W. Lewallen, private correspondence.
[9]Amidon Associates, 12033 Otsego St, North Hollywood, CA 91607.

A Servo-Controlled Antenna Tuner

By John Svoboda, W6MIT
517 Santa Clara Way
San Mateo, CA 94403

The tuner to be described uses the same L network as many others. But it does work, seemingly by magic, and it sure beats going out into the rain to change bands. It will work with any end-fed antenna. The antenna I am currently using is a 53-foot vertical with an 80-meter trap above a 160-meter loading coil positioned 39 feet from the bottom. This antenna works from 160 through 20 meters when used with the servo-controlled antenna tuner.

The two principle components of the tuner are available from either swap meets or from Fair Radio.[1] One is the motor-driven inductor assembly from a Collins 180L remote tuner, priced around $30. The other is a motor-driven vacuum-variable capacitor from the same tuner, priced at about $70. You can buy the entire tuner for around $110. Another important item is the enclosure. In my case, I found a brand new steel Hoffman electrical control box, 12 × 24 × 10 inches, at my local surplus store. It has a waterproof hinged door on the front. These are also made in fiberglass, which probably would have been easier to work with. I have since found several suitable enclosures at swap meets.

Fig 1 shows the RF circuit of the tuner. L1 and C1 are the motor-driven components mentioned above. Relay K1 switches in L2, and is used for matching impedances of less than 50 Ω (160 and 75 meters). It should be able to handle a reasonable amount of current. The voltage rating need be only moderate. Both relays are off for high-impedance matches. K2 is activated for impedances greater than 50 ohms (20 meters) and is a high-voltage vacuum type. It must be well insulated since it is connected to the RF output. Both relays should have the same coil voltages to simplify the interconnections. Also included are an RF ammeter and an RF-sampling transformer and detector. The detector provides remote current sensing for use during tune-up. The component layout should minimize lead length. Leads should be no. 10 wire or copper strap to reduce stray inductance. L2 was salvaged from a BC-375 10-MHz tuning unit.

Three basic control methods are available to the designer. Using manual control, one

[1]Notes appear on page 181.

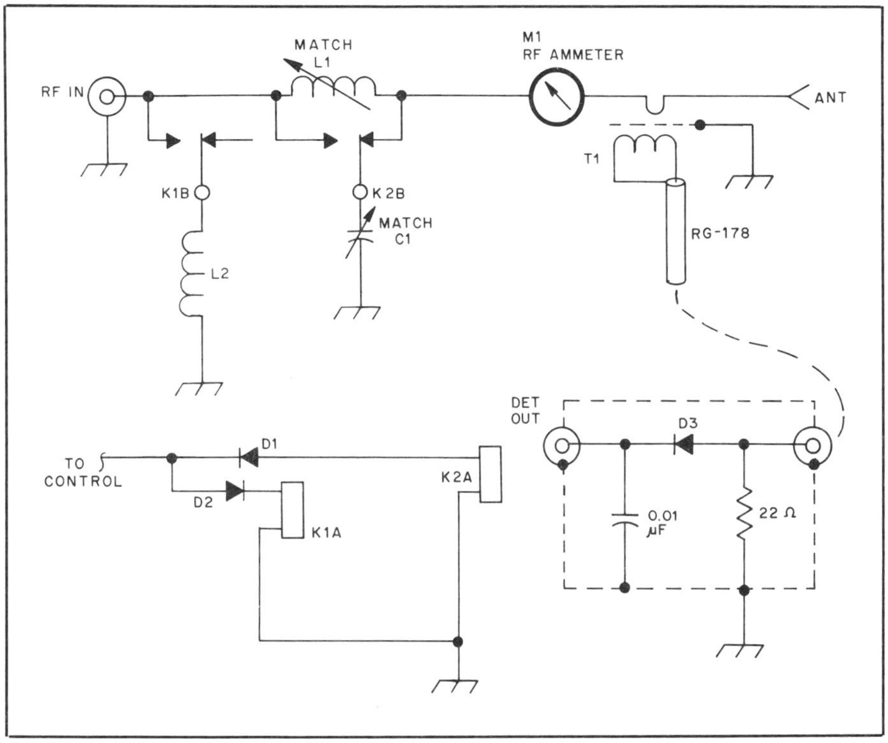

Fig 1—Circuit diagram of the tuner.

C1—7-1000 pF, 3-kV motor-driven vacuum-variable capacitor from a Collins 180L remote tuner.
D1, D2—Silicon power diodes.
D3—Germanium small-signal diode such as 1N34A.
K1—SPDT relay with a 24-V dc coil.
K2—SPDT high-voltage vacuum relay with a 24-V dc coil.

L1—Approx 40 µH motor-driven inductor assembly from a Collins 180L remote tuner.
L2—Inductor of 1.5 to 2 µH. The one used in this project is from a BC-375 10-MHz tuning unit.
M1—RF ammeter, 0-5 amperes.
T1—See Fig 2.

must hold a control switch until the device arrives at the desired position and then bring the next device to its position. This must be repeated for each adjustment, and is slow at best. Fully automatic tuners that are capable of matching over a broad range tend to be rather complicated, and are beyond the scope of this article. Semiautomatic control makes things easier since preset controls can be adjusted and null circuits will bring the variables to their desired settings at top speed. This is the basis of the tuner described in this paper. It uses dc motors, servo amplifiers and components that minimize the problems

associated with construction of gearing systems in a home workshop.

Modifications of L1 and C1

Remove the rotary switch from L1 and the limit switches from C1. Remove the power resistor from both units. I suggest that you do some experimenting by running these assemblies with a variable low-voltage supply (12 volts maximum) so that you get some idea of exactly what happens during operation. In actual operation the motors will see only about half the supply voltage. This will also allow you the opportunity to learn something about the "dead band" in servo systems. Dead band, in this case, is the range through which the signal from the control pots can be varied without initiating a response from the motors. Your experiments should yield stall voltage and current as well as start voltage and current. Take care during your experiments to prevent damage to the components; do not allow them to slam against the stops. Note also that on the low-capacitance end of the travel of C1, it takes a great deal of power to drive it the full range. I made a bracket and attached a spring to neutralize some of this force, reducing the power required to drive it to the extremes.

In the department of "more confusing to write about than to do" is the task of adding the position pot to C1. The last external gear of the drive train turns somewhat less than one turn when running from one extreme to the other. Therefore, you will need to find a slightly smaller gear to mount on the pot—such that it will produce somewhat less than one turn. The exact size is not important so long as they mesh. With a suitable gear on hand, fabrication of the mounting bracket should pose no great problem. Finding the pots and the needed hardware may be a problem, but a few surplus stores should yield all you need.

Positioning Potentiometers

Positioning potentiometers are special pots with a high degree of resetability and linearity, and are basically free-turning 360° degree controls. They have little or no dead space, and go from full resistance to zero in 2 or 3° at the end of their travel. Ordinary pots operate over about 270°, and the beginning and end are dead for about 10 to 20°. They are not suitable for these applications. The resistance value of the position pots is not terribly important as long as the maximum bridge-unbalance voltage does not exceed the servo-amplifier limits, in this case 7 volts. The 2.5-kΩ pots used here are about the minimum acceptable value. For other values simply scale the divider resistance values to suit. The higher the pot resistance, the more Q1 and Q2 (Fig 3) will affect the circuit. Pots with a value of 10-kΩ would probably be a good bet.

Most precision pots have a flange type mount. They are attached to the panel with a ring that allows the pot to be rotated to exact angular position and then locked in place with eccentric washers that fit into a groove on the side of the unit. This permits placing the dead space exactly where you want it. Although mounting seems difficult, I found that I was able to do an adequate job with a fly cutter on the drill press. Figure out the required counterbore, cut that first, then reset the cutter for the through-hole.

The pots used in the controller should be good quality 10-turn types with counter dials. The resistance values should be close to those in the tuner. In my case I used 2.5-kΩ pots in the tuner and 2-kΩ in the

controller. Standard 270° linear pots will work at the sacrifice of some resetability.

Construction

The entire tuner is built on a 19-inch rack panel 8-3/4 inches high and 1/8 inch thick. The panel is then mounted to the inside of the box on short standoffs. This permits removing the tuner for maintenance without dismounting the box, and also makes it a lot easier to work on. With the coil assembly attached to the panel, an additional hole was drilled in line with the shaft of the old switch assembly. The position potentiometer was then mounted on the back of the panel. It must be oriented to establish the dead space outside the area traversed from start to end of the coil.

A small Plexiglas® panel supports the RF ammeter. It in turn is mounted on fiberglass-rod standoffs at a height such that the lead from the coil and the lead to the feedthrough are kept short. The RF sampling transformer is actually mounted within the dome of the feedthrough insulator. See Fig 2.

The circuitry is housed in a closed chassis at the bottom of the mounting plate. Terminal strips are provided for all pots, motors and relay connections. A connector is also provided for the interconnection between the tuner and the wall of the box. All interconnection wiring is shielded and located on the far side of the base plate where possible. Control and power leads

exit the box via a surplus water-resistant AN connector. RF IN and DETECTOR OUT lines are via UHF feedthrough connectors. Grounding bolts are located at each end of the box for convenience. Copper ground straps from the bolts go to the base plate via the cabinet standoffs to assure a low-loss path to ground. Putting the holes in the steel Hoffman box is a demanding task. The output dome insulator required a 4-inch diam hole. I laid out and drilled pilot holes and then visited an electrician friend who has a hydraulic punch set. I'd still be filing today without that punch. The LOCAL C & L controls permit working on the unit without the control box.

Electronics

The basis of the servo amplifier, shown in Fig 3, is the National LM2878 IC, a stereo amplifier chip designed to provide 5 watts per channel. In this configuration it is operating as a high-gain bridging dc amplifier. This arrangement permits zero crossing from a single 29-volt supply. The trade-off is that the motors see only about half the supply voltage. Fortunately, the motors run quite well at this level.

The amplifier however, is operating near maximum dissipation. Therefore, a first-rate heat sink is required. The mounting tab, which is grounded, is clamped between two aluminum bars $3/8 \times 3/8$ inch and $3/8 \times 1/2$ inch, which are attached to a base plate of 0.09-inch thick stock that makes

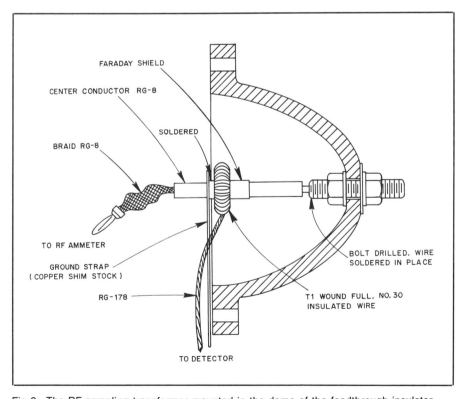

Fig 2—The RF sampling transformer mounted in the dome of the feedthrough insulator. The secondary of T1 is made by winding enough no. 30 wire on an insulated Amidon T-68-2 core to cover it with a single layer winding. The Faraday shield is a tube made from copper shim stock.

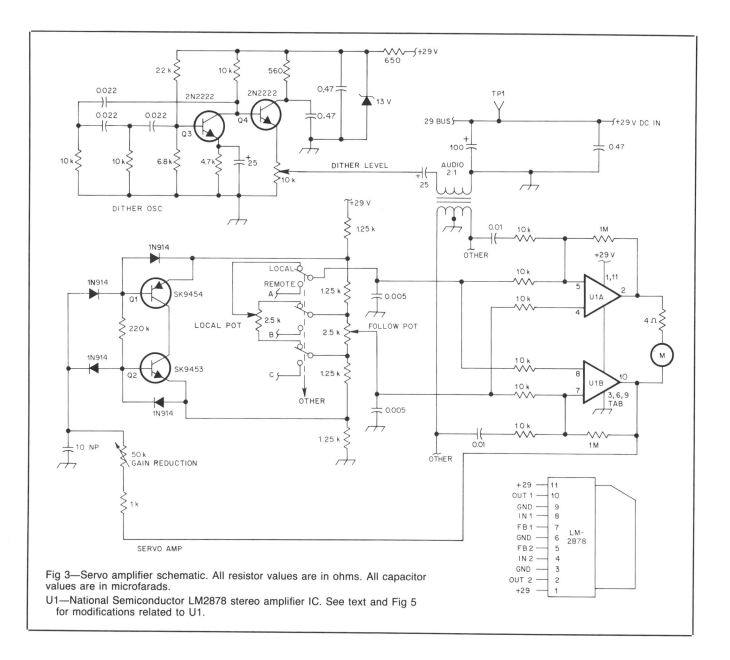

Fig 3—Servo amplifier schematic. All resistor values are in ohms. All capacitor values are in microfarads.

U1—National Semiconductor LM2878 stereo amplifier IC. See text and Fig 5 for modifications related to U1.

a base for the electronics assembly. Over-heating of the amplifiers results in lower gain, producing broader nulls and reduced resetability. This does not occur in normal operation but can and has happened during initial setup demonstrations to friends. Allowing the amplifiers to cool will restore normal operation.

The IC pins were bent at a right angle (as in a DIP) and plugged into the circuit card. The complete assembly is attached to the chassis box. The terminal strip and associated parts are mounted on the back wall of the chassis box.

Q1 and Q2 reduce bridge sensitivity at low amplifier output. The effect is to ease making small adjustments near null. The RCA complimentary pair, types SK9453 and SK9454, are used in this unit. The higher gain SK9455 and SK9456 might be interesting to try.

Dither—Reducing the Dead Band

Dither is a process rather like shaking an alarm clock to try to get it started. The shaking overcomes the mechanical resistance. In this case, ac is fed into the amplifier to encourage the motor to move a little closer to null. The frequency of this signal should, by definition, be several times the system frequency. Probably any frequency between 300 and 2000 Hz would satisfy the requirement. During my experiments I found no optimum and eventually settled on 300 Hz. The peak-to-peak amplitude of the signal should be somewhat greater than the dead band as determined earlier. The effect of the gain-reduction circuit and dither should bring the system closer to null or the preset values. The contribution made by dither alone to the overall system performance is a bit questionable. On the other hand, the parts are not expensive and

it can only help. In as much as these drive assemblies have a lot of mechanical resistance (C1) and slack (L1), perfect resetability will not occur. Some touch-up will always be required to reach a 1:1 SWR.

The Control Unit

The control unit, Fig 4, is mounted in an LMB cabinet and contains two power supplies, the inductor and capacitor (L and C) PRE-SET controls, limit controls, the remote RF ammeter and miscellaneous controls. The LOAD switch provides power for a relay to select the dummy load and the AUX switch provides power to select an alternate antenna, etc.

The first power supply section supplies dc to the tuner. Relay K3 is controlled by an NE555 timer and turns the regulator on and off. Pressing START turns on the line for 12 to 15 seconds. This is more than

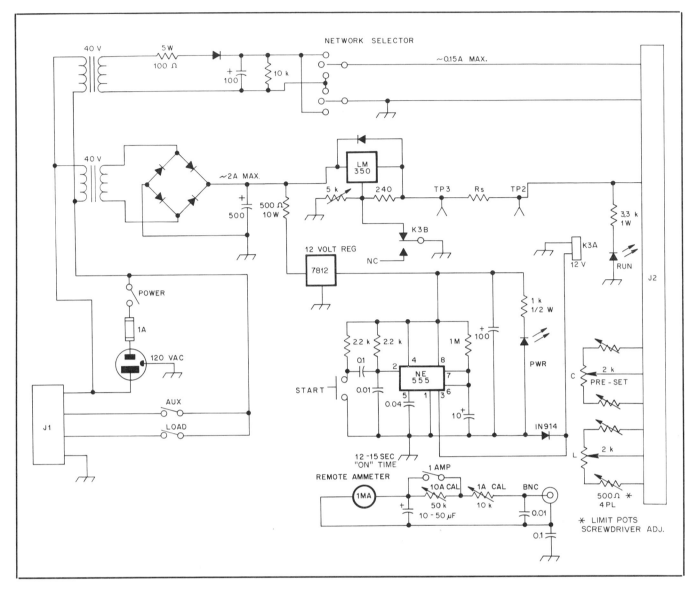

Fig 4—Tuner controller-unit schematic. All resistor values are in ohms and all capacitor values are in microfarads. K3 coil resistance must be greater than 60 ohms and the pull-in current must be less than 200 mA.

enough time for the coil to run from one end to the other. The timer offers two benefits. First, it shuts down the servo amplifiers when they are not needed, thus reducing dissipation. Second, it reduces filtering and shielding problems that arise during high-power operation. Tuning is done only at low RF levels. Final voltage adjustment to the tuner is done after installation. Test points should show 24-26 volts with the servos at null.

A second, poorly regulated power supply operates the relays in the tuner. It provides plenty of closing voltage but drops to a reasonable holding voltage. The control wire is Belden rotator wire with two large and seven smaller wires. By setting up the power supplies and regulator as described, control-cable length can largely be ignored.

The remote meter is fed with a separate shielded wire. It enters the control chassis

with an insulated BNC connector. The return is dc blocked to prevent motor and relay current from influencing the meter.

Checking the Servo Amplifiers

To verify operation of the servo amplifiers, I suggest that you substitute a motor similar to the type in the assemblies or that you temporarily remove the motor from C1. Connect it and adjust the LOCAL pot until the motor stops. Continue turning the pot slowly and the motor should start again, but in the reverse direction. After you have checked out both amplifiers, reinstall the motor. Before continuing, be sure to double check all wiring, as an open in the bridge circuit can cause the system to malfunction. Ensure that the LOCAL-RMT (normal) switch is set to LOCAL and that the L and C controls are set near the center of their ranges. Reduce operating voltage as much as possible and connect power.

The motors should drive toward a null and stop. If they simply go to one end (take care hitting the end stop), it probably means that you will need to reverse the motor leads. Try again, and the unit should drive to the null. If minimum dial setting now occurs at maximum component value, reverse the outer connections of the pot and the two motor leads. After you are satisfied with basic operation then increase voltage to the normal 29 volts.

Adjustment of the limit pots in the control unit is fairly straightforward. Set the switch to RMT (normal) and the limit pots to about midrange. Check system operation, then set what would appear a safe value near minimum resistance and run to the limit of the device. After several trials with different pot settings you should be able to come very near the end without hitting the mechanical stop.

Calibration of the remote ammeter can

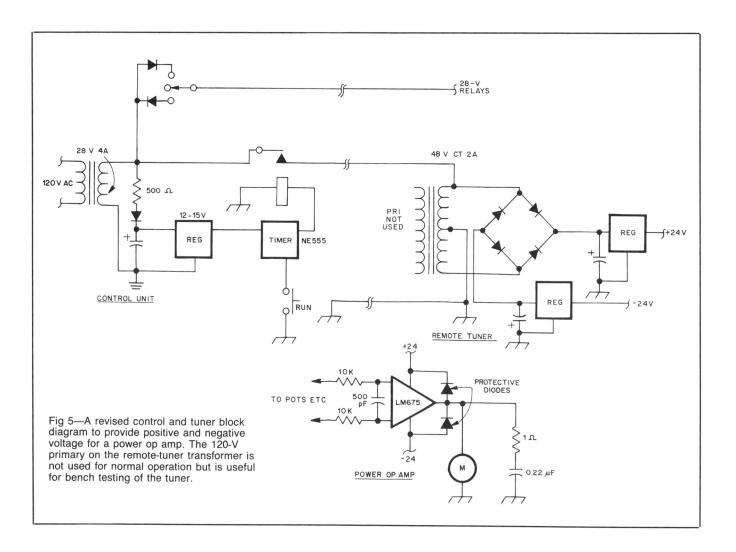

Fig 5—A revised control and tuner block diagram to provide positive and negative voltage for a power op amp. The 120-V primary on the remote-tuner transformer is not used for normal operation but is useful for bench testing of the tuner.

be done any time after the system is installed and operating. Tune up normally, then go out and note the RF ammeter reading. Assuming the range switch is set for the 1-A range, adjust that pot. For improved accuracy, it is best to set the 10-A range at high power. Note the reading and adjust accordingly. If the meter is too "lively" on SSB, increase the value of the shunt capacitor.

Operation

I tune up into a dummy load, then switch to the network and bring it into adjustment. Minimum SWR and maximum antenna current should occur at the same setting if all is well. For your initial band settings, log your values and then back off the tuner to some other setting. Now reset your logged value and run the tuner to it. You should be right on. If not, record the value and try again. The average of several trials should be quite good. Allow the amplifiers to cool between (extended) trials to ensure sharp nulls. Dial settings on 160 meters will probably always require some touch-up as bandwidth may be rather narrow. A simple

Fig 6—The position-reading potentiometer for C1 is in the foreground. The electronics-to-cabinet connector is seen in the square opening in the side of the electronics chassis to the left. The detector enclosure is a small box to the extreme right and rear. The object just in front of the detector is a static bleed resistor left over from an earlier version. The spring and bolt standing above the position pot are to balance tension on C1, as mentioned in text.

Fig 7—This view shows details of the homemade high-voltage relay. A vacuum switching element was available surplus. It was mounted in a fiberglass tube and bonded to a fiberglass base plate. A ¼-inch fiberglass rod was bonded to the switch arm and runs through to the base where it is activated by a stripped-down relay assembly. Contact between the clamp and switch body is via copper shim stock exiting one of the four slots (one visible in the photo). A wide copper strap connects the shield of the RG-8 atop L1 to the panel below.

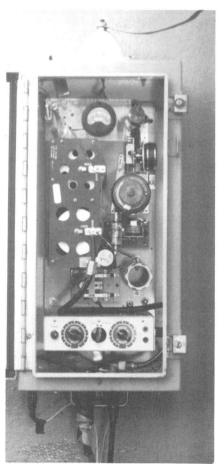

←

Fig 8—The tuner installed in the weatherproof Hoffman cabinet. L1 occupies most of the space. The tuner can be controlled with the LOCAL controls (bottom of cabinet) as well as from the remote unit. Bags of silica gel reduce humidity buildup in the cabinet.

calibration chart will now get you tuned up in short order.

After operating the tuner and the vertical antenna for several years, I find that the speed and convenience has made band changing a pleasure. The hum of the tuning motors doing their thing has made the difference.

Addendum

Some time has passed now since I built the unit just described and wrote the foregoing material. Probably the availability of parts has changed. One noteworthy change has occurred, namely the LM2878 has been discontinued by National Semiconductor. This is unfortunate since this complex chip handles a reasonable amount of power and the price, $2.95, was attractive. I have not been able to locate a direct replacement. There are several things to consider when looking for a substitute. The power rating should be 5 watts or more per channel. The maximum operating voltage should be around 30 volts. There are many audio chips on the market, but most are designed for operation in the 12- to 15-volt region with a maximum rating of 20 volts. This

Fig 9—The position-reading pot for L1 is in the foreground. A small flexible coupling connects this pot to an existing shaft on L1. Direct-current wiring can be seen under the panel. Note: L1 assembly is raised slightly on thick washers to accommodate the pot-to-shaft interface.

Fig 10—The remote control unit is mounted in a small LMB cabinet. The meter reads antenna current at 1 or 10 amperes full scale. Standard 270° pots with calibrated knobs were used here. These were later replaced with ten-turn pots and counter dials to improve repeatability. The ANT. CUR.-SERVO meter switch is not shown on the schematic since it served only passing interest. What shows as a white line between the cabinet and the feet is a T-shaped calibration chart retained by an elastic band. It provides dial settings for frequencies of interest.

results in less than 10 volts to the motors and the results are not too satisfactory. The chips should have differential inputs.

My overall design approach has been to minimize the mechanical complexity as well as the electronics. Keeping it simple generally improves the reliability. A discrete amplifier design is certainly possible, however it will require at least four power ICs per channel, plus numerous other parts. Power operational amplifiers (op amps) are probably the best bet. Unfortunately, because of limited markets, they tend to be expensive—$25 and up. Burr-Brown and National Semiconductor make a variety of these.[2] Fortunately, National makes the LM675 which is affordable.[3] I have not had a chance to try it but expect to for my next version.

The most practical circuit configuration requires a plus and minus supply. This can be achieved without adding another wire to the remote unit by modifying the power system as shown in Fig 5. Moving the rectifier and regulators to the tuner will also improve overall performance. The unused primary winding of the tuner supply will be very handy when bench testing the unit.

Although it may not be possible to make an exact duplicate of the control circuit, the the mechanical and RF system probably won't require any modifications. Mechani-

Fig 11—This interior view of the electronics chassis shows the servo amplifiers adjacent to the heat-sink bar near the center of the enclosure. At top left is a gain-reduction adjustment pot. Center left is the adjustment for dither amplitude and below that is the second gain-reduction pot. To the right of the heat sink is the dither oscillator.

cal details and components are illustrated in Figs 6 to 11.

As with most ham projects, they are never really finished—there is always another improvement just around the corner. This unit is actually model 3. If you happen to locate a great deal on power op amps, please drop me a line and I will begin work on model 4.

Notes

[1]Fair Radio Sales Co, Inc, Box 1105, 1016 E Eureka St, Lima, OH 45802, tel 419-227-6573.
[2]Burr-Brown Corp (corporate headquarters) Dept G, PO Box 11400, Tucson, AZ 85734, tel 602-746-1111.
[3]Available from Digi-Key Corp, 701 Brooks Ave S, PO Box 677, Thief River Falls, MN 56701. Price class as of this writing: $6.

Remotely Controlled Antenna Coupler

By Richard Z. Plasencia, WØRPV

PO Box 1195
Cedar Rapids, IA 52406

The seemingly endless low at the end of the previous sunspot cycle and the resultant lack of openings in the upper HF bands have given me ample opportunity to optimize my antenna system for the lower bands. I settled on a vertical antenna for use on 160 through 40 meters for this location back in 1984. Although the vertical has provided superior performance compared to the various horizontal antennas used in the past, the vertical has unique problems. The two most important of these problems are that the feed point must be directly over a ground screen or radial system, and that the length of the antenna feed line is measured in inches. Add to these the complications caused by large feed-point impedance changes because of multiband operation, and the inescapable conclusion is that one needs an antenna coupler at the base of the vertical as part of the antenna system. One of my better couplers was the "ammunition-box" model shown in my December 1985 *QST* article.[1]

The First Coupler

In the ammunition-box coupler, band changing was accomplished by moving alligator clips to different coil or capacitor taps. The ammunition-box coupler was good simply because the lid was easy to

---

[1]R. Z. Plasencia, "Computer-Aided Two Band Vertical Antenna Design," *QST*, Dec 1985.

open and close when changing bands, even with mittens on during Iowa's harsh winter. Whatever invective appropriately describes our winters, one thing is true, they spark creativity. I must confess it took two winters to spark my creativity. To change bands I had to go through two rooms, put on a parka, mittens, and rubber boots, then go through the garage and one more door to reach the back yard. Now I was ready to trudge through 75 feet of back yard usually covered with drifted snow, locate the coupler by where the vertical disappeared into the snow, and clear the snow away before opening the easy-to-open lid. That was fun. Try it a few times with the parka over your pajamas in near blizzard conditions and see if that doesn't cause a spark of creativity.

Remote Control

While thawing out after my return from one of these treks I resolved to design a remotely tuned antenna coupler. I placed some constraints on the design. This had to be a relatively inexpensive design, easily reproduced by the average ham. After all, there would be little challenge or fun in this project for me if I used the resources made available to me by my gracious employer. No special tooling or components could be used. The original coupler was destroyed while being cleaned to be photographed for this paper. So I am presenting version 2 of the coupler, a much

simplified and more reliable unit than the prototype.

This article is not meant to be a step-by-step construction article to duplicate my antenna coupler. This type unit is amply described in various reference sources available to most hams. However, if you study the photographs (Figs 1 through 8) and the schematics (Figs 9 and 10) you should be able to make a good copy. My intent is to pass along some design criteria and a method of building a high quality capacitor.

The electrical design is our old standby, the L network. An L network has a minimum of components and can be used to match almost any antenna load encountered by hams. The main components of an L network are a coil and a capacitor. If the coupler is to be used with a 1-kilowatt transmitter, these two circuit elements will have to be capable of handling large currents and high voltages. Standard coil stock in the 2- to 3-inch size with 8 to 10 turns per inch will handle the current. The variable capacitor will need a rating of at least 10,000 volts to cope with the voltages on the antenna when the operating frequency is such that the antenna is operating as a half-wave radiator. The maximum capacitance is dictated by several design parameters such as L/C ratio (related to the required impedance-transformation ratio), desired range of adjustment, and lowest band to be covered.

Fig 1—General view of coupler. The white cylinder in the foreground is PVC pipe used as the maximum-capacitance travel limit. The variable capacitor is at half maximum position.

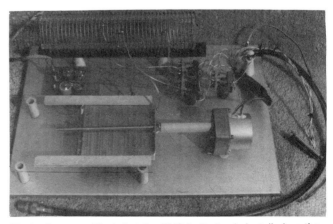

Fig 2—Same as Fig 1, but the PVC limit stop is installed on the motor shaft and the capacitor is at maximum capacitance.

Fig 4—Weather cover showing feed-through connector with caulking material to waterproof the joint.

Fig 3—Same as Fig 2 but with the capacitor at minimum setting where the movable plate would bear against weather cover. This mechanism is used to create the minimum-capacitance limit stop.

The Variable Capacitor Solution

Often a very low minimum capacitance is more important than a very large maximum. If you read the above to mean a vacuum variable is required, go to the head of the class. If you also understand vacuum variable equals lots of money, you know your capacitors. It is apparent from the above discussion that the single most expensive and difficult to procure component will be the variable capacitor. One more item: Conventional capacitors are unsuited because of moisture shorting the plates, and corrosion of the rotary contacts causing noise in the receiver. The ammunition-box coupler used a small tapped inductor in series with the fixed capacitor to provide a means of trimming the fixed capacitor to the exact required value for antenna matching. Using a motor-driven roller inductor in the new coupler meant using moving contacts, a source of noise and eventual circuit failure because of corrosion. Therefore, a variable inductor was ruled out both as the L component and as a means of trimming circuit capacitance.

We will make our own tuning capacitor. For our purposes it will have characteristics similar to the finest vacuum variable except for the price. Take a good look at the photographs of the coupler. The variable capacitor is the device on four pillars to the left of the motor. It consists of two aluminum plates separated by a piece of window glass. The lower plate is epoxied to the glass. The upper plate is free to move in a wooden track epoxied to the upper surface of the glass. The motor is reversible and moves the upper capacitor plate by rotating a threaded rod in a wing nut pinned to a tab on the capacitor plate. The four pillars are cut from PVC pipe to insulate the capacitor from the chassis and to elevate it into alignment with the motor shaft. The PVC supports are also epoxied to the glass and chassis.

One source of single-weight glass is dime-store picture frames. Threaded rods, wing nuts, small nails or brads, PVC pipe, and epoxy are found at hardware stores. The wooden track for the upper capacitor plate is made from a single wooden paint stirrer obtained free at the hardware store. A source of aluminum sheet is cooking utensils such as cookie sheets or pie plates. That takes care of the logistics.

Here are the design criteria so you can make a capacitor suited to your needs. For reference look in the section on capacitance in your *ARRL Handbook for the Radio Amateur*, the single most important reference book in your shack. There you will find a table of dielectric constants and breakdown voltages and the formula for calculating capacitance. That formula is

$$C = \frac{0.224KA}{d} (n - 1)$$

where
  C = capacitance, pF
  K = dielectric constant of material between plates
  A = area of one side of one plate in square inches
  d = separation of plate surfaces in inches
  n = number of plates

If the plates in one group do not have the same area as the plates in the other, use the area of the smaller plates.

Fig 5—The method of assembly of wing nut to movable capacitor plate. The PC board with the four ceramic capacitors is the series capacitor to electrically shorten the antenna to 40-meter resonance. The banana plug leading out of the picture goes to the antenna lead and the hot end of the coil.

Fig 6—Details of the lower capacitor plate and support stand-off insulators. The wooden track for the upper plate is made by splitting a wooden paint stirrer with a knife into one narrow and one wide strip. The narrow strip is cemented to the glass; the wide strip is cemented on top and overhangs the movable plate, creating a slotted track. Since the wood is supported by the glass plate, its insulating qualities are of no importance.

Notice the influence that the dielectric constant, K, has on the numerical results of the formula. Increasing K increases the total capacitance. Common window glass has the highest K factor in the table. This means that the capacitor plate area can be eight times smaller than a similar capacitor using air as a dielectric, further simplifying our project. I have been using K = 8 in my calculations, and checking prototype capacitors with a capacitance bridge verifies this assumption.

Single-weight glass is commonly used in picture framing. This type of glass is about half the thickness of ordinary window glass. I used pieces measuring approximately 0.063 inch thick, or 63 mils. According to the table this will yield a breakdown

Fig 7—Detail of relays, coaxial cable connection, and ground lug. The diode going to relay no. 1 is inside the plastic tubing that protects the motor-lead splices.

Fig 8—A view of the control box for the system. Red LEDs at the upper right area of the panel indicate when travel limits of the variable capacitor are reached.

voltage rating of $200 \times 63 = 12,600$ volts. If you assume a rating of 250 volts per mil of thickness then the capacitor will have a breakdown rating of 15,750 volts.

My requirements were for a maximum capacitance of 550 pF. I designed for 575 pF because the formula is for a single dielectric. Since I cannot achieve perfect metal-to-glass contact, some air intervenes. This has the effect of lowering the total capacitance from the calculated value. The minimum capacitance is the circuit stray capacitance, as the capacitor plates can be widely separated. The capacitor has the characteristics shown in Table 1.

The Mechanism

The principle of operation is simple. A reversible motor turns a threaded rod that engages a wing nut attached to the movable capacitor plate. This plate is grounded through a strap to the chassis. This simplifies the electrical design in that no insulation is required for the capacitor shaft. The lower capacitor plate is connected to the circuit through a strap. Bear in mind that the lower plate also acts as a capacitor in relation to the chassis. For this reason it must be elevated a sufficient distance to reduce the stray capacitance thus created to an acceptable level. I could have mounted the capacitor on its side and reduced the stray capacitance to that provided by the leads alone, but the ensuing mechanical complexity was not worth the effort.

Threaded rod of ¼-inch diameter is made with a pitch of 20 threads to the inch. This means that a nut threaded on the rod will move a distance of one inch for 20 turns of the rod. Keep this fact in mind when you select the motor shaft speed. My motor came from a vending machine and the shaft turns 90 r/min. The total travel for my capacitor plate from maximum to minimum capacitance is 4½ inches, which means the motor must run for approximately one minute to run the capacitor out and one more minute to run it in all the way. While this may seem slow, it does allow me to fine tune with ease.

An improvement would be to use a faster motor and variable motor voltage to control the tuning rate. You can also change the shape of the capacitor plates to suit the

capacitance change to the rate of travel.

If you have a computer, you can use a simple program written in standard BASIC to calculate the capacitor design parameters. A program listing is given as Program 1, and a sample program run is given in Table 2. [This program is available on diskette for the IBM PC and compatibles; see information on an early page of this book.—*Ed.*]

Construction of the coupler follows normal procedures and is adaptable to materials at hand. My chassis is an aluminum cookie sheet. All components are attached to this by screws and internal-tooth washers. Don't forget these washers. After the aluminum ages a bit it will be covered with aluminum oxide—one of the best insulators available. The teeth in the washers bite into the aluminum, assuring a long-lasting reliable contact.

The PVC stand-offs for the capacitor are epoxied both to the glass plate and the chassis. The motor is held to the chassis by two screws underneath the chassis that go into holes drilled and tapped into the gear case. This is more secure than a strap around the gear case. A spacer was used under the motor to bring the drive system into alignment. The relays are surplus 24-volt dc units, mounted on insulating spacers. Only one set of contacts is used. The relays were made to control large motor loads, and the contacts are rated at 10 amperes each. Wiring to the relay movable contacts uses hook-up wire to provide the required flexibility. The fixed-contact connections to the coil use no. 12 solid wire. The coax RF input is by means of a short cable terminated in a stand-off insulator and laced to a large ground lug. The ground lug has the coax braid and the ground-lead braid soldered to it. The ground lead is attached to the antenna system radial ground.

Keep It Dry

The cover to protect the coupler from the weather is a surplus parts tray. A plastic dish pan can be used instead. The cover has only one hole in it, and that is for the antenna connection. The antenna lead—the braid seen originating from the PC board with the four fixed capacitors—is attached to the cover feed-through with a

---

Table 1

Home-Built Variable Capacitor With Glass Dielectric

Size of glass plate = 8.5 × 5.5 inches
Size of metal plates = 4.5 × 4.5 inches
Breakdown voltage = 12,000 volts (average value from tests)
Maximum capacitance = 542 pF
Minimum capacitance = 16 pF
Power factor = negligible, about the same as air
Cost of materials = $2.87 (the picture frame cost $1.25)
Time to build = 2 hours

Table 2

Sample Run of Program 1

```
RUN
DIELECTRIC CONSTANT =? 8
AREA OF ONE PLATE IN SQUARE INCHES =? 20.25
SEPARATION OF PLATES IN INCHES =? 0.063
TOTAL NUMBER OF PLATES =? 2
MAXIMUM CAPACITANCE IS : 576 pF
RUN PROGRAM AGAIN? Y/N :? N
Ok
```

screw. The cover is lowered over the chassis and secured by a bead of silicone caulking material. The cover extends beyond the end of the chassis to which the cables are attached, leaving a gap. This serves a dual purpose. One, it allows the cables to reach outside without cutting into the cover and compromising its rain-protecting qualities. Two, it allows circulation of air to prevent harmful condensation because of weather changes.

The multiple-pin connector for the control wires is mated to the cable coming from the shack, and the connectors are pushed up into the coupler to protect them from rain. Notice the unused pins on the control connector. This is from the original coupler where several extra wires were needed for capacitor-travel limit sensing and capacitor position. The chassis sits on a concrete block at the base of the antenna.

The control unit sits on a bookshelf above my operating position. The case is from a surplus computer disk drive. Inside is a simple 24-V dc power supply rated at 2 amperes. A double-pole rotary switch controls power on/off and band changing. A green LED in the upper right of the panel indicates power on. A double-pole double-throw spring center-off rotary switch controls the capacitor motor. Red LEDs in the upper right of the panel indicate limit of capacitor travel. I could have used a standard toggle switch instead of a fancy rotary, but I had both types in the junk box and I do enjoy my luxuries.

Improvements

Here is what I learned from the prototype coupler: KISS. Keep it simple, stupid. I had micro switches at either end of the capacitor to sense limits of travel. They were hard to keep adjusted and ultimately failed because of corrosion. The original capacitor drive system used flexible couplings and springs to take up excess motor torque. The couplings would slip, and the springs work hardened and broke. The new model does away with all these problems by not using limit switches or flexible couplings.

I originally thought it important to know how much capacitance was in the circuit. I read this by using a dial cord attached to the movable capacitor plate that went around a plastic pulley with a 5-inch circumference. The other end of the dial cord was kept taut by a long coil spring running back under the capacitor. The pulley rotated a potentiometer whose resistance was read by a milliammeter on the control panel. The circumference of the pulley was made longer than the length of the capacitor-plate travel to accommodate the fact that most potentiometers only rotate 330 degrees. This still provided an accurate linear position readout. Mute testimony to the usefulness of this complicated feature is seen in the fact that I left it out of the new improved coupler. I now rely entirely on my Bird wattmeter for tuning indica-

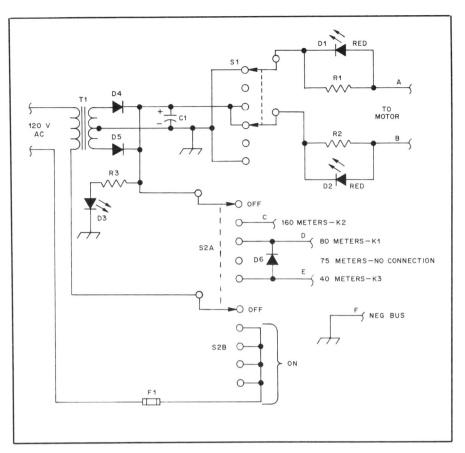

Fig 9—Schematic diagram of the power supply and control unit.
C1—150-μF, 50-V capacitor.
D1, D2, D3—Radio Shack 276-1655 assorted LEDs.
D4, D5, D6—Radio Shack 276-1611 diodes, 50 V, 6 A.
F1—¼-A fuse.
R1, R2—3 Ω, 2 W.
R3—1.2 kΩ, 1 W.
S1—DPDT, center off.
S2—Double-pole, 5-position wafer.
T1—Power transformer; secondary 48 V, 2 A, center tapped.

tion. If I run the capacitor toward MAX and the SWR starts to go up, I reverse the switch position to MIN. Egad, how simple.

In the accompanying photographs you will notice a short length of PVC pipe covering the capacitor-drive screw between the motor and the glass support/insulator plate. When the motor rotates to move the capacitor plate to the maximum-capacitance position, the plate will jam against the end of the PVC pipe stopping further motion. In the other direction the capacitor plate will jam against the plastic weather cover. The results are positive limit stops—crude but effective and maintenance free.

How does the operator know the limit has been reached? Think a bit. What happens to motor current when the motor stalls? Right, it goes up! We sense the excess current condition with a red LED bypassed by a 3-ohm resistor in each of the two leads to the motor. The operating current is about 300 mA; under this condition not enough voltage is developed across the 3-ohm resistor to light the LED. When the motor stalls, the current increases to over 500 mA; the LED will now light. Two LEDs are used because the polarity of the

motor control leads changes with rotation direction. A single no. 51 bulb used in automobiles works very well, but I changed to LEDs to get that modern solid-state look.

I eliminated the other source of trouble, slipping couplings, by not using them. The threaded shaft is pinned to the motor shaft by drilling through the coupling and motor shaft and using a small nail as a pin. This is cut almost flush and riveted into place by a couple of well-applied hammer blows. The threaded-shaft to motor-coupling interface is handled the same way. Some system flexibility is preserved by the method of attaching the drive nut to the capacitor plate. A wing nut is used. The extreme ends of the nut are drilled with the same drill bit used on the motor end of the shaft. The movable capacitor plate is drilled to pass the threaded rod and, on either side, slots are made to accept the wings of the wing nut. (Use two or three small holes and a small Swiss file to connect the holes, thereby forming the slots.) Use two small nails for cotter pins to secure the wing nut to the capacitor plate. If you have done the job as intended, the nut will

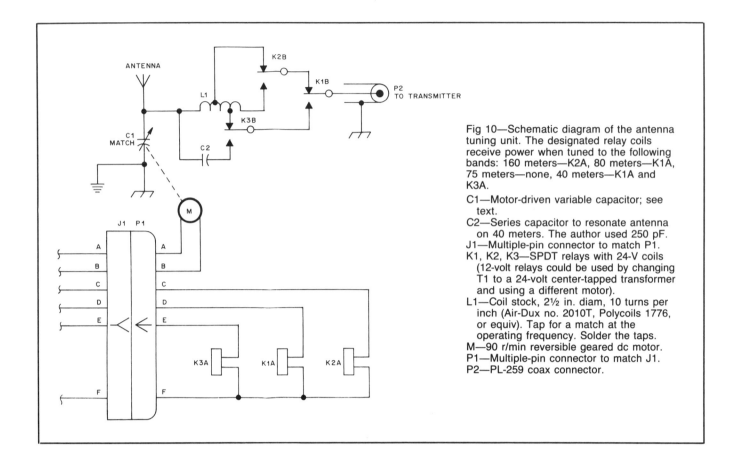

Fig 10—Schematic diagram of the antenna tuning unit. The designated relay coils receive power when tuned to the following bands: 160 meters—K2A, 80 meters—K1A, 75 meters—none, 40 meters—K1A and K3A.

C1—Motor-driven variable capacitor; see text.
C2—Series capacitor to resonate antenna on 40 meters. The author used 250 pF.
J1—Multiple-pin connector to match P1.
K1, K2, K3—SPDT relays with 24-V coils (12-volt relays could be used by changing T1 to a 24-volt center-tapped transformer and using a different motor).
L1—Coil stock, 2½ in. diam, 10 turns per inch (Air-Dux no. 2010T, Polycoils 1776, or equiv). Tap for a match at the operating frequency. Solder the taps.
M—90 r/min reversible geared dc motor.
P1—Multiple-pin connector to match J1.
P2—PL-259 coax connector.

be free to move in all axes, except rotational, in a manner similar to a universal joint.

Here are some further notes on the variable capacitor that may be of general interest. I have another version in which the plates are fixed; one plate is the side of the metal equipment enclosure, and the glass is moved in and out. This type of capacitor avoids movable contacts and the inductance of connecting leads, but it provides only a 6-to-1 capacitance range. Dual movable plates can also be used by placing another sheet of glass on the other side of the fixed plate and yoking another movable plate to the original with a long U-shaped strap. This scheme doubles the maximum capacitance.

Program 1

Calculates capacitor design parameters. [The ARRL-supplied disk filename for this program is PLASENCI.BAS.—*Ed.*]

```
10  REM CAPACITOR CALCULATOR
20  INPUT "DIELECTRIC CONSTANT =";K
30  INPUT "AREA OF ONE PLATE IN SQUARE INCHES =";A
40  INPUT "SEPARATION OF PLATES IN INCHES =";D
50  INPUT "TOTAL NUMBER OF PLATES =";N
60  C=[.224*[(K*A)/D)]*(N-1)
70  PRINT "MAXIMUM CAPACITANCE IS : ";C;"pF"
80  INPUT "RUN PROGRAM AGAIN? Y/N :";A$
90  IF A$="Y" GOTO 20
100 IF A$="y" GOTO 20
110 IF A$="N" GOTO 130
120 IF A$="n" GOTO 130
130 END
```

Phase-Shift Design of Pi, T and L Networks

By Robert F. White, W6PY

130 Heather Lane
Palo Alto, CA 94303

Pi, T and L networks are widely used in radio systems to match impedances and to provide selectivity. Pi and T networks comprise one inductor and two capacitors (or the reverse), while L networks comprise one inductor and one capacitor. Assuming an ideal system in which the impedances to be matched are pure resistances and the network elements pure reactances, there are, for any specific pair of resistances, infinite mathematically correct pi nets, each with a unique combination of elements and characteristics. For each such pi net there is an equivalent T net. The pi and T nets come in two categories: conventional and unconventional. There is a unique L net which can be treated as either a degenerate pi or a degenerate T net. Finally, every one of the foregoing networks can be configured in either a low-pass or a high-pass form.

Many articles on network design have appeared in the literature over the years. Mostly they have focused on conventional low-pass pi and pi-L nets, with system Q as the parameter of choice. Wingfield and Whyman give especially good treatments.[1,2] The methods are precise and straightforward but they lack generality and do not provide much insight into the relationships among network types.

System Q, an important system characteristic which affects selectivity and (in real systems) system losses, has some limitations as a computational aid. Another system characteristic, its phase shift, is an ideal mathematical tool and in some cases may be of interest *per se*. Each of the infinite network possibilities described earlier has a specific phase shift. Everitt suggested a way in which ideal lossless networks could be designed for such a specific phase shift.[3] He left the matter there because, he said, the phase shift was of little interest.

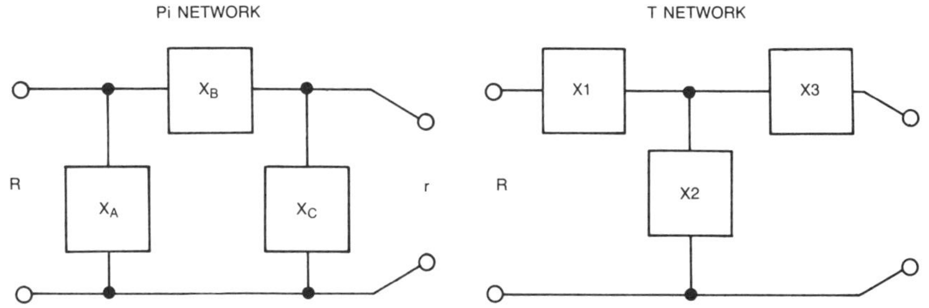

A Normalized, Trigonometric Design Process

By pursuing Everitt's suggestion, adding ideas of my own plus some borrowed from Whyman, Wingfield and others, and by tinkering with it over a number of years I've arrived at the phase-shift approach which is the subject of this paper. It has complete generality and is in a form which facilitates analysis as well as synthesis of

[1]Notes appear on page 196.

Fig 1—Basic normalized phase-shift model and design algorithm. The pi network shown at the left is equivalent to the T network at the right.

such networks. The method gives both the magnitudes (in reactive ohms) and the signs (+ for inductive, − for capacitive) of the network reactances. It is ideally suited for programmable-calculator or computer implementations.

Fig 1 provides the model and the basic algorithm for the normalized trigonometric phase-shift design method. For a given resistance pair (R,r) one can in principle design an ideal matched system to have any desired phase shift between (but not including) 0 and 180°. Practical considerations impose some limitations in real systems.

Fig 1 provides all the ingredients for synthesizing all possible pi and T networks to match a given R,r pair. The only variable in the process is the choice of a suitable value of phase shift, P. Once a P has been selected, the calculation of the network elements and the system Q is straightforward.

The phase shift can be positive for low-pass networks or negative for high-pass networks. It is easily seen from Fig 1 that the only effect of changing the sign of P will be to change the signs of A, B, C and the six X values, without affecting their magnitudes. To simplify discussion and avoid confusion I'll focus on positive P and low-pass networks from here on, except where otherwise noted.

The method and the algorithm of Fig 1 are very easy to explain and to understand if the phase shift, P, is regarded as a primary system objective rather than merely a useful mathematical tool. (Just such an application is discussed briefly near the very end of this paper).

To illustrate the simplicity of the method, suppose for example that we wanted to construct a network to match 50 ohms to 200 ohms with a phase shift of + 5.768°, another network with a P of

+ 154.50°, and a third with P of + 60.00°.

From Fig 1, calculate $N = \sqrt{50 \times 200} = 100$, and $T = 200/50 = 4$. With $T = 4$ and and with each P in turn, calculate A, B and C (to 4 decimal places). The results are shown in Table 1. From these results we can deduce:

1) The systems are all low-pass (Bs are +),

2) The first system is unconventional (A and C have unlike signs),

3) The second system is conventional (A and C have like signs), and

4) The third system is an L network (C is nearly infinite). [The author's result for C in this network is an approximation, probably the result of internal calculator rounding. The denominator value when solving for C is actually zero, and a divide-by-zero error will occur in many calculators.—*Ed.*]

By using the appropriate equations from Fig 1, the ohmic pi and T reactances for the three networks can be calculated. These results are shown in the lower portion of Table 1. See Fig 2A for the L-network configuration, Fig 2B for the conventional pi and T configurations, and Fig 2C for the unconventional pi and T configurations.

Choosing a Suitable P

Since P is the only variable in the whole process, it needs to be carefully considered. The chosen P should produce a network of the desired category and should provide an appropriate value of Q—something which so far has not been mentioned. Following is a menu for the choices.

L network:

$$P_m = \arccos \sqrt{\frac{1}{T}}$$

$$Q_m = \sqrt{T - 1}$$

Conventional pi or T network:

$$P_m < P_c < 180°$$

$$Q_c > Q_m$$

Unconventional pi or T network:

$$P_m > P_u > 0°$$

$$Q_u > Q_m$$

The L net doesn't require or allow any choice. Since T is fixed, P_m and Q_m are also fixed. In the case of our example, $T = 4$, $P_m = 60°$, and $Q_m = 1.732$.

With the fixed-choice L net out of the way we can tackle the more complicated problem of selecting an appropriate P for the 3-element nets. Values of P (P_c) between P_m and 180° will produce the conventional low-pass configurations of Fig. 2B. Q_c will increase monotonically from Q_m to infinity as P_c increases from P_m to 180°.

Values of P (P_u) between P_m and 0° will produce unconventional low-pass configurations as in Fig 2C. Q_u will increase monotonically from Q_m to infinity as P_u

Table 1

Results of Calculations for R = 200, r = 50

P	A	B	C
5.768°	− 0.2000	+ 0.1005	+ 0.1015
154.50°	− 0.2966	+ 0.4305	− 0.1535
60.00°	− 1.1547	+ 0.8660	+ 4.3 × 10¹¹

X_A	X_B	X_C	X_1	X_2	X_3
− 20.00	+ 10.05	+ 10.15	− 984.94	− 955.02	+ 500.03
− 29.66	+ 43.05	− 15.35	+ 651.59	− 232.28	+ 337.11
− 115.47	+ 86.60	4.3 × ¹³	− 000.00	− 115.47	+ 86.60

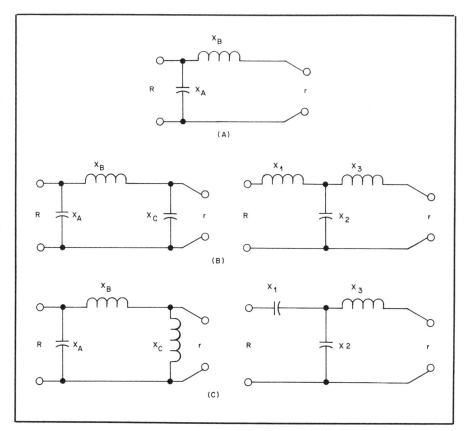

Fig 2—The five possible low-pass configurations. For high-pass configurations, change inductors to capacitors and vice versa. At A, the L net. At B, conventional pi and T networks. At C, unconventional pi and T networks. The following equations apply for these networks.

For A: $P_m = \arccos \sqrt{\frac{1}{T}}$

For B: $P_m < P_c < 180°$

For C: $P_m > P_u > 0°$

decreases from P_m to 0°.

T, P and Q Relationships

These three fundamental system characteristics are so related that if two are known, the third is calculable. Since T is fixed by the initial conditions, we can calculate Q as a function of T and P, or calculate P as a function of T and Q. In each case there is a slight complication.

When T and P are known there is only one possible Q, but the calculations are different for a P_u than for a P_c, as given

by Eqs 6 and 7 below. For *any* P, the applicable Q is the larger of Q_c and Q_u, so the simple answer is to calculate both and pick the larger.

In our example, with a T of 4,

P of 5.768° gives $Q_c = 5.08$ and $Q_u = 10.00$

P of 154.50° gives $Q_c = 10.00$ and $Q_u = 6.74$

P of 60.00° gives both Q_c and $Q_u = 1.732$

When T and Q are the known quantities,

there are two complications. The first is that Q must not be less than $\sqrt{T - 1}$ (Q_m). The second is that for any value of Q greater than Q_m there are two possible values of P, one in the conventional range and the other in the unconventional range. P_c and P_u in terms of T and Q ($Q > Q_m$) are given by Eqs 5c and 5u below. In our example, the desired Q was 10. Eqs 5c and 5u give P_c as 154.50°, and P_u as 5.768°.

Consider now this set of numbered equations.

Pre-choice:

Normalization factor:
$$N = \sqrt{R \times r} \qquad \text{(Eq 1)}$$

Transformation ratio:
$$T = \frac{R}{r} \qquad \text{(Eq 2)}$$

L-net phase shift, the minimum for P_c, the maximum for P_u:
$$P_m = \arccos\sqrt{\frac{1}{T}} \qquad \text{(Eq 3)}$$

L-net Q, the minimum Q for a given T:
$$Q_m = \sqrt{T - 1} \qquad \text{(Eq 4)}$$

Exact Q:

The P_c for any $Q > Q_m$:
$$P_c = \arccos \psi$$
where $\psi =$
$$\frac{2(\sqrt{T} + \sqrt{1/T}) - Q\sqrt{Q^2 - T + 2 - 1/T}}{Q^2 + 4}$$
$$\text{(Eq 5c)}$$

The P_u for any $Q > Q_m$:
$$P_u = \arccos \frac{\sqrt{T} + Q\sqrt{Q^2 - T + 1}}{Q^2 + 1}$$
$$\text{(Eq 5u)}$$

Post-choice:

The Q_c for a $P > P_m$:
$$Q_c = \left| \frac{\sqrt{T} + \sqrt{1/T} - 2\cos P_c}{\sin P_c} \right| \qquad \text{(Eq 6)}$$

The Q_u for a $P < P_m$:
$$Q_u = \left| \frac{\sqrt{T} - \cos P_u}{\sin P_u} \right| \qquad \text{(Eq 7)}$$

Note: The Q for *any* P is the larger of Q_c and Q_u. If Q_c is larger, the system is conventional; if Q_u is larger, the system is unconventional; if Q_c and Q_u are equal, it's an L net.

Eqs 1 and 2 in effect convert R and r into the normalizable identities $R = N\sqrt{T}$ and $r = N\sqrt{1/T}$. Once N and T have been calculated, R and r drop out of the picture

completely and N plays only a very minor role as a multiplying constant in the reactance equations. The *transformation ratio* T, on the other hand, is the key element in the whole process. In the normalized form R is represented by $\sqrt{T}$, r by $\sqrt{1/T}$, and $\sqrt{R \times r}$ by 1.

Eqs 3 and 4, which depend only on T and hence are constants for a particular application, give directly the phase shift and the Q for the unique L-net solution. P_m also provides useful information for P selection, and Q_m provides similar information for Q selection.

Eqs 5c and 5u provide an "exact Q" design option by means of which one can specify any Q ($> Q_m$) and use it with T to calculate the corresponding P. Eq 5c will generate a conventional solution and Eq 5u an unconventional one. Eqs 5c and 5u are alternates, used one at a time, and only with the exact-Q option.

Eqs 6 and 7 give Q_c and Q_u. For *any* value of P, the larger of Q_c and Q_u gives the applicable Q. Eqs 6 and 7 are redundant when the Q option is used, but can provide a useful check.

After these seven calculations have been made, the design is essentially complete. N, T, P, and Q are known, and all that remains is to calculate the network reactances as a function of N, T, and P. This second set of equations may be used for that purpose.

Low-pass pi reactances:
$$X_A = \frac{N \sin P}{\sqrt{1/T} \cos P - 1} \qquad \text{(Eq 8)}$$

$$X_B = N \sin P \qquad \text{(Eq 9)}$$

$$X_C = \frac{N \sin P}{\sqrt{T} \cos P - 1} \qquad \text{(Eq 10)}$$

Low-pass T reactances:
$$X1 = \frac{-N^2}{X_C} \qquad \text{(Eq 11)}$$

$$X2 = \frac{-N^2}{X_B} \qquad \text{(Eq 12)}$$

$$X3 = \frac{-N^2}{X_A} \qquad \text{(Eq 13)}$$

System type indicator:
$$SGN = Q_c - Q_u \qquad \text{(Eq 14)}$$

Note: SGN is positive for conventional systems, negative for unconventional, and zero for an L net.

Automating the Process

The two sets of equations constitute a blueprint for automating the calculation process. In my case it is by the TI-59 calculator program of List 1. My interest in these networks is almost entirely theoretical, so the program is broken into four subroutines which can be used indepen-

dently for system analysis or as a chained sequence to construct networks for specific applications. The subroutines share a coordinated bank of data registers; the equation numbers and the corresponding data-register numbers have been made the same. Although the program in List 1 is for a specific machine, it should be easily adaptable to other calculators or to computers. I've programmed it on a Casio fx4000P machine and the results are identical. However the program on the vintage TI-59 machine is easier to describe.

The Q-Design Option

The List 1 program is set up to favor the exact-Q option for two reasons. One reason is that it becomes easy to compare this phase-shift method with the Q-based methods in the current literature. A more compelling reason is that Q as a system characteristic is already a familiar concept, and a rationale for choosing Q is well established. In theory, Q can be assigned any value from Q_m to infinity, but practical considerations impose an upper limit. If harmonic attenuation as well as impedance matching is desired, the consensus seems to be that Q should be at least 10 but no more than about 20. For simple impedance matching, Q values fairly close to Q_m will keep real system losses to a minimum and real system performance closer to that of an ideal system than would be the case for higher values of Q.

List 1 shows how to use each subroutine individually but not how to chain them to "Q-design" for a particular application. Here's how: Store R in 21 and r in 22, and press A to do the pre-choice equations. When the machine stops, Q_m will appear in the display. Choose any value of Q greater than Q_m, key it, and store it in 23. Then press B for a conventional solution or press C for an unconventional one. When P_c or P_u appears in the display, values for N, T, and P will all be duly stored and D can be pressed to run the main program. Subsequent runs for the same R,r pair can be made without repeating SBR-A. With R = 200, r = 50, and Q = 10, the A,B sequence will produce a P_c of 154.50° and an A,C, sequence a P_u of 5.768° as in the previously described example.

An Interactive P-Design Option

SBR-D, the main program of List 1, is essentially complete in itself and can be used in an interactive manner to empirically find a suitable value of P for a particular application. The procedure is to manually calculate and store N and T, then use the program itself to evaluate different values of P until a suitable one is found. It takes only a few seconds to run the program and only a few iterations are needed. Indeed, with the help of precalculated information in Tables 2 and 3, which I'll get to shortly, the initial choice for P may well be the final one.

List 1

The Main TI-59 Program

SBR-A: Pre-stored inputs, R in 21, r in 22.
Press A.

LBL A (Sbr A starts)

(Eq 1) = N   Sto 01
(Eq 2) = T   Sto 02
(Eq 3) = P_m Sto 03
(Eq 4) = Q_m Sto 04

R/S (Sbr A stops with Q_m displayed)

SBR-B and SBR-C: Pre-stored inputs, T in 02, Q (> Q_m) in 23.
Press B for P_c or C for P_u

LBL B (Sbr B starts)

(Eq 5c) = P_c Sto 05

R/S (Sbr B stops with P_c displayed)

LBL C (Sbr C starts)

(Eq 5u) = P_u Sto 05

R/S (Sbr C stops with P_u displayed)

SBR-D: Pre-stored inputs, N in 01, T in 02, P in 05. Press D.

LBL D (Sbr D starts)

(Eq 6) = Q_c Sto 06
(Eq 7) = Q_u Sto 07
(Eq 8) = X_A Sto 08
(Eq 9) = X_B Sto 09
(Eq 10) = X_C Sto 10
(Eq 11) = X1   Sto 11
(Eq 12) = X2   Sto 12
(Eq 13) = X3   Sto 13
(Eq 14) = SGN

R/S (Sbr D stops with SGN in display).

The simple program shown completely in List 2 is remarkable in that any value of P, except for zero or *exact* multiples of 180°, will produce a mathematically correct set of network elements. The program will also identify the category, conventional if SGN is positive, unconventional if SGN is negative, or L net if SGN is zero. Q_c is applicable for positive SGN, Q_u for negative SGN, and either for an L net. The applicable Q will be the larger of Q_c and Q_u as calculated from Eqs 6 and 7.

Without recourse to the tables, here's how we might tackle our example problem using the List 2 program. Our objective: a conventional system to match 200 to 50 ohms with a Q of about 10.

1) Set the machine to display three decimal places.

2) Manually calculate N = 100 and store it in 01, then manually calculate T = 4 and store it in 02.

3) For the first run calculate $P_m = 60°$, store it in 05, and press D. In about 7 seconds, 0.000 will appear on the TI-59 display, indicating an L net. Q_c and Q_u will both show 1.732.

4) Now do successive runs with P values of 90, 120, 150, 160, 155, 154, and 154.5° producing Q_c values of 2.5, 4.041, 8.464, 12.804, 10.205, 9.804, and 10.000 respectively.

List 2

SBR-D as a Stand-Alone Program

SBR-D: Pre-stored inputs, N in 01, T in 02, P in 05. Press D.

LBL D (SBR-D starts)

$|(\sqrt{T} + \sqrt{1/T} - 2\cos P)/\sin P| = Q_c$ Sto 06
$|(\sqrt{T} - \cos P)/\sin P| = Q_u$ Sto 07
$N\sin P/(\sqrt{1/T}\cos P - 1) = X_A$ Sto 08
$N\sin P = X_B$ Sto 09
$N\sin P/(\sqrt{T}\cos P - 1) = X_C$ Sto 10
$-N^2/X_C = X1$ Sto 11
$-N^2 X_B = X2$ Sto 12
$-N^2 X_A = X3$ Sto 13
$Q_c - Q_u = SGN$

R/S (SBR-D stops with SGN in display)

The whole process, including the manual calculations and entries, takes no more than a couple of minutes, even less on the Casio calculator which does the calculation almost instantaneously.

A similar sequence to find the P_u for an unconventional solution with a Q of about 10 might be: P_u values of 30, 20, 10, 5, 5.5, 6, and 5.75° giving respective Q_u values of 2.268, 3.100, 5.846, 11.517, 10.481, 9.619 and 10.031.

I used these rather long sequences to show how easily one could converge on an exact or near-exact Q by iteration. For an approximate Q, two or three iterations should suffice, particularly if Table 2 or Table 3 is used for the initial selection.

The P-Picker Tables

Tables 2 and 3 provide a complete

mapping of the T, P and Q relationships for transformation ratios from 1 to about 250 and for Q values ranging from Q_m to 20. This covers all ranges of practical interest.

For any of the 20 specific T lines and the six Q columns, you can read off directly the values of P for each such Q. In our example with T = 4, we can immediately find 60° for P_m, 5.768° for an unconventional system with a Q of 10, and 154.50° for a conventional system with the same Q. We could also deduce that any P between 2.87° and 167.14° would have a Q of 20 or less, and also that the minimum Q is 1.732 and the Q at 90° is 2.5.

For values of T which fall between the lines or for values of Q which fall between columns, a simple and rather rough interpolation will give a good initial value. There is really no need to be overly precise in selecting an initial P.

For example, consider a system with a T of 70 for which a P_c is desired to give a Q of about 12. Using the closest T line, 65 in this case, and picking a P_c about halfway between 115° and 140° should give a good initial value. Actual calculations with T = 70 and P = 127.5° produce a Q_c of 12.2, very close to the objective. A caveat: Some of the very low P_u values in Table 3 may require impractical large or small network elements.

For easy reference, Table 4 lists the results of six different calculations for the example problem. Because N in this example is 100, the ohmic X values can be converted to the equivalent normalized values by moving the decimal point two places to the left.

So far in both the discussions and the

Table 2

P, T, and Q Relationships for Conventional Configurations

$P_m < P_c < 180°$; $Q_c > Q_m$

P_{min}	T	Q_{min}	Q_c at 90°	P_c for $Q_c = 5$	P_c for $Q_c = 10$	P_c for $Q_c = 15$	P_c for $Q_c = 20$
0	1	0	2.00	136.40°	157.38°	164.81°	168.58°
26.565°	1¼	0.5	2.01	136.25°	157.31°	164.76°	168.54°
35.264°	1½	0.71	2.04	135.92°	157.14°	164.65°	168.46°
45.000°	2	1	2.12	135.00°	156.68°	164.35°	168.23°
54.736°	3	$\sqrt{2}$	2.31	132.80°	155.60°	163.63°	167.69°
60.000°	4	$\sqrt{3}$	2.50	130.54°	154.50°	162.90°	167.14°
63.435°	5	2	2.68	128.31°	153.43°	162.19°	166.62°
71.565°	10	3	3.48	117.96°	148.75°	159.12°	164.32°
75.964°	17	4	4.37	104.04°	143.34°	155.64°	161.74°
78.690°	26	5	5.30	78.69°	137.41°	151.92°	159.01°
80.538°	37	6	6.25		130.91°	148.02°	156.18°
81.870°	50	7	7.21		123.68°	143.94°	153.26°
82.875°	65	8	8.19		115.30°	139.66°	150.26°
83.660°	82	9	9.17		104.69°	135.13°	147.16°
84.289°	101	10	10.15		84.29°	130.29°	143.96°
84.806°	122	11	11.14			125.02°	140.65°
85.236°	145	12	12.12			119.16°	137.19°
85.601°	170	13	13.12			112.33°	133.56°
85.914°	197	14	14.11			103.62°	129.71°
86.186°	226	15	15.10			86.19°	125.59°

Table 3
P, T, and Q Relationships for Unconventional Configurations

$P_m > P_u > 0°$; $Q_u > Q_m$

P_{max}	T	Q_{min}	P_u for $Q_u = \sqrt{T}$	P_u for $Q_u = 5$	P_u for $Q_u = 10$	P_u for $Q_u = 15$	P_u for $Q_u = 20$
26.565°	1¼	0.5	6.38°	1.356°	0.677°	0.451°	0.338°
35.264°	1½	0.71	11.54°	2.588°	1.289°	0.859°	0.644°
45.000°	2	1	19.47°	4.792°	2.379°	1.584°	1.187°
54.736°	3	$\sqrt{2}$	30.00°	8.548°	4.214°	2.802°	2.100°
60.000°	4	$\sqrt{3}$	36.87°	11.784°	5.768°	3.831°	2.870°
63.435°	5	2	41.81°	14.700°	7.145°	4.740°	3.549°
71.565°	10	3	54.90°	27.019°	12.630°	8.329°	6.224°
75.964°	17	4	62.73°	42.650°	18.511°	12.104°	9.020°
78.690°	26	5	67.81°	78.690°	24.778°	16.013°	11.889°
80.538°	37	6	71.33°		31.537°	20.053°	14.821°
81.870°	50	7	73.90°		39.006°	24.244°	17.815°
82.875°	65	8	75.86°		47.632°	28.618°	20.879°
83.660°	82	9	77.40°		58.585°	33.225°	24.023°
84.289°	101	10	78.64°		84.289°	38.138°	27.261°
84.806°	122	11	79.65°			43.470°	30.613°
85.236°	145	12	80.51°			49.411°	34.103°
85.601°	170	13	81.23°			56.333°	37.763°
85.914°	197	14	81.85°			65.195°	41.637°
86.186°	226	15	82.39°			86.186°	45.791°

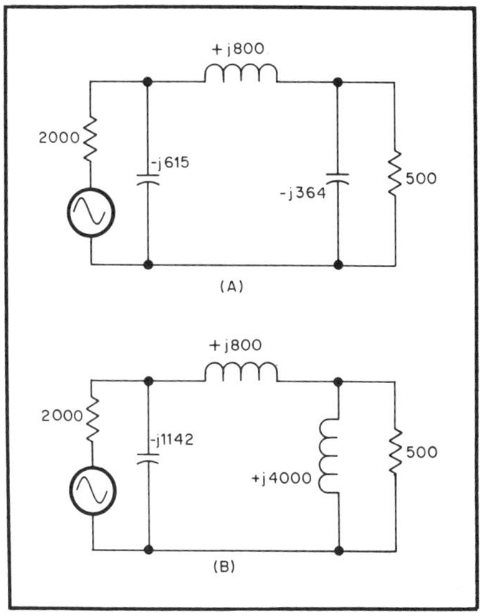

Fig 3—The two possibilities of Everitt's example, from his Figs 120(a) and 120(b); see text and note 3. The network at A results when positive signs are used before the radicals (Ev. 36 through 38, given in the text of this paper), and that at B from negative signs before the radical.

Table 4
Example Calculations

$R = 200$, $r = 50$, $N = 100$, $T = 4$, $P_m = 60°$, $Q_m = 1.732$

P			60°		90°	
Q	10	2		2		10
P	5.768°	36.87°		72.89°		154.50°
Q_c	5.08	1.50	1.732	2.00	2.50	10.00
Q_u	10.00	2.00	1.732	1.78	2.00	6.74
X_A	−20.00	−100.00	−115.47	−112.06	−100	−29.66
X_B	10.05	60.00	86.60	95.57	100	43.05
X_C	10.15	100.00	4.33×10^{13}	−232.29	−100	−15.35
$X1$	−984.89	−100.00	0.00	43.05	100	651.58
$X2$	−994.96	−166.67	−115.47	−104.63	−100	−232.28
$X3$	500.00	100.00	86.60	89.24	100	337.10
SGN	Neg	Neg	Zero	Pos	Pos	Pos

mathematical treatment, equal attention has been given to the conventional and the unconventional configurations. From a practical viewpoint, interest is overwhelmingly in the conventional configurations. Indeed, the current literature doesn't appear to acknowledge that the unconventional configurations even exist.

A discussion of the classical pi-network equations which Everitt did pursue and develop may help to put things into perspective. Whyman and Wingfield cited these equations and used portions of them, but not the complete set, in their articles.[4]

Everitt's Pi-Network Equations

Using my terminology (ie, R and r instead of R1 and R2) the Everitt equations are:

$$|X_B| < \sqrt{R \times r} \qquad \text{(Ev. 38)}$$

$$X_A = \frac{-R \times X_B}{R \pm \sqrt{R \times r - X_B^2}} \qquad \text{(Ev. 37)}$$

$$X_C = \frac{-r \times X_B}{r \pm \sqrt{R \times r - X_B^2}} \qquad \text{(Ev. 36)}$$

The Everitt approach was to arbitrarily select an X_B less than $\sqrt{R \times r}$, and then use Ev. 37 and Ev. 36 to calculate X_A and X_C. He pointed out that for any X_B there are two possible X_A, X_C pairs, corresponding to the plus or minus signs in front of the radicals in the two equations. He warned that if the positive sign is used for X_A it must also be used for X_C, and vice versa.

To illustrate the two possibilities, Everitt used an example: a generator of 2000 ohms to be matched against a 500-ohm line. He arbitrarily chose a value of 800 ohms for X_B (the maximum would be 1000 ohms), then calculated two X_A, X_C pairs using Ev. 37 and Ev. 36, producing the networks of Fig 3.

After pointing out that either network would have an input impedance of 2000 ohms and would absorb maximum power from the generator, he went on to say: "The network of Fig 120(a) will attenuate the harmonics, *while Fig 120(b) will not*, and therefore Fig 120(a) is usually the more desirable." (Emphasis added.)

The italicized portion of that statement is inaccurate. Both networks attenuate the harmonics, but the attenuation is substantially higher for the conventional network, and hence Everitt's conclusion that Fig 120(a) is usually the more desirable is valid.

Things would have been much more interesting had Everitt chosen 866 ohms for X_B. He would then have found, using the minus signs for the radicals, values of −1154.7 ohms for X_A, and about ten million ohms for X_C. Indeed, had he chosen 866.0254038 and made the calculations *exactly*, he would have found X_C to be about 4.8×10^{14}, with X_A unchanged at −1154.7 ohms. Clearly an X_B of 866 ± about an ohm would make X_C so large that it could be replaced by an open circuit, thus producing an L net. Using my terminology, the value of X_B to produce an L net can be calculated as

$$X_B = N \sqrt{\frac{T - 1}{T}}$$

In principle, X_C should become infinite for this X_B, but usually when done on a calculator it is some huge number which can be either positive or negative.

A further point of interest: Had Everitt chosen an X_B between 866 and 1000, X_C

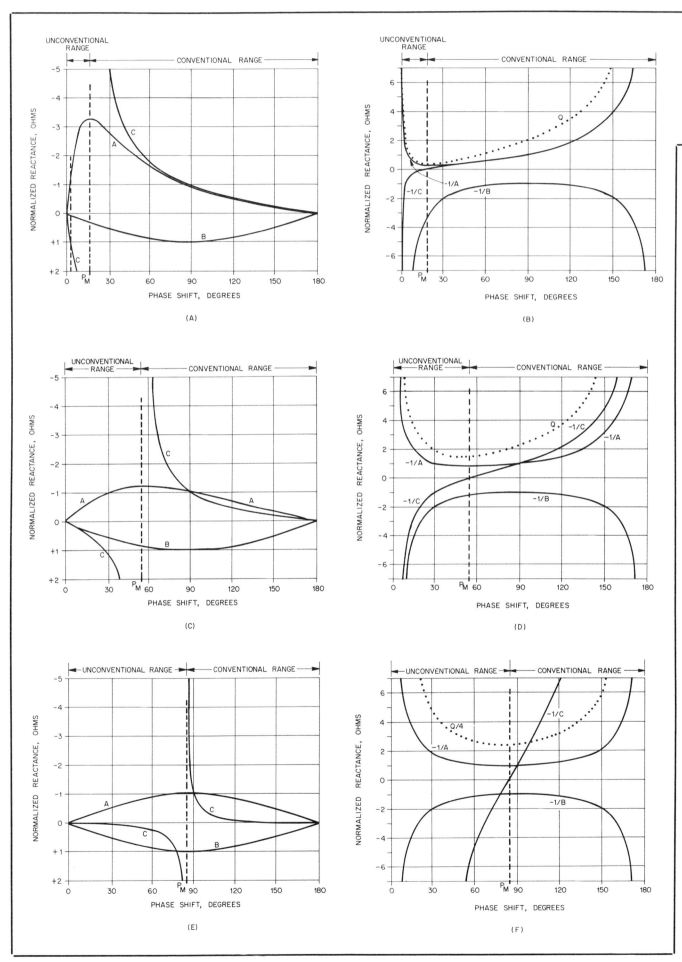

as calculated using the minus sign before the radical would have come out negative, in which case Everitt's Fig 120(b)—my Fig 3B—would have both X_A and X_C as capacitive, the conventional configuration. Only for X_B values less than 866 ohms would the configuration be as actually shown in Fig 3B.

Everitt simply ignored these possibilities and went on to other things, no doubt a wise decision since a full explanation would have been complicated. In phase-shift terms the explanations are fairly easy. Ev. 37 and Ev. 36 with plus signs before the radicals correspond to phase shifts between 90 and 180°, the configuration always being conventional. With minus signs before the radicals they correspond to phase shifts between 0 and 90°, the configuration being conventional for phase-shifts between P_m and 90°, and unconventional for those between P_m and 0°. Exactly P_m of course gives the L-net solution.

A Bit of System Analysis

The Everitt example provides a good basis for some system analysis. Using the phase-shift method with R = 2000 and r = 500 gives these pre-choice numbers.

N = 1000, T = 4, P_m = 60°,
$Q_m = \sqrt{3} = 1.732$

When Everitt chose 800 ohms for X_B he was indirectly choosing two values of P, one in the first quadrant and the other in the second. By using a variation of Eq 9 with X_B = 800 ohms and N = 1000, we can calculate

$$P1 = \arcsin \frac{X_B}{N} = 53.13°$$

From this,

$$P2 = 180 - 53.13 = 126.87°$$

And indeed, if we run the SBR-D TI-59 program with N = 1000, T = 4 and P = 53.13, we get the following.

$X_A = -1143$, $X_B = 800$, $X_C = 4000$, $Q_u = 1.75$

Note that Everitt made an error in X_A; his method also gives -1143)

An SBR-D run with P = 126.87° gives -615, 800 and -364 for the reactance

values, identical to Everitt's numbers. The Q (Q_c) is 4.625.

Even though he didn't mention Q at all, we see that Everitt was indirectly choosing a Q_c of 4.625 for Ev. 120(a) and a Q_u of 1.75 for Ev. 120(b).

Normalized Calculations using SBR-D

The algorithm of SBR-D, unlike that of Fig 1, does not explicitly give the normalized elements. By the simple expedient of setting N = 1 (ie, storing a "1" in 01), SBR-D can be used to produce A, B, C, $-1/C$, $-1/B$, and $-1/A$ instead of the six ohmic reactances. SBR-D was used in this manner to derive the six graphs of Fig 4. A and B of Fig 4 depict the behavior of the normalized elements and the Q for a low value of T (1.1), C and D for a medium value of T (3), and E and F for a high value of T (100).

These graphs are not meant to be used as calculation aids, but rather to provide insights into the way the various systems and elements are interrelated. Much useful information can be gleaned from the graphs, but that is peripheral to the main thrust of this paper. I won't attempt to elaborate except for a couple of points. One is the special nature of the network elements at P_m and at 90°. P_m has already been discussed in detail. At 90°, the derived networks have a special characteristic similar to that of a quarter-wave section of transmission line. The nets are symmetrical, with B = 1 and A = C = -1. A system with 90° phase shift will match any R pair whose product is equal to N^2, regardless of the value of T. However, the Q will depend on T according to the following relationship.

$$Q_{90} = \sqrt{T} + \sqrt{1/T} \qquad \text{(Eq 15)}$$

A second point is the behavior of the normalized elements for the special case when R = r and T = 1. If in Fig 4A we imagine T being reduced below 1.1, P_m will move toward zero and the peak of the curve for A will increase in magnitude. In the limit, as T approaches 1, P_m and Q_m will approach zero, while A and C will be equal and will approach $-\infty$. At 90° A and C will be -1, and at 180° they will go to zero. Thus, the entire range from 0 to 180° is conventional, which accounts for Table 3 not having a line for T = 1. The T = 1 line in Table 2 is useful for networks intended to provide selectivity or a specific phase shift without changing the impedance.

The Unconventional Networks

The unconventional versions of the pi and T networks seem to have received little or no attention in the literature in the half-century since Everitt briefly described and dismissed them as inferior to the conventional types. Do they have any real usefulness, or are they merely of academic interest?

I've just recently stumbled across a reference which indicates that they are useful and have been used, but perhaps have not been identified as belonging to the pi and T family. The reference is a circuit in the 1986 *ARRL Handbook*, Fig 84D of Chapter 2, shown here as Fig 5A.[5] A statement in the text accompanying the *Handbook* circuit says, "The circuit of Fig 84D has never been given any special name but is quite popular in both antenna and transistor matching applications."

When redrawn as in Fig 5B, the circuit is seen to be identical to the high-pass version of an unconventional pi net. The component relationships are R = R2, r = R1, $X_A = X_L$, $X_B = X_{C1}$, and $X_C = X_{C2}$.

Perhaps there are other circuits in use that would fit into the—"unconventional" pi- and T-network family. All such circuits could be easily designed using the phase-shift approach.

Frequency Response, Unconventional Networks

At the risk of confusing matters, I have to mention that the unconventional networks are not truly "low-pass" or "high-pass" circuits except in a limited sense. For values of Q_u close to the minimum Q, the responses of the unconventional nets approximate those of conventional nets with the same Q, which are true low pass or high pass. But for considerably larger Q values, the unconventional circuits have a rather symmetrical band-pass characteristic. Fig 89 and the accompanying text in the 1986 *ARRL Handbook* give an indication of the responses of the circuit of Fig 84D for several conditions.[6] The responses show a "high-pass" system. A mirror image of the response curve would show how the low-pass responses look.

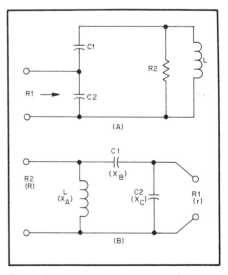

Fig 5—At A, the "no special name" circuit of *The ARRL Handbook.* At B its equivalent which is seen to be an unconventional high-pass pi network.

To give these orphan systems an identifying name it is convenient, at least for the purposes of this paper, to continue to identify them as unconventional low-pass designs, where P_u and X_B are positive, and unconventional high-pass where P_u and X_B are negative. Perhaps a less cumbersome description can be found.

Matching Complex Impedances

The simplicity and generality of the approach described so far result from the assumptions that the impedances to be matched are purely resistive and that the network elements are purely reactive. Things get somewhat more complicated but still manageable if we allow one or both of the impedances to be matched to be complex, but continue to treat the network elements as pure reactances. Fortunately there is a fairly simple approach that will yield good results in most situations of practical interest, provided high-quality inductors and capacitors are used, the operational Q is not excessive, and the frequencies involved are in the HF range.

The approach matches a generator with a purely resistive impedance, R_g, to a load with a complex impedance. The load impedance in series form is given by $R_s + jX_s$ and in parallel-equivalent form is given by

$$R_p = \frac{R_s^2 + X_s^2}{R_s} = R_s + \frac{X_s^2}{R_s}$$
(Eq 16)

$$X_p = \frac{R_s^2 + X_s^2}{X_s} = X_s + \frac{R_s^2}{X_s}$$
(Eq 17)

The method is to design a basic network to match R_g to R_s if a T net is desired, or R_g to R_p if a pi net is desired, and then modify the network element facing R_s or R_p so that it absorbs or cancels the associated reactive part (X_s or X_p).

Note that when using this method, one must choose the type of network before doing the network calculations. A T net requires R_s and X_s for the load impedance, while a pi net requires R_p and X_p. Since R_p and R_s will always be different (R_p larger), the pi and T nets will no longer be interchangeable as they are in the ideal case. It is also desirable, though not absolutely necessary, to decide in advance between low-pass and high-pass designs.

There is another new element in the overall picture—the direction of transmission. When taking all the new elements into account, there are four possible configurations as depicted by Cases 1 through 4 in Fig 6.

The equations for modifying the network elements in the four cases are:

Case 1:

$$X3' + X_s = X3; \quad X3' = X3 - X_s$$
(Eq 18)

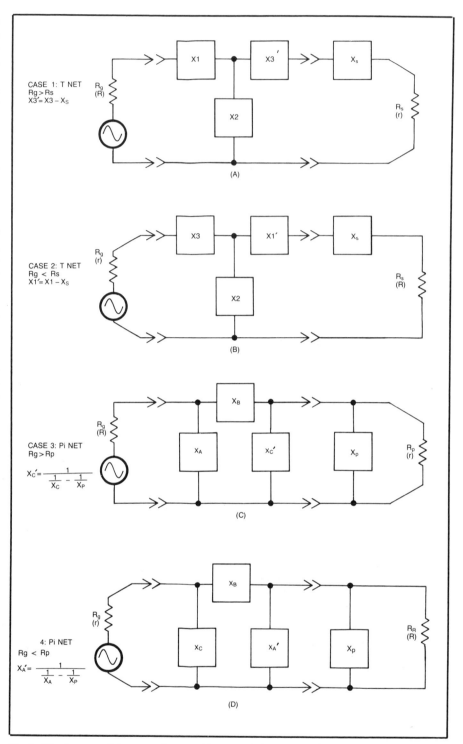

Fig 6—The four possible cases for matching a resistive source, R_g, to a complex load.

Case 2:

$$X1' + X_s = X1; \quad X1' = X1 - X_s$$
(Eq 19)

Case 3:

$$\frac{1}{X_C'} + \frac{1}{X_p} = \frac{1}{X_C}; \quad X_C' = \frac{1}{\frac{1}{X_C} - \frac{1}{X_p}}$$
(Eq 20)

Case 4:

$$\frac{1}{X_A'} + \frac{1}{X_p} = \frac{1}{X_A}; \quad X_A' = \frac{1}{\frac{1}{X_A} - \frac{1}{X_p}}$$
(Eq 21)

Note that careful attention must be given to the signs of all the X values. In some cases, modifying an element may even change its sign. Note also that following the rule that X_A and X1 always face the larger resistance, the network elements in Cases 2 and 4 are reversed from their usual positions.

The calculation process is best described by way of example. The one I'll use is a generator impedance R_g = 50 ohms and a complex load impedance $Z = 200 - j100$ ohms.

First we'll look for a conventional low-pass pi net with a Q of 10, second a conventional T network with the same Q, third an unconventional T net with the same Q, and finally a low-pass L net, which curiously enough is the most complicated of all.

Pi Network with Complex Load

For the pi net (see Fig 7) we must convert Z to the parallel-equivalent form using Eqs 18 and 19, giving R_p = 250 and X_p = −500 ohms. So the basic network will have R = 250 and r = 50. Using SBR-A or by direct calculation, we find N = 111.8, T = 5, and from Table 2 we find that for Q = 10 we need a P_c of 153.43°. With N, T and P_c as inputs, SBR-D or an equivalent calculation gives the following basic network elements.

$$X_A = -35.72; X_B = 50.01;$$
$$X_C = -16.67; Q = 10$$

Since R_g is less than R_p the category is Case 4, so we must use Eq 21 to calculate $X_A' = -38.47$. See Fig 7 for the resulting matched system.

T Network with Complex Load

For the T network (see Fig 8) the initial data are R_g = 50, R_s = 200 and X_s = −100. With an R of 200 and r of 50 we find N = 100, T = 4, and from Table 2 we read a P_c of 154.50° for a Q of 10. The basic network elements are then

$$X1 = 651.6; X2 = -232.3; X3 = 337.1;$$
$$Q = 10$$

Since R_g is less than R_s the category is Case 2 and we must use Eq 19 to calculate X1 ′ = 751.6. See Fig 8 for the resulting matched system.

Unconventional T-network with Complex Load

This is just like the preceding case except that we use Table 3 instead of Table 2 and find that P_u is 5.768° instead of 154.50°. The basic network elements are then

$$X1 = -985.0; X2 = -995.0; X3 = 500;$$
$$Q = 10$$

The category is Case 2 so we use Eq 19 to calculate X1 ′ = −885.0. See Fig 9 for the resulting matched system.

Low-Pass L Net with Complex Load

As indicated earlier, the L net is a bit more complicated. There are three pos-

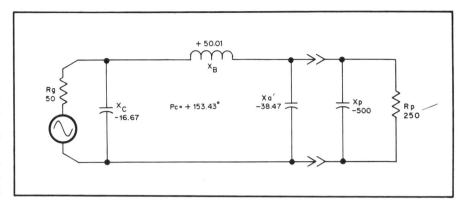

Fig 7—Low-pass conventional pi network with complex load.

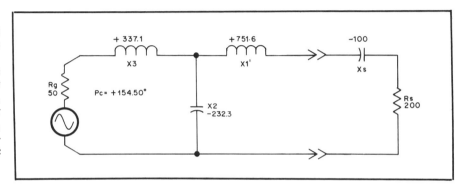

Fig 8—Low-pass conventional T network with complex load.

sibilities, depending on the relationship of R_g to R_s and R_p. For R_g less than R_s use Case 4 to get a 2-element net. For R_g greater than R_p use Case 1 to get a 2-element net. If R_g is greater than R_s but less than R_p, either Case 1 or Case 4 can be used. (In Cases 2 and 3 the 2-element basic nets require three elements when the load-facing element is modified).

In our example R_g is less than R_s so we must use the Case 4 pi net approach but with $P = P_m$. We must also match R_g to R_p, 250 ohms. N now equals 111.8 and T = 5. For this T we find in Table 2 that P_m = 63.435°. The resulting basic elements are

$$X_A = -125; X_B = 100; X_C = \infty; Q = 2$$

Using Eq 21 we get $X_A' = -166.67$. See Fig 10 for the matched system.

What If?

What if Z were $200 + j100$ instead of $200 - j100$? The process would be exactly the same except that X_s and X_p would change to inductors of +200 and +500 ohms, respectively. X_A' would change to −33.33 in Fig 7 and to −100 in Fig 10. X1 ′ would change to +551.6 in Fig 8 and to −1085 in Fig 9.

What if Z were $200 + j100$ and we made all P angles negative? In that case the magnitudes of all the P values and all reactances in Figs 7 through 10 would be unchanged,

but all plus signs would become minus and vice versa. Every inductor would become a capacitor, every capacitor an inductor, and the resulting networks would all be high pass instead of low pass.

To simplify discussion and avoid confusion, the calculation processes described earlier all assume low-pass networks and positive P angles, with high-pass equivalents to be derived by merely changing the signs of all the network reactances. This sign changing can be avoided by simply using negative values of P with SBR-D or the equivalent calculation when the objective is high-pass networks.

Q Values of the Modified Systems

When a modified network X and the load X it faces have the same sign (as in Figs 7, 9 and 10), the modified system Q and the basic system Q are the same.

When these two X values have different signs (as in Fig 8), the modified system Q is greater than the basic system Q by

$$|X_s/R_s|.$$

Phase Shift as a System Objective

In some situations the phase shift itself may be an important system objective. A case in point in the fairly recent amateur literature appears in an article by Forrest Gehrke.[7] In his fourth of a series of articles on vertical phased arrays Gehrke shows, among other things, how symmetrical (T = 1 in my terminology) "coaxial-equivalent" pi or T circuits could be used instead of coaxial delay lines to establish precise phase relationships among the currents in the driven elements of such systems.

Gehrke uses a specific length of coax between each antenna element and the junction point where the feeders are paralleled. He then uses an L net in each line to transform the rotated complex impedance as seen at the input to each coax to the previously calculated pure resistance needed for combining with the other feeders. Finally, with the current phase of one of the lines as a reference, he uses a symmetrical pi or T net in each other line to bring all current phases into synchronism without changing the already established resistive impedances.

Gehrke points out several advantages in using these discrete networks rather than coaxial "delay lines" to bring about this phase synchronization. The discrete networks can be designed for any desired characteristic impedance, and for leading (high-pass nets) as well as lagging (low-pass nets) phase shifts. Coax, on the other hand, is available in only a few specific impedances and can provide only lagging phase shifts unless rather long lines are used. A disadvantage of the discrete circuits is that they function over a narrower frequency range than coax, but Gehrke shows how this disadvantage can be minimized by using two or more networks in tandem.

Gehrke's equations for calculating the pi or T nets are identical to mine with T set equal to 1. The pi and T nets in his approach always match pure resistances. Additional flexibility can be achieved by using nets designed to do impedance transforming in addition to the phase-shifting function. This, however, is only a small part of a very complex process, and is far beyond the scope of this paper.

It is clear, nevertheless, that phase shift is a useful and significant system characteristic worthy of more attention than Everitt gave it.

Validating the Phase-Shift Approach

The validity of solutions produced by the phase-shift algorithm can be tested by experiment, by comparison with known results where available, or by mathematical analysis. Experimental validation of the many possible variations would be an awesome task, but in any case I have no facilities for such testing.

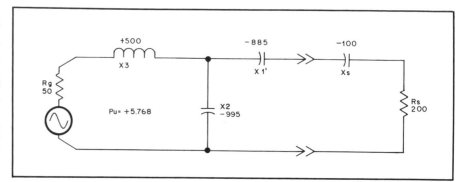

Fig 9—Low-pass unconventional T network with complex load.

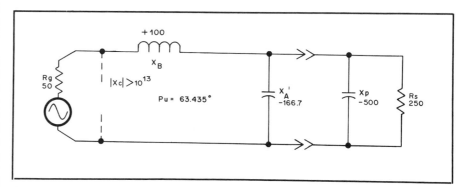

Fig 10—Low-pass L network with complex load.

The Wingfield, Whyman and Gehrke articles[8] provide many tabulated examples for conventional pi- and L-net configurations, and in all cases the phase-shift method exactly duplicates their solutions. But except for the single example in Everitt's Fig 120(b) I find nothing in the literature about the unconventional networks and very little about T networks.

By far the most comprehensive and useful test is by mathematical analysis, in my case by another TI-59 program, a "ladder analysis" program adapted from an HP-25 program in an article by Leonard H. Anderson.[9] My program checks the voltage phase shift, the current phase shift and the system Q, in addition to checking the impedance match. This ladder program confirms the validity of the phase-shift method for all possible configurations. The ladder program can also be used to check the response of a network at frequencies other than the design frequency.

In closing I'd like to repeat the cautions about the care which must be given to the signs of the X values and the orientation of the network elements. The asymmetrical "impedance tapered" shape for the network models and the use of R and r for the resistances are mnemonic aids which I find useful in keeping the orientation straight. Others may not need such mental props.

Notes

[1]E. W. Whyman, "Pi-Network Design and Analysis," *Ham Radio*, Sep 1977, pp 30-39.
[2]E. A. Wingfield, "New and Improved Formulas for the Design of Pi and Pi-L Networks," *QST*, Aug 1983, pp. 23-29.
[3]W. L. Everitt, *Communication Engineering*, 1st ed. (New York: McGraw-Hill Book Co Inc, 1932), Fig 127 and text, p 247; Eqs 36, 37 and 38, p 236; Fig 120, p 237.
[4]See notes 1 and 2.
[5]M. Wilson, Ed., *The ARRL 1986 Handbook* (Newington: The American Radio Relay League, 1985), Fig 84D, p 2-49. (The same circuit appears as Fig 80D in years immediately prior to 1986, and as Fig 84D in 1987 through 1990 editions of The Handbook.)
[6]See note 5; Fig 89 in recent editions is presented as Fig 84 in years immediately prior to 1986.
[7]F. Gehrke, "Vertical Phased Arrays, Part 4," *Ham Radio*, Oct 1983, pp 34-45.
[8]See notes 1, 2 and 7.
[9]L. A. Anderson, "Calculator-Aided Circuit Analysis," *Ham Radio*, Oct 1977, pp 38-46.

Solar Activity
and
Ionospheric Effects

Sunspots, Flares and HF Propagation

By Richard W. Miller, VE3CIE
RR 1
Hillsburgh, ON NØB 1ZØ, Canada

In December 1982, Bob Rose, K6GKU, introduced the MINIMUF computer program to the Amateur Radio world.[1] MINIMUF provides a prediction of maximum usable frequency (MUF) between two locations taking into account time of day, season, solar activity, and transmission path geometry. Since that time many commercial and home-programmed versions have been produced for a variety of computers. MINIMUF type programs are now found in the shacks of most amateurs who are interested in various aspects of HF propagation. The level of long-term solar activity is accounted for by using the 10-cm solar radio flux which is well correlated with the sunspot number.[2] This allows for the prediction of the MUF for various times in the approximately 11-year solar cycle. However, it does not account for variations in the day-to-day level of solar activity produced by solar flares, disappearing filaments and coronal holes and their effects on HF propagation. In this paper we examine some of the ways in which amateur operators can include the variations caused by solar flares in their forecasts of propagation conditions.

The Ionosphere

The normal ionosphere consists of three layers which are important to HF communications; the D, E, and F layers. The F layer sometimes splits into two layers, the F1 and F2. The parameter of significance is the maximum electron density of the layer, which is related to the maximum frequency that can be reflected by the layer. If the electron density of a layer is sufficiently low, the signal is transmitted through the layer. If the density is sufficiently high, the signal is reflected. However, if the density is high enough, the signal may be absorbed. See Fig 1. The portion of the HF spectrum usable for communications is limited on the high frequency side by the maximum electron density of the F layer and on the low frequency side by signal absorption in the D layer. The useful frequency range lies between the maximum usable frequency, MUF, and the lowest usable frequency,

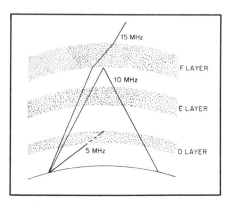

Fig 1—The MUF lies between 5 and 15 MHz. Frequencies above the MUF pass through the ionosphere. Frequencies below the LUF are absorbed.

LUF. Since absorption varies inversely with the square of the frequency, one should use a frequency as close as possible to the MUF. When the ionosphere is relatively well behaved under normal (quiet) solar conditions, programs such as MINIMUF can predict the MUF within an accuracy of a few percent.[3] However, during solar active conditions, solar flares may cause ionospheric storms, geomagnetic storms, and radio noise storms. The effects of these phenomena are to lower MUF values and increase absorption and noise. These changes can be so intense as to cause a complete radio blackout along certain transmission paths. These sporadic changes are not accounted for in MINIMUF type programs.

Solar Flares and Their Effects

Solar flares are responsible for a variety of disturbances which affect HF propagation, as indicated in Fig 2 and Table 1. These effects may occur almost immediately (within 8 minutes) if they result from electromagnetic radiation such as ultraviolet, X-rays or radio emissions. Cosmic ray particles (mainly protons) require 15 minutes to several hours to reach the earth, while

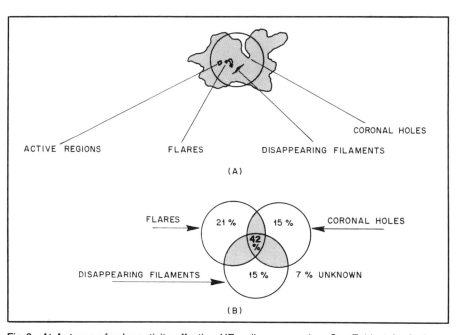

Fig 2—At A, types of solar activity affecting HF radio propagation. See Table 1 for further information. The drawing at B indicates the percent of geomagnetic disturbances by cause from June 1, 1976, to December 31, 1983 (*After Joselyn, 1984—see note 12*).

Table 1

Types of Solar Activity and Effects

Active Regions

10-cm solar flux—used in MINIMUF type
    programs
Flux levels and occurrence of active
    regions increase with peak of sunspot
    cycle
Increased F layer ionization
Increased D layer absorption
Higher MUFs with higher flux levels
Sun-earth transit time 8.3 minutes

Flares

X-rays, light, radio emissions
    Sudden ionospheric disturbances (SIDs)
    Radio noise storms
    F layer increase
    E layer increase
    Sun-earth transit time 8.3 minutes

Cosmic ray particles
    Polar cap absorption (PCA)
    Sun-earth transit time ¼ to several hours

Solar wind particles
    Geomagnetic storms
    Aurora
    Ionospheric storms
    Average sun-earth transit time 48 hours
7% of flares produce geomagnetic disturbances

Disappearing Filaments

Number of filaments which disappear
    increases at sunspot cycle peak
Geomagnetic storms
Aurora
Ionospheric storms
Average sun-earth transit time 134 hours
15% of disappearing filaments produce
    geomagnetic disturbances

Coronal Holes

Recur at 27 day intervals (the rotational
    period of the sun). Once established, they
    persist up to about 6-8 rotations
Geomagnetic storms
Aurora
Ionospheric storms
Average sun-earth transit time 72 hours
43% of coronal holes produce geomagnetic
    disturbances

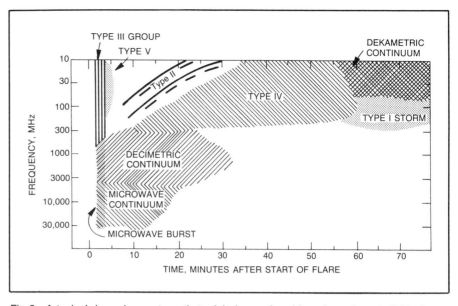

Fig 3—A typical dynamic spectrum that might be produced by a large flare. Individual flares exhibit many variations to this spectrum. Type II emissions begin at about 300 MHz and sweep slowly to 10 MHz; they are loosely associated with major flares and indicate a major disturbance in the solar wind. Type IV emissions are broadbanded bursts between 300-30 MHz. They occur with some major flares and begin 10 to 20 minutes after the flare maximum; they can last for several hours.

solar wind disturbances require about 48 hours, on average, for their effects to become apparent.

Ultraviolet and X-ray emissions from solar flares are responsible for a class of disturbances called sudden ionospheric disturbances (SIDs). Of these, the SID of primary importance to the amateur is the shortwave fade-out (SWF). SWFs can be quite intense, resulting in a complete blackout of signals on sunlit paths. SWFs are caused by increased D-layer absorption of signals. Solar radio emissions occur at

wavelengths ranging from centimeters to decameters. These emissions may occur as storms, bursts or sweeps. They are classified according to the frequencies effected and their duration, as shown in Fig 3.

Cosmic ray particles may cause polar cap absorption events (PCA). PCAs occur with major flares. They begin after a few hours and peak after one or two days. PCAs result in anomalous absorption of polar path HF signals. Disturbances which result from shock waves propagating in the solar wind result in geomagnetic and ionospheric

Table 2

Effects of Solar Flares on HF Propagation

(See note 13.)

	Shortwave Fade-out	Polar Cap Absorption	Ionospheric Storm	Polar Blackout
Duration	Few minutes to several hours	Several minutes to several days	Several hours to several days	Several hours to several days
Greatest Effect	Day side	Polar regions, day or night	Polar regions, mid-latitudes day or night	Polar regions, mid-latitudes day or night
Least Effect	Night side	Mid-latitude and equatorial regions	Low latitude and equatorial regions	Low latitude and equatorial regions
Best Bands	10-15 m	10 m	80-160 m	None
Worst Bands	20-160 m	15-160 m	10-40 m	10-160 m
Remedial Action	Use night paths; use higher frequencies on day paths.	Use low latitude and equatorial paths; use higher frequencies on high latitude and polar paths.	Use low latitude and equatorial paths; use lower frequencies on high latitude and polar paths.	Use low latitude and equatorial paths.

storms when the shock waves interact with the earth's magnetic field. (Note: These shock waves do not always reach the earth, and as a result geomagnetic storms are extremely difficult to predict even though a major flare may have occurred.)

Geomagnetic storms are worldwide disturbances in the earth's magnetic field, apart from the normal diurnal variations. Ionospheric storms, which occur in connection with geomagnetic storms, are disturbances in the F region of the ionosphere. They begin in the auroral regions and spread slowly toward the equator. They result in the reduction of the MUF due to decreased F layer ionization. This may result in a partial or total radio blackout along certain paths. Geomagnetic storms also cause increased auroral activity that is useful for extended-range VHF propagation.[4]

Solar flares are classified according to their increasing intensity of X-ray emission as C, M, and X class flares.[5] C flares have little effect on HF propagation, while M flares may cause minor events. X flares may result in SWFs of major proportion with total radio blackout. When accompanied by type II or type IV radio emissions, extreme radio noise may occur. The emission of protons (cosmic ray particles) will result in a PCA event. X class flares can produce massive solar-wind shock waves and hence intense ionospheric storms. The effects of solar flare activity and resulting geomagnetic and ionospheric disturbances are summarized in Table 2.

Monitoring Solar Flare Activity

Amateurs who are interested in monitoring solar flare activity may obtain information from the Space Environment Services Center (SESC), Boulder, CO, either as GEOALERT messages transmitted via radio station WWV or the SESC computer bulletin board, tel 303-497-5000.[6] However, amateurs with small telescopes may monitor potential solar flare activity directly by observing sunspots.

Observations of sunspots in white light are easily made with a small telescope. (I use a refracting telescope of 60 mm aperture.) Safe methods of solar observation are described by Richard Hill.[7] **_WARNING: Never look at the sun without proper precautions. Retinal damage can result in nearly instantaneous blindness!_**

In 1966, Patrick S. McIntosh of SESC introduced a modification to the long-established Zurich classification of sunspots.[8] The Zurich classes had already been correlated with solar flares,[9] but the McIntosh system incorporated further structural and dynamic characteristics of sunspot groups which were noted by flare forecasters to enhance the correlation. Evaluation of the relationship between sunspots as classified by McIntosh and X-ray flare activity indicates that the McIntosh system effectively distinguishes between flare-active and non-flaring sunspots.

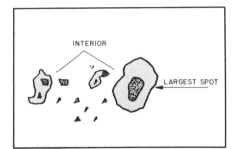

Fig 4—Components of the McIntosh sunspot classification scheme. See text and Fig 5 for additional information.

The McIntosh sunspot classification consists of three components as shown in Fig 4. In this classification scheme,

Sunspot group type = D + a + i

where

D = modified Zurich class
a = type of largest spot
i = interior distribution of spots between leader and follower spots

The complete classification is illustrated in Fig 5 and Table 3. There are 63 possible

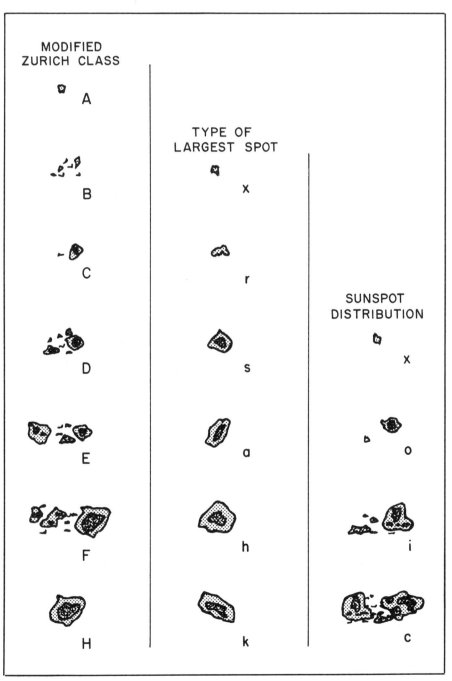

Fig 5—McIntosh sunspot classification. The classification covers three categories: modified Zurich class, type of largest spot, and sunspot distribution, defined as indicated in Table 3.

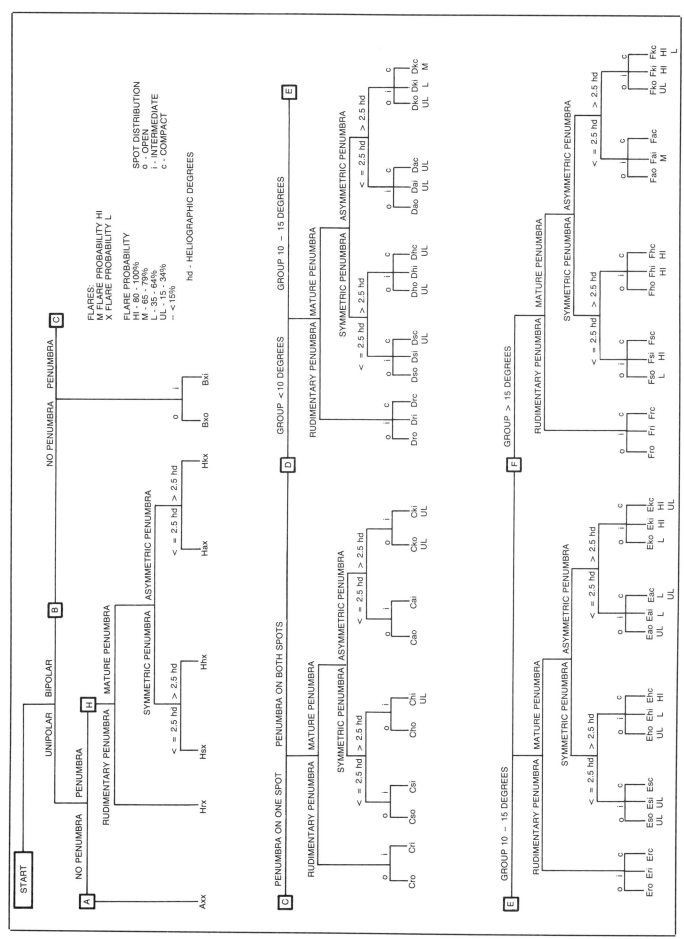

Fig 6—Author's key for allocating sunspot groups to McIntosh classes.

Table 3

Definitions for McIntosh Sunspot Classification

See Fig 5 for drawings of the various classifications.

Abbreviations and symbols in this table (not part of McIntosh classification)
hd—heliographic degree
=—equal to
>—greater than
<—less than

Modified Zurich class:

A—Unipolar group, no penumbra, early or final stage of evolution
B—Bipolar group, without penumbra on any spots
C—Bipolar group with penumbra on one end of group, usually the largest of leader umbrae
D—Bipolar group, penumbra on spots at both ends, length < 10 hd
E—Bipolar group, penumbra on spots at both ends, length 10-15 hd inclusive
F—Bipolar group, penumbra on spots at both ends, length > 15 hd
H—Unipolar group with penumbra, usually evolved from a larger group

Type of largest spot:

x—No penumbra
r—Rudimentary penumbra partially surrounding largest spot
s—Small, symmetric, north to south diameter < = 2.5 hd
a—Small, asymmetric, north to south diameter < = 2.5 hd
h—Large, symmetric, north to south diameter > = 2.5 hd
k—Large asymmetric, north to south diameter > = 2.5 hd

Sunspot distribution:

x—Undefined for unipolar groups
o—Open, few spots between leader and follower, weak magnetic field gradient
i—Intermediate, many spots between leader and follower, none with penumbra
c—Compact, many strong spots between leader and follower, at least one has penumbra

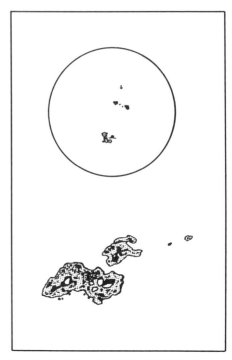

Fig 7—SESC region 5060, July 2, 1988, 1314 UT from a 60 mm f/17 refracting telescope, by the author.

categories to which sunspot groups may be allocated. In order to determine the classification, an observer must note the following features: (1) group polarity, (2) group length, (3) penumbra on the largest spot, (4) maturity and shape of the penumbra, (5) penumbra size on the largest spot, and (6) spot distribution within the group.

Once the McIntosh classification has been determined for the sunspot group, it can be related to the potential for X-ray flare production as determined from a statistical analysis of previous activity of similar spot groups.[10] I have prepared "A Guide to Sunspot Classification and Solar Flare Prediction," which contains a key to allocating sunspot groups to McIntosh classes and the results of an analysis of 12,411 sunspot groups and 1485 associated flares. The key, included here as Fig 6, can be used to assist in the determination of the McIntosh class.

To use the key, select a sunspot group for classification and begin at START. Proceed through each successive division of the key, determining the presence or absence of the appropriate feature in the selected group. When moving along a vertical path in the key, you will encounter a horizontal line at which you have a choice to move either left or right, depending on the feature present. Move in the direction indicated for the desired feature. This will allow you to correctly determine the McIntosh classification of the observed group. In the key, below each class designation, the probability of flare occurence greater than 15% is indicated according to the following classifications.

VL—very low, 15-34%
L—low, 35-64%
M—moderate, 65-79%
HI—high, 80-100%

Where two probabilities are indicated, the upper is for M class flares and the lower for X class flares. If there is only one probability, it is for M class flares. For example, a sunspot group classed as Ekc will have a high probability of producing

M flares and a very low probability of producing X flares. Fig 7 illustrates a drawing of an Fkc sunspot group, designated SESC 5060, which I observed on July 2, 1988 at 1314 UT. This group ¼ rotated onto the solar disk on June 25, 1988 as a Dkc group. It grew quickly to Fkc within one day and maintained this configuration until July 4, when it decayed to Ekc and finally to Eki on July 6. It crossed the west limb of the disk on July 8. During the disk transit, this group produced 42 C class flares and 13 M flares.[11] (Note the key of Fig 6 indicates high probability of M flares and low probability of X flares for an Fkc group.) The group produced no X flares. Intervals of minor and major geomagnetic storms occurred June 29-July 2 at high latitudes as a result of flares from this group and several others on the sun's surface. The Anchorage (Alaska) K index reached storm levels of 5 and 6 on June 30. As well, several shortwave fade-outs and type II and IV radio noise storms occurred during the passage of SESC 5060.

Summary

Programs such as MINIMUF are useful for predicting propagation conditions during various phases of the sunspot cycle. However, their predictions for the MUF are based on average conditions when the ionosphere is well behaved. Day-to-day variations in solar activity are not taken into account. In order to consider these effects, amateur operators can monitor solar activity. Amateurs may do their own monitoring by directly observing sunspots on the solar disk. This information can be used to make changes to planned operating frequencies or to pass priority traffic before bands close. Amateurs who take into account the effects of solar flares on propagation conditions are able to make more effective use of MUF predictions.

Notes

[1] R. B. Rose, "MINIMUF: A Simplified MUF-Prediction Program for Microcomputers," *QST*, Dec 1982, pp 36-38.

[2] A. E. Covington, "Solar Noise Observations on 10.7 cm," *Proc IRE*, no. 36, pp 454-457, 1948.

[3] R. B. Rose and J. N. Martin, "MINIMUF-3: A Simplified HF Prediction Algorithm," *Technical Report 186*, Naval Ocean Systems Center, 1978.

[4] R. W. Miller, "Radio Aurora," *QST*, Jan 1985, pp 14-18

[5] D. M. Baker, "Flare Classification Based on X-ray Intensity," *AIAA Paper No. 70-1370*, 1970.

[6] SESC, "Descriptive Text," *Contents of Preliminary Report and Forecast of Solar Geophysical Data*, US Department of Commerce, 1988, p 16.

[7] R. Hill, "The Handbook for the White Light Observation of Solar Phenomena," *The Association of Lunar and Planetary Observers, Solar Section*. Also *Observe - A Program for Observing and Photographing the Sun*, The Astronomical League.

[8] P. S. McIntosh, "Flare Forecasting Based On Sunspot Classification," in Solar-Terrestrial Predictions, P. A. Simon et al, Eds., *Proceedings of A Workshop at Meudon, France*, US Department of Commerce, 1986, p 357

[9] M. Waldmeir, *Ergebnisse und Probleme der Sonnenforschung*, 2nd ed. (Leipzig: Geest u Portig, 1955).

[10] K. Kildahl, "Frequency of Class M and X Flares by Sunspot Class (1969-1976)," R. Donnelly, Ed., *Solar-Terrestrial Predictions Proceedings*, Vol 3, US Department of Commerce, 1980, p C-166.

[11] SESC, "Preliminary Report and Forecast of Solar Geophysical Data," *SESC PRF 671*, Jul 12, 1988, US Department of Commerce.

[12] J. Joselyn, "SESC Methods for Short-Term Geomagnetic Predictions," in Solar-Terrestrial Predictions, P. A. Simon et al, Eds., *Proceedings of a Workshop at Meudon, France*, US Department of Commerce, 1986, p 404.

[13] G. Jacobs and T. J. Cohen, *The Shortwave Propagation Handbook*, 2nd ed (Hicksville, NY: CQ Publishing, Inc, 1982).

Visible Phenomena of the Ionosphere

By Bradley Wells, KR7L
1290 Puget Dr East
Port Orchard, WA 98366

Many amateurs have an intense personal interest in the ionosphere because of its effect on their communications activities. Hardly a QSO goes by without a conversational reference to propagation conditions. Without a doubt, hams are a most attentive audience for the WWV forecasts at 18 minutes past each hour. Many facets of our hobby would not exist without these refractive layers of the atmosphere. We would be essentially limited to line-of-sight communications, regardless of operating frequency or power level.

Continuing investigations have shown the ionosphere to be a region of dynamic activity. It forms a boundary zone in which the atmosphere interacts with the earth's magnetosphere. Atmospheric sciences and geophysics can provide us with an understanding of the physical and chemical processes operative in the ionosphere.

Scientists divide the earth's atmosphere into various regions, each with its own particular characteristics. See Fig 1. While each region is defined by altitude, it is important to realize that these various zones grade into each other. Atmospheric characteristics do not abruptly change across sharply defined boundaries.

The lowest region of the atmosphere is the troposphere, where the temperature decreases with increasing altitude. (Refer to Fig 1.) It is here that most of the world's weather takes place. The weather reports and photographs presented by television meteorologists are manifestations of activity in this region.

At about 10 miles altitude we enter the stratosphere, where temperature increases with altitude. It is here that we find the ozone layer, the subject of much recent investigation. Here too, primary cosmic rays are intercepted when they impact air molecules. This gives rise to showers of less energetic and less damaging particles.

Above the stratosphere, beginning at an altitude of 30 miles, we move into the realm of the mesosphere. Temperature again drops with altitude, reaching a low of −225 °F at a height of 50 miles. The D layer of the ionosphere forms the transition between the mesosphere and the higher thermosphere. Above this lowest layer of the ionosphere, the temperature climbs with altitude, reaching some 3600 °F at 180 miles. This is the region of the upper

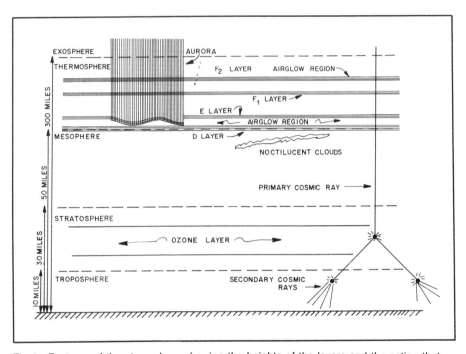

Fig 1—Features of the atmosphere showing the heights of the layers and the action that takes place at different levels.

ionosphere, airglow and aurora.

Above 300 miles we find the exosphere, which has no real upper boundary. The atmosphere becomes progressively thinner, finally merging with the magnetosphere and solar wind. While it is convenient to speak of these various atmospheric regions, none is sharply defined, but grade one into the other.

Active Regions

The territory of primary concern to Amateur Radio operators, and to many other users of the radio spectrum, is that region extending from the middle mesosphere to the upper reaches of the thermosphere. It is here we find the ionosphere and a variety of atmospheric features which can affect radio communications locally and worldwide.

Surprisingly, one of the first features encountered at this altitude is reminiscent of activity within the troposphere. Clouds may form in the mesosphere above the high-latitude regions of the earth. However, they are unlike anything found in the lower atmosphere. These clouds are

visible only after dark, when the sun is well below the horizon. For this reason they are called noctilucent clouds. In addition, they are normally visible only during the summer months of the Northern and Southern Hemispheres. They form white veils, somewhat resembling cirrus clouds, at altitudes of 50 miles above the surface of the earth in that boundary region called the mesopause. Amateurs more commonly know this region as the D layer of the ionosphere.

At this altitude, these clouds are above 99.9% of the atmosphere. The mesopause is the coldest region of the earth's atmosphere, where summer temperatures may drop to −225 °F. It is interesting to note that the lowest temperatures recorded in the mesopause have occurred during the summer months, rather than winter.

A few sounding rockets have been fired through these clouds to collect samples of material. Analysis of this particulate matter indicates that a significant proportion is of extraterrestrial origin. Many of these recovered particles are considerably larger than the maximum size of terrestrial dust or volcanic ash which could have been carried to this height. It is possible these clouds are made up of microscopic dust particles produced when meteors burn up within the upper atmosphere. The 1908 Tunguska Event in Siberia (the fall of a meteorite or comet estimated to weigh 1000 tons) caused the immediate appearance of very striking noctilucent clouds in the Northern Hemisphere. Some scientists feel that these dust particles are merely the nuclei for ice crystals. However, a difficulty with this theory is the almost total lack of water vapor at this altitude.

It is possible that we are able to perceive this material only where its accumulation becomes thick enough to reflect the sun's light. This observation poses another question. Why is there such an uneven distribution of these particles throughout the upper reaches of the atmosphere? These clouds form and move through the lower reaches of the ionosphere, but there has been little investigation of their possible interaction with ongoing activity in the D layer.

The density and temperature of the atmosphere at altitudes above 50 miles remained relatively unknown until the advent of artificial satellites. Prior to this time, infrequent measurements were made through the use of sounding rockets or the computation of meteor trajectories. Gradual changes in the orbits of satellites allowed scientists to calculate densities in the upper atmosphere.

These computations revealed that less than 1/10,000 of our atmosphere exists above 50 miles. In spite of this very low density, most of the incoming solar radiation in the short ultraviolet (UV) range is absorbed here. As a result of this absorption, daytime temperatures may soar to 3600°F around noon and drop to 1000°F during the night. In reality, these temperatures have little correlation with our everyday experiences. Temperature is a measure of the average random velocity of gas molecules, but the low densities involved here reduce this to an almost abstract concept. The individual gas molecules are moving at high speeds, but since so few are involved, there is a negligible heat content. In this rarefied atmosphere, with its gas temperature of 3600°F, a person would freeze to death in the shade (disregarding, for the moment, the lack of oxygen to support life).

Early Research

In 1902, Leon Philippe Teisserenc de Bott presented the French Academy of Sciences with the results of his three-year study of the atmosphere. Teisserenc de Bott, an amateur meteorologist, had launched some 236 balloons to probe the upper atmosphere. He began his investigations using kites tethered with thin steel cables. Much to the relief of local residents, he was forced to discontinue this practice since the cables repeatedly broke and would

deposit several miles of wire across the city of Paris. In addition to the obvious effects of draping a steel cable over power lines, Teisserenc de Bott holds the dubious honor of having shorted, on several occasions, the telegraph wires which were carrying the press accounts of the Alfred Dreyfus trial.

The result of his balloon experiments demonstrated that the atmosphere did not continue to get colder with altitude, but leveled off at about 7 miles. He assumed that temperatures remained constant above this altitude. Without a temperature differential, Teisserenc de Bott hypothesized the upper atmosphere would be devoid of convection currents, and atmospheric gases would settle into distinct layers based on their molecular weight.

While later investigation proved this incorrect, it does occur farther up in the thermosphere. Here, the short UV radiation of the sun breaks molecules into their constituent atoms. Then these atoms tend to sort into layers determined by their atomic weight. Nitrogen and oxygen settle below the lighter gases of helium and hydrogen because of the almost total absence of generalized atmospheric motion in this region of the atmosphere.

Atomic oxygen, combined with smaller proportions of nitrogen and nitrogen-oxide ions, is the principal constituent of the lower thermosphere. Between 200 and 600 miles altitude, helium becomes the primary gas of the atmosphere. Above 600 miles, the prevalent gas is hydrogen, which becomes increasingly rarefied until it merges with the earth's magnetosphere.

As we now know, the atmosphere has electromagnetic layers in addition to its chemical layers. The explanation for this phenomenon was first advanced by the English physicist Oliver Heaviside, several

months after the transatlantic success of Guglielmo Marconi. He stated, "There may possibly be a sufficient conducting layer in the upper air. If so, the waves will, so to speak, catch on to it more or less. Then the guidance will be by the sea on one side and the upper layer on the other." Heaviside's explanation was also arrived at independently by the American electrical engineer Arthur Edwin Kennelly.

In this region of the mesosphere, X-rays and UV radiation from the sun ionize atoms and molecules. Free electrons may then be caught up in the electromagnetic field of a passing radio wave. They first absorb and then radiate energy, and the effect of this interaction is to slightly bend the radio wave. The more free electrons present, the more the wave front is bent. At a critical density, the wave front will be bent enough to return it to the earth's surface.

There is also a direct relationship between electron density and the highest frequency of radio wave affected. Since ionization is a function of incoming solar radiation and atmospheric density, the number of free electrons increases with altitude. Maximum density occurs around 185 miles. This is not an even progression, but occurs in steps which give rise to the various layers of the ionosphere. These are the familiar D, E and F layers, which affect radio communications worldwide.

Since the extent of these layers can only be approximated, it is more convenient and correct to refer to them as regions: the D region at altitudes below 55 miles, the E region between 55 and 100 miles, and the F region above 100 miles. The F region is subdivided into two layers, the F1 and F2. The uppermost (F2) layer, beginning around 155 miles, has the highest concen-

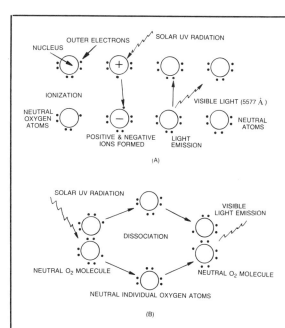

Fig 2—Mechanisms of airglow emission. At A, ionization is caused by UV radiation and X-rays from the sun. Recombintion emits light. At B, dissociation of gas molecules by solar radiation produces light upon recombination.

tration of free electrons.

Light-Producing Regions

In addition to its effects on radio waves, the ionosphere makes its presence known in another way. Simply put, it glows. If you have ever been outside on a truly dark night, far from the effects of urban light pollution, you know that it is not completely dark. As your eyes become dark-adapted, there is often enough light to read letters ½ inch high. Under these conditions, individual stars appear sharp and brilliant, while the Milky Way stretches out in a glowing band across the sky.

It was originally thought that this residual light was the contribution of the 6000 stars visible to the naked eye. We now know that starlight accounts for only 15% of this illumination, and the rest comes from the ionosphere. This diffuse layer of light, called airglow, is faintest at the zenith and strongest at about 10 degrees above the horizon, which makes it an atmospheric, rather than stellar, phenomenon.

Two types of reaction take place within the ionosphere. The first is simple ionization of atoms and molecules, as shown in Fig 2A. UV radiation and X-rays from the sun separate atoms into ions and free electrons. The recombination of these particles causes the emission of light. Airglow, because of its faintness, often appears colorless to the eye, but is composed of a variety of discrete wavelengths.

When an atom is ionized, some of its remaining electrons may be raised to higher energy levels without actually leaving the atom. When these electrons spontaneously drop to lower energy levels, they emit radiation. Each type of atom has its own unique emission spectrum. Some of this radiation may be visible light, depending upon an atom's electron structure and energy levels.

Characteristic emission lines found in airglow spectra are green, 5577 angstroms (Å) and red; 6300, 6364 and 6391Å. These are the lines of atomic oxygen. Green is the most prominent color at altitudes of 60 to 75 miles and the red lines originate around 100 miles. The intensity of 5577Å emission is greatest around midnight, while the red lines are strongest during the early evening and around dawn.

In addition to the emission lines of gases, the spectral lines of metallic sodium, lithium, calcium and magnesium have also been identified. The ratio of sodium to potassium in airglow emission is similar to the ratio of these elements in seawater. Additionally, the abundance of sodium appears highest in winter and lowest in summer. The conclusion is that these elements are carried high into the atmosphere by a very slow, large-scale circulation corresponding with the seasons of the year. The source of other metallic ions is believed to be the result of meteors burning up within the atmosphere.

Dissociation, the second mechanism generating airglow, is more chemical. Here, UV radiation dissociates molecules of gas (primarily oxygen) into individual atoms. Free oxygen can easily combine with other atoms, primarily nitrogen and sodium. The result of this recombination is the emission of visible light. The process is called chemiluminescence and occurs as shown in Fig 2B.

Airglow is not only visible from the ground, but can be seen from orbiting satellites. Identified by John Glenn, the first American to orbit the earth, it has been observed and photographed on many subsequent manned missions. Pictures of this phenomenon show it as colored bands which follow the curvature of the earth's surface and extend above the lower atmosphere.

The ionosphere reaches almost to the top of the thermosphere. At sea level, an air molecule can travel only 3/1,000,000 inch before colliding with another molecule. In the upper reaches of the thermosphere, molecules travel an average of six miles before colliding. This increase in the mean free path is responsible for the relative longevity of the F layer when contrasted to the D layer.

At the altitude of the F layer, deionization proceeds at a very low rate because of

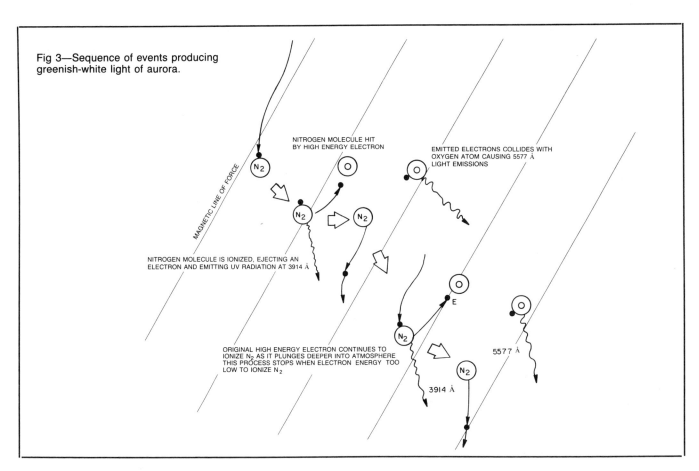

Fig 3—Sequence of events producing greenish-white light of aurora.

the low frequency of recombination of ions and electrons. Since the rate of collision is so low at this altitude, the lighter gas atoms move relatively unimpeded and can be accelerated to escape velocities. Thus, there is a gradual loss of hydrogen and helium out of the atmosphere and into space. However, there is no net loss of these elements. Helium is injected into the atmosphere as a by-product of radioactive decay, while photochemical breakdown of water releases molecular hydrogen.

Aurora

Perhaps the most spectacular phenomenon occurring within the ionosphere is the aurora borealis and its southern counterpart, aurora australis. The light of the aurora is generated by the interaction of solar particles and the atoms of the upper atmosphere. Ions and electrons of the solar wind can become imprisoned within the Van Allen belts. They gradually leak out of this magnetic trap where it closely approaches the earth's surface. These charged particles spiral along the magnetic lines of force and form a glowing ring around the earth's north and south magnetic poles.

An auroral display normally becomes visible a few hours after sunset and disappears with the sunrise. The activity of an aurora does not cease during daylight hours. Its light is simply lost in the overwhelming luminosity of the sun. From the ground, an aurora may be seen as convoluted sheets of light or a glowing arch directly over the observer. Actually, both of these forms are one and the same, but appear visually different because of the effects of perspective. Only from high in space, looking down over the poles, can its true circular shape be seen and photographed. Astronauts in low earth orbit have described their passage through aurora as similar to flying between columns of flaming gas.

One early significant discovery was that the aurora follows the 11-year sunspot cycle. During the years of sunspot maxima, auroras are more numerous and brilliant, while during the sunspot minima they are fewer and smaller. The years from 1645 to 1715 are often referred to as the Little Ice Age, a time of worldwide sub-normal temperatures. These decades also encompass the Maunder Minimum, named after E. Walter Maunder, the British astronomer who first pointed out this irregularity. During this 70-year period there were few, if any, sunspots on the solar surface. Auroras were so uncommon during the Maunder Minimum that an entire generation never witnessed their beauty.

How do incoming particles, spilling out of the Van Allen belts, produce auroral light? The mechanism is similar to that which produces airglow. Spectroscopic

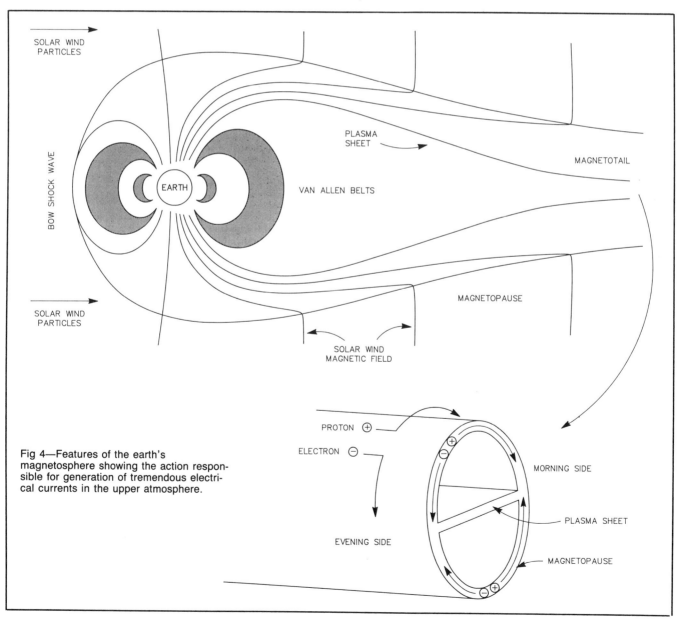

Fig 4—Features of the earth's magnetosphere showing the action responsible for generation of tremendous electrical currents in the upper atmosphere.

207

studies have revealed the most common light to be a whitish-green color with a wavelength of 5577 Å. This is the familiar emission line of atomic oxygen, produced as shown in Fig 3. This form of oxygen can also emit light at a wavelength of 6300 Å. This color forms the "blood red" aurora so much feared in medieval times as the precursor to war, famine and pestilence. If an aurora becomes very active, the bottom of the curtain is often tinted crimson red, a visible light emission by molecular nitrogen.

Auroral emission is triggered by high-speed electrons passing through a tenuous medium. A neon light works in much the same manner. In the upper atmosphere an energetic electron impinges upon a nitrogen atom, ionizing it. The ionized nitrogen atom emits light in the ultraviolet region at a wavelength of 3914 Å. The electron ejected from the nitrogen, as it was ionized, can hit an oxygen atom and excite it. The oxygen emits its characteristic white-green light at a wavelength of 5577 Å as it returns to its lowest energy state. Since the original incoming electron is very energetic, it continues to ionize large numbers of nitrogen atoms as it passes through the upper atmosphere. When its energy drops to the point of being unable to ionize nitrogen, it can still excite these same atoms to cause the crimson red emission often seen at the bottom of auroral curtains. In addition to emission at visible and UV wavelengths, an aurora also emits radiation in the infrared and radio portions of the electromagnetic spectrum.

The Earth as a Generator

Auroras represent an electrical discharge phenomenon involving an enormous flow of current at a power level of one trillion watts. The annual energy release is 9000 billion kilowatt hours, or nine times the annual consumption of electrical energy in the United States. The source of this energy is provided by the solar wind as it cuts through the lines of force of the earth's magnetic field. This is illustrated in Fig 4.

In essence, the earth's magnetosphere functions as a tremendous generator with currents of 1,000,000 amperes flowing across a potential difference of 100,000 volts. As with any generator, there must be two terminals. In the case of the magnetosphere, the positive terminal is located on the morning side and the negative terminal on the evening side. The protons in the solar wind tend to move to the morning side of the magnetopause (the boundary layer of the magnetosphere), while the electrons migrate to the evening side. The interaction of the earth's magnetic field lines with those of the solar wind provide the path for current flow between the two "terminals" of the magnetopause. Electrons moving along these field lines provide the current for upper-atmosphere electrical discharge, which we see as aurora.

Auroras are not limited to the earth's environment. Any planet possessing a magnetic field and atmosphere can generate these phenomena. Auroras above the magnetic poles of Jupiter were photographed by the Voyager spacecraft in 1977. Strong radio emission from Jupiter's magnetosphere and auroras are detectable at a frequency range of 18-22 MHz by relatively simple equipment.

The impact of auroras on radio communications is varied, depending on frequency, path and geography. A large aurora manifests itself in several ways. Incoming electrons and UV emission greatly strengthen the D layer, which results in increased absorption of HF radio energy. Thus, activity on the high-frequency bands is shut down for those living in the high latitudes. Polar paths between stations in mid-latitudes are also disrupted.

At the higher frequencies of the VHF spectrum, the auroral curtains act like imperfect reflectors. VHF stations can work greater than normal distances during an auroral disturbance by pointing their antennas in a generally northern direction, rather than the traditional great-circle path. Since the reflection characteristics of the curtains are constantly changing in response to shifts in electron density, the operator using a very high gain, narrow beamwidth antenna may be at a disadvantage when compared to those stations using more modest antenna systems.

Aurora, airglow and the ionosphere are all features of the upper atmosphere. They are interrelated and affected by activity within the earth's magnetosphere. The electrical activity of the magnetosphere is controlled by events on the surface of the sun. Over the past several decades it has become clear that none of these phenomena can be studied or explained in isolation. Each is affected by the other.

Answers have become more involved for those of us asking questions about the ionosphere. However, with this complexity comes order and, with order, comes knowledge. While we still have a long way to go, each passing year brings us closer to the goal of completely understanding the ionosphere. The achievement of this objective would give us the ability to predict accurately which DX can be worked during a certain hour on any given day of the year.

Bibliography

S. I. Akasofu, "Aurora Borealis—The Amazing Northern Lights," *Alaska Geographic*; Vol 6, No. 2, 1979.

S. I. Akasofu, "The Aurora," *American Scientist*, Sep-Oct 1981.

O. E. Allen, *Atmosphere* (Virginia: Time-Life Books, 1983).

I. Asimov, *Asimov's New Guide to Science* (New York: Basic Books, Inc, 1984).

I. Asimov, *The Universe* (New York: Walker and Company, 1980).

F. W. Cole, *Introduction to Meteorology* (New York: John Wiley and Sons, 1980).

H. Hellman, *Light and Electricity in the Atmosphere* (New York: Holiday House, 1968).

J. H. Jackson and J. H. Baumert, *Pictorial Guide to the Planets*, 3rd ed. (New York: Harper and Row, 1981).

G. Jacobs and T. J. Cohen, *The Shortwave Propagation Handbook* (New York: CQ Publishing, Inc, 1982).

J. A. Ratcliffe, *Sun, Earth and Radio: An Introduction to the Ionosphere and Magnetosphere* (New York: McGraw-Hill, 1970).

R. K. Soberman, "Noctilucent Clouds," *Scientific American*, Jun 1963.

J. A. Van Allen, "Interplanetary Particles and Fields," *Scientific American*, Sep 1975.

L. B. Young, *Earth's Aura* (New York: Alfred A. Knopf, 1977).

Notes

Notes

Notes

Notes

ARRL MEMBERS

This proof of purchase
may be used as a $1.20
credit on your next ARRL
purchase. Limit one
coupon per new membership, renewal or publi-
cation ordered from ARRL Headquarters. No
other coupon may be used with this coupon.
Validate by entering your membership number—the first 7 digits on your QST label—
below:

FEEDBACK

Please use this form to give us your comments on this book and what you'd like to see in future editions.

License class:

☐ Novice ☐ Technician ☐ Technician with HF privileges
☐ General ☐ Advanced ☐ Extra

Name

_____ Call sign _____

Address _____

City, State/Province, ZIP/Postal Code _____

Daytime Phone ( ) _____ Age _____

If licensed, how long? _____ ARRL member? ☐ Yes ☐ No

Other hobbies _____

_____

Occupation _____

From _____

_____

_____

EDITOR, ANTENNA COMPENDIUM, VOL 2
AMERICAN RADIO RELAY LEAGUE
225 MAIN ST
NEWINGTON CT 06111